The Cerambycidae of North America, Part VII, No. 2: Taxonomy and Classification of the Subfamily Lamiinae, Tribes Acanthocinini through Hemilophini

E. Gorton Linsley and John A. Chemsak

UNIVERSITY OF CALIFORNIA PRESS
Berkeley • Los Angeles • London

UNIVERSITY OF CALIFORNIA PUBLICATIONS IN ENTOMOLOGY

Editorial Board: Howell V. Daly, John Doyen,
Henry Hespenheide, Robert S. Lane, John Pinto, Jerry A. Powell,
Arthur Shapiro, Phillip S. Ward, John Smiley

Volume 114
Issue Date: January 1995

UNIVERSITY OF CALIFORNIA PRESS
BERKELEY AND LOS ANGELES, CALIFORNIA

UNIVERSITY OF CALIFORNIA PRESS, LTD.
LONDON, ENGLAND

© 1995 BY THE REGENTS OF THE UNIVERSITY OF CALIFORNIA
PRINTED IN THE UNITED STATES OF AMERICA

Library of Congress Cataloging-in-Publication Data
(Revised for pt. 7, no. 2)
Linsley, E. Gorton (Earle Gorton), 1910–
    The Cerambycidae of North America.
    (University of California publications in entomology, v. 18–)
    Pt.   : By E. Gorton Linsley and John A. Chemsak.
    Includes bibliographical references.
    Contents: pt. 1. Introduction. — pt. 2. Taxonomy and classification of the Parandrinae, Prioninae, Spondylinae, and Aseminae. — [etc.] — pt. 7, no. 2. Taxonomy and classification of the subfamily Lamiinae, tribes Acanthocinini through Hemilophini.
    1. Cerambycidae—North America. 2. Insects—North America. I. Chemsak, John A.   II. Title.   III. Series: University of California publications in entomology; v. 18, etc.
QL461.C17 vol. 18, etc.   595.7'008 s 61-63700
ISBN 0-520-09556-1 (pt. 6, no. 2)   [595.76'48]
ISBN 0-520-09795-5 (pt. 7, no. 2)

The paper used in this publication meets the minimum requirements of American National Standard for Information Sciences—Permanence of Paper for Printed Library Materials, ANSI Z39.48-1984.

Contents

Abstract, x

Acknowledgments, xii

INTRODUCTION   1

TRIBE ACANTHOCININI   2
   Alcidion, 5
      A. umbraticum, 6
   Glaucotes, 8
      G. yuccivorus, 9
   Urographis, 11
      U. despectus, 12
      U. fasciatus, 13
      U. triangulifer, 16
   Trichocanonura, 18
      T. linearis, 19
   Eutrichillus, 20
      E. biguttatus, 21
      E. neomexicanus, 22
      E. canescens, 24
      E. pini, 25
   Dectes, 26
      D. sayi, 28
      D. texanus, 29
   Valenus, 32
      V. inornatus, 32
   Sternidius, 34
      S. variegatus, 35
   Acanthocinus, 37
      A. nodosus, 40
      A. obsoletus, 41
      A. pusillus, 43
      A. angulosus, 44
      A. obliquus, 46
      A. spectabilis, 50
      A. princeps, 52
      A. leechi, 53

Sternidocinus, 55
    S. barbarus, 56
Astylopsis, 58
    A. arcuata, 60
    A. sexguttata, 62
    A. macula, 64
    A. perplexa, 66
    A. collaris, 68
Hyperplatys, 69
    H. femoralis, 71
    H. maculata, 71
    H. californica, 73
    H. aspersa, 75
Lepturges, 77
    L. regularis, 79
    L. megalops, 80
    L. infilatus, 81
    L. yucca, 82
    L. vogti, 83
    L. symmetricus, 84
    L. pictus, 85
    L. confluens, 86
    L. angulatus, 87
Liopinus, 88
    L. wilti, 92
    L. chemsaki, 93
    L. punctatus, 94
    L. misellus, 95
    L. imitans, 96
    L. incognitus, 97
    L. decorus, 98
    L. centralis, 99
    L. mimeticus, 100
    L. alpha, 102
Nyssodrysina, 106
    N. haldemani, 107
Urgleptes, 109
    U. foveatocollis, 110
    U. celtis, 112
    U. facetus, 113
    U. querci, 115
    U. signatus, 117
Coenopoeus, 118
    C. palmeri, 119
Styloleptus, 121
    S. biustus, 122

Pseudastylopsis, 124
  P. nebulosus, 125
  P. nelsoni, 127
    P. n. australis, 128
  P. pini, 128
Trichastylopsis, 130
  T. albidus, 131
Astylidius, 132
  A. parvus, 133
Lagocheirus, 134
  L. undatus, 136
  L. araneiformis, 137
    L. a. stroheckeri, 138
Leptostylopsis, 139
  L. luteus, 140
  L. terraecolor, 141
  L. planidorsus, 142
  L. albofasciatus, 144
  L. argentatus, 145
Leptostylus, 146
  L. asperatus, 148
  L. transversus, 148
  L. gibbulosus, 153

TRIBE CYRTININI   156
  Cyrtinus, 156
    C. beckeri, 157
    C. pygmaeus, 158

TRIBE SAPERDINI   161
  Saperda, 162
    S. obliqua, 165
    S. calcarata, 168
    S. populnea, 172
      S. p. moesta, 173
      S. p. tulari, 174
    S. horni, 175
    S. mutica, 177
    S. candida, 178
    S. puncticollis, 181
    S. lateralis, 182
    S. imitans, 185
    S. tridentata, 186
    S. vestita, 190
    S. discoidea, 192
    S. cretata, 196
    S. fayi, 197
    S. inornata, 198

TRIBE PHYTOECIINI   201
  Mecas, 202
    Mecas (Dylobolus), 205
      M. (D.) rotundicollis, 205
    Mecas (Mecas), 208
      M. (M.) marginella, 208
      M. (M.) confusa, 209
      M. (M.) femoralis, 211
      M. (M.) cineracea, 212
      M. (M.) linsleyi, 212
      M. (M.) pergrata, 214
      M. (M.) bicallosa, 215
      M. (M.) menthae, 216
      M. (M.) cana, 218
        M. (M.) c. cana, 219
        M. (M.) c. saturnina, 219
  Oberea, 220
    O. gracilis, 223
    O. ruficollis, 224
    O. quadricallosa, 226
    O. schaumi, 228
    O. flavipes, 230
    O. oculaticollis, 231
    O. ocellata, 232
    O. delongi, 234
    O. affinis, 235
    O. perspicillata, 236
    O. tripunctata, 241
    O. myops, 244
    O. praelonga, 246
  Incertae sedis, 247
    O. bimaculata, 247
    O. mairei, 248

TRIBE TETRAOPINI   249
  Tetraopes, 250
    T. melanurus, 254
    T. linsleyi, 256
    T. discoideus, 258
    T. annulatus, 261
    T. pilosus, 263
    T. tetrophthalmus, 265
    T. texanus, 268
    T. mandibularis, 270
    T. quinquemaculatus, 271
    T. sublaevis, 272

  T. thermophilus, 274
  T. basalis, 275
  T. femoratus, 277
 Phaea, 280
  P. canescens, 282
  P. monostigma, 283

TRIBE HEMILOPHINI   286
 Cathetopteron, 287
  C. amoena, 287
 Hemierana, 288
  H. marginata, 289
   H. m. marginata, 290
   H. m. ardens, 291
   H. m. suturalis, 292

Abstract

This volume concludes the taxonomy and classification of the family Cerambycidae of America north of Mexico. This part includes the remainder of the subfamily Lamiinae, tribes Acanthocinini, Cyrtinini, Saperdini, Phytoeciini, Tetraopini, and Hemilophini. The 32 genera and 138 species are all fully described with keys included to separate all taxa. Complete synonymical bibliographies are presented.

New taxa are: *Liopinus*, n. gen. (type species = *Lamia alpha* Say); *Pseudastylopsis nelsoni*, n. sp. from Arizona; and *P. nelsoni australis*, n. subsp. from Mexico. Type species designations are proposed for the genera *Urographis* Horn (*Cerambyx fasciatus* Degeer); *Lepturges* (*L. elegantulus* Bates); and *Myrmolamia* (*M. opacicollis* Bates). New synonymies include: *Ceratographis* Gahan = *Eutrichillus* Bates; *Astyleiopus* Dillon = *Sternidius* LeConte; *Graphisurus* Kirby, *Canonura* Casey, *Tylocerina* Casey, and *Neacanthocinus* Dillon all equal *Acanthocinus* Megerle; *Maculurges* Dillon = *Lepturges* Bates; *Eutrichillus canescens nelsoni* Dillon = *E. canescens* Dillon; *Dectes alticola* Casey, *D. thoracicus* Casey, *D. aridus* Casey, and *D. texanus murinus* Dillon all equal *D. texanus* LeConte; *Graphisurus punctatus* Casey = *Acanthocinus obsoletus* (Olivier); *Graphisurus pacificus* Casey, *G. obtusus* Casey, and *G. obliquus chihuahuae* Casey all equal *Acanthocinus obliquus* (LeConte); *Lepturges minutus* Champlain and Knull, *Urgleptes kissingeri* Dillon, and *U. knulli* Dillon = *U. foveatocollis* (Hamilton); *Liopus minuens* Leng and Hamilton = *Styloleptus biustus* (LeConte); *Astylidius versutus* Casey and *A. versutus downiei* Casey = *A. parvus* (LeConte); *Lagocheirus texensis* Dillon, *L. undatus mariorum* Dillon, *L. zimmermani* Dillon, and *L. zimmermani aukena* Dillon all equal *L. undatus* (Voet); *Leptostylus transversus dakotensis* Dillon, *L. transversus dietrichi* Dillon, and *L. transversus floridellus* Dillon = *L. transversus* (Gyllenhal); *Leptostylus knulli* Fisher = *Astylopsis arcuata* (LeConte); *Liopus schwarzi* Hamilton and *Leiopus texanus* Casey = *Liopinus mimeticus* (Casey); *Lamia fascicularis* Harris, *Liopus rusticus* LeConte, *Liopus floridanus* Hamilton, and *Sternidius vittatus* Dillon all equal *Liopinus alpha* (Say); *Saperda bipunctata* Hopping = *S. candida* Fabricius; *Saperda shoemakeri* Davis = *S. fayi* Bland; *Oberea canadensis* Fisher = *O. myops* Haldeman. New combinations proposed: *Urographis despectus* (LeConte); *U. fasciatus* (Degeer); *U. triangulifer* (Haldeman); *Eutrichillus biguttatus* (LeConte); *Astylopsis arcuata* (LeConte); *A. perplexa* (Haldeman); *Lepturges regularis* (LeConte); *Liopinus wilti* (Horn); *L. chemsaki* (Lewis); *L. punctatus* (Haldeman); *L. misellus* (LeConte); *L.*

imitans (Knull); *L. incognitus* (Lewis); *L. decorus* (Fall); *L. centralis* (LeConte); *L. mimeticus* (Casey); and *L. alpha* (Say). Two species, *Oberea bimaculata* (Olivier) and *O. mairei* Chevrolat, both originally described from France, are placed under *Incertae sedis*.

Thirty-nine species are illustrated, and distribution maps are provided for 15 species.

Acknowledgments

Acknowledgments to the various individuals and institutions that have contributed toward making this study possible have appeared in previous parts. We wish especially to express our gratitude to the Essig Museum of Entomology and the Department of Entomological Sciences, University of California at Berkeley, for providing the means to complete this study.

Christina Jordan should be added to the list of artists who produced the outstanding illustrations.

INTRODUCTION

This is the second number of the seventh part of a series of studies comprising a monograph of the Cerambycidae of America north of Mexico. The previous parts have appeared in the University of California Publications in Entomology as follows: Part I, Volume 18, 97 pp., 1961; II, Volume 19, 102 pp., 1962; III, Volume 20, 188 pp., 1962; IV, Volume 21, 165 pp., 1963; V, Volume 22, 197 pp., 1964; VI(1), Volume 69, 138 pp., 1972; VI(2), Volume 80, 186 pp., 1976; VII(1), Volume 102, 258 pp., 1984. Part VIII will include the bibliography, complete index, and host plant index.

TRIBE ACANTHOCININI THOMSON

Thomson, 1860, Class. ceramb., p. 6 (Acanthocinitae).
Thomson, 1864, Syst. ceramb., p. 23 (Acanthocinitae).
Lacordaire, 1872, Genera des coléoptères, 9(2): 757 (Acanthocinides).
LeConte, 1873, Smithson. Misc. Coll., 11(265): 337.
Horn, 1880, Trans. Amer. Entomol. Soc., 8: 116.
LeConte and Horn, 1883, Smithson. Misc. Coll., 507: 322.
Leng and Hamilton, 1896, Trans. Amer. Entomol. Soc., 23: 113.
Blatchley, 1910, Coleoptera in Indiana, p. 1069 (Acanthoderini, part).
Casey, 1913, Memoirs on the Coleoptera, 4: 303.
Bradley, 1930, Man. Genera Beetles, p. 143.
Chagnon, 1933-40, Coleop. Prov. Quebec, p. 271 (Acanthoderini, part).
Doane et al., 1936, For. Ins., p. 188.
Knull, 1946, Ohio Biol. Surv. Bull., 39: 244.
Duffy, 1953, Mon. British Timber Beetles, p. 271.
Dillon, 1956, Ann. Entomol. Soc. Amer., 49: 134.
Duffy, 1960, Mon. Neotrop. Timber Beetles, p. 235.
Dillon and Dillon, 1961, Man. Beetles East. North America, p. 637.
Chagnon and Robert, 1962, Prin. Coleop. Prov. Quebec, p. 271 (Acanthoderini, part).
Arnett, 1962, Beetles U.S., 103: 871.
Bayer and Shenefelt, 1969, Univ. Wisc. Res. Bull., 275: 26.
Hatch, 1971, Univ. Wash. Pub. Biol., 16: 150.
Villiers, 1980, Ann. Soc. Entomol. Fr., (n.s.) 16: 90.
Rice and Enns, 1981, Trans. Mo. Acad. Sci., 15: 97.

Form small to large. Head with front large, quadrate; labrum large, base coriaceous; palpi slender, apical segments pointed; eyes finely faceted, emarginate, lower lobes subquadrate; antennae slender, usually elongate, scape elongate, slender, barely thickened apically, lacking a cicatrix, segments often fimbriate beneath. Pronotum transverse; apex usually narrowly impressed at apex; sides usually with lateral tubercles; prosternum narrow, front coxal cavities closed behind, rounded or feebly angulate laterally; mesosternum with coxal cavities closed to epimeron. Elytra often basally bigibbose, often costate, tufted tubercles often present; erect setae occasionally present. Legs with femora clavate; middle tibiae with an external sinus; tarsi with claws divaricate. Abdomen normally segmented, fifth segment often elongate in females.

The slender, elongate antennal scape, clavate femora, divaricate tarsal claws, and the rounded precoxal cavities readily distinguish this tribe.

The Acanthocinini are abundantly represented in the Neotropical region and comprise about one-third of the species of Lamiinae in North America, where 24 genera are known.

KEY TO THE NORTH AMERICAN GENERA OF ACANTHOCININI

1	Elytra with long or short erect setae or flying hairs	2
	Elytra with appressed pubescence or tufted tubercles	8
2(1)	Pronotum with obtuse tubercles at sides	3
	Pronotum with acute tubercles or spines at sides	5
3(2)	Elytra with apices rounded to emarginate	4
	Elytra with outer margins of apices produced into broad spines; pronotum with small, obtuse lateral tubercles behind middle. Form convex, broad	*Alcidion*
4(3)	Prothorax with intercoxal process angulate at sides. Elytra with apices rounded. Abdomen with last segment not elongate in females	*Glaucotes*
	Prothorax with intercoxal process not angulate at sides. Elytra with apices truncate to emarginate. Abdomen with last segment elongate in females	*Urographis*
5(2)	Elytra with small tufted tubercles, at least on basal crests	6
	Elytra lacking tufted tubercles	7
6(5)	Pronotum with sides broadly tuberculate behind middle. Males with front tarsi broad, fringed laterally. Females with last abdominal segment greatly elongate	*Trichocanonura*
	Pronotum with sides acutely tuberculate behind middle. Males with front tarsi slender. Females with last abdominal segment moderately elongate	*Eutrichillus*
7(5)	Elytra with pubescence dense, appressed, suberect setae short. Body cylindrical	*Dectes*
	Elytra with pubescence very fine, short, not obscuring surface. Body subdepressed	*Valenus*
8(1)	Pronotum with acutely spined tubercles at sides	9
	Pronotum with lateral tubercles rounded or obtuse	17
9(8)	Pronotum with lateral spines a little behind middle	10
	Pronotum with lateral spines near base	13
10(9)	Pronotum with basal transverse impression extending below tubercles onto sides	11
	Pronotum with basal impression extending only to bases of lateral tubercles, not onto sides	*Sternidius*

11(10)	Antennae with segments three to five at most with several suberect hairs beneath, at most one-third longer than body. Abdomen of females not elongate 12
	Antennae with segments three to five densely fimbriate beneath, at least 1-1/2 times as long as body. Abdomen with last segment elongate in females .... *Acanthocinus*
12(11)	Pronotum with disk convex, not callused. Antennae with scape subequal in length to third segment. Front tibiae with short apical spurs .... *Sternidocinus*
	Pronotum with five obtuse discal calluses. Antennae with scape shorter than third segment. Tibial spurs normal................................. *Astylopsis*
13(9)	Elytra with epipleura rounded 14
	Elytra with epipleura vertical, delimited by a lateral carina. Elytra usually with small round spots..... *Hyperplatys*
14(13)	Pronotum with basal sulcus extending onto sides 15
	Pronotum with basal sulcus confined to disk, not extending beyond bases of lateral tubercules 16
15(14)	Mesosternum with intercoxal process as broad as prosternal process, about one-sixth as broad as coxal cavities; lateral tubercules placed at or near extreme base. Pronotum lacking discal calluses ..... *Urgleptes*
	Mesosternal process much broader than prosternal; lateral tubercles placed before base. Pronotum with three discal calluses .... *Liopinus*
16(14)	Mesosternum with intercoxal process about two-thirds as broad as coxal cavities, much broader than prosternal process. Body ovoid .... *Nyssodrysina*
	Mesosternal process about one-sixth as broad as coxal cavities, only slightly broader than prosternal process. Body small, subdepressed .... *Lepturges*
17(8)	Prothorax with intercoxal process less than half as broad as coxal cavities 18
	Prothorax with intercoxal process more than half as broad as coxal cavities 20
18(17)	Elytra with costae and small tufted tubercles. Form small to moderate-sized 19
	Elytra lacking costae and tufted tubercles. Antennae of males with sixth segment apically produced. Form large, robust .... *Coenopoeus*
19(18)	Pronotum with obtuse tumid tubercles at sides just before basal impression; disk with three calluses; mesosternal process about as broad as coxal cavities .... *Styloleptus*

	Pronotum with obtuse lateral tubercles at middle; disk with five calluses; mesosternal process about two-thirds as broad as coxal cavities .... *Pseudastylopsis*
20(17)	Pronotum with three discal calluses 21
	Pronotum with five or more discal calluses 22
21(20)	Legs with numerous long flying hairs. Prosternal process two-thirds as broad as coxal cavity, mesosternal process about as broad as coxal cavity, abruptly declivous anteriorly .... *Trichastylopsis*
	Legs lacking long flying hairs. Prosternal process half as broad as coxal cavity, mesosternal process as broad as coxal cavity, arcuate anteriorly .... *Astylidius*
22(20)	Form small to moderate-sized. Pronotum with lateral tubercules small to moderate-sized. Antennae and front tarsi of males not modified....................... 23
	Form large, robust. Pronotum with robust lateral tubercles. Antennae of males with an apical spur on sixth segment. Front tarsi of males broadened and strongly fringed .... *Lagocheirus*
23(22)	Mesosternum with intercoxal process as broad as or slightly less broad than coxal cavities; prosternal process about half to two-thirds as broad as coxal cavities 24
	Mesosternal process broader than coxal cavities; prosternal process about two-thirds as broad as coxal cavities............................. *Leptostylopsis*
24(23)	Pronotum with discal calluses obtuse, shallow area between calluses densely punctate; mesosternal process about four-fifths as broad as coxal cavities .. *Astylopsis*
	Pronotum with discal calluses prominent, punctures between calluses not distinct; mesosternal process as broad as coxal cavities .... *Leptostylus*

Genus *Alcidion* Sturm

Alcidion Sturm, 1843 (not White, 1855; not Thomson, 1864), Cat. Käfer-Samml. Jacob Sturm, p. 254; Monne, 1977, Rev. Brasil. Biol., 37: 698; Villiers, 1980, Ann. Soc. Entomol. Fr., (n.s.) 16: 90, 94.

Probatius White, 1855, Cat. Col. British Museum, 8: 389; Thomson, 1860, Class. ceramb., p. 16; Thomson, 1864, Syst. ceramb., pp. 27, 355; Bates, 1864, Ann. Mag. Nat. Hist., (3) 13:47; Lacordaire, 1872, Genera des coléoptères, 9(2):781; Bates, 1881, Biol. Centr.-Amer., Coleoptera, 5: 175; Dillon, 1962, Coleop. Bull., 16: 32 (synonymy); de Zayas, 1975, Rev. Fam. Ceramb., p. 271. (Type species: *Acanthocinus humeralis* Perty, Monne designation, 1977.)

Hirsutographis Dillon, 1956, Ann. Entomol. Soc. Amer., 49: 207; Arnett, 1962, Beetles U.S., 103: 872; Dillon, 1962, Coleop. Bull., 16: 32 (synonymy). (Type species: *Hirsutographis pulchra* Dillon, by original designation.)

Form small to moderate-sized, convex; pubescence fine, appressed, elytra with long, suberect setae. Head with front moderately convex, quadrate; mandibles feebly arcuate; genae slightly convergent, about two-thirds as long as lower eye lobes; eyes moderate-sized, deeply emarginate, lower lobes longer than broad, upper lobes small, separated by slightly more than their width; antennal tubercles prominent, broadly divergent from bases; antennae eleven-segmented, slender, about 1-1/2 times as long as body, segments two to about seven with a few short, erect setae beneath and at apices, third segment subequal to first, fourth shorter than third, fifth shorter than fourth. Pronotum broader than long, sides with a small, obtuse tubercle behind middle; disk convex, base broadly impressed, sulcus extending below tubercles, apex narrowly impressed; prosternum narrow, intercoxal process almost half as broad as coxal cavity, cavities closed behind; mesosternal process a little broader than prosternal process; metasternum with episternum narrow, subparallel. Elytra slightly less than twice as long as broad; base vaguely, shallowly gibbose, disk not impressed; pubescence short, appressed, with moderately long, suberect setae sparsely interspersed, each seta arising out of a puncture; apices obliquely emarginate truncate, outer margins produced into broad spines. Legs stout, femora clavate; tarsi moderate, first segment of hind pair about as long as following two together, third segment cleft to base. Abdomen normally segmented.

Type species. Acanthocinus humeralis Perty (monobasic).

The convex, rather broad form, presence of erect setae on the elytra, and proportions of the intercoxal processes readily separate this genus from other North American Acanthocinini. The above generic description is based on the single species, *Alcidion umbraticum*, occurring in the North American fauna, and may not be adequate for the other species, most of which are Neotropical.

Alcidion umbraticum (Jacquelin du Val)
(Figure 1)

Probatius umbraticus Jacquelin du Val, 1857, in Sagra, Hist. Cuba, 7: 272, pl. 10, fig. 10; Chevrolet, 1862, Ann. Soc. Entomol. Fr., (4) 2: 249; Gundlach, 1891, Contr. Entomol. Cuba., p. 210, pl. 3; Leng and Hamilton, 1896, Trans. Amer. Entomol. Soc., 23: 182; Wolcott, 1948, J. Agr. Univ. Puerto Rico, 32: 346; Knull, 1954, Ohio J. Sci., 54: 130; Dillon, 1962, Coleop. Bull., 16: 32; de Zayas, 1975, Rev. Fam. Ceramb., p. 271, pl. 23, fig. d.
Alcidion umbraticum: Monne, 1977, Rev. Brasil. Biol., 37: 698.
Hirsutographis pulchra Dillon, 1956, Ann. Entomol. Soc. Amer., 49: 207; Dillon, 1962, Coleop. Bull., 16: 32 (synonymy).

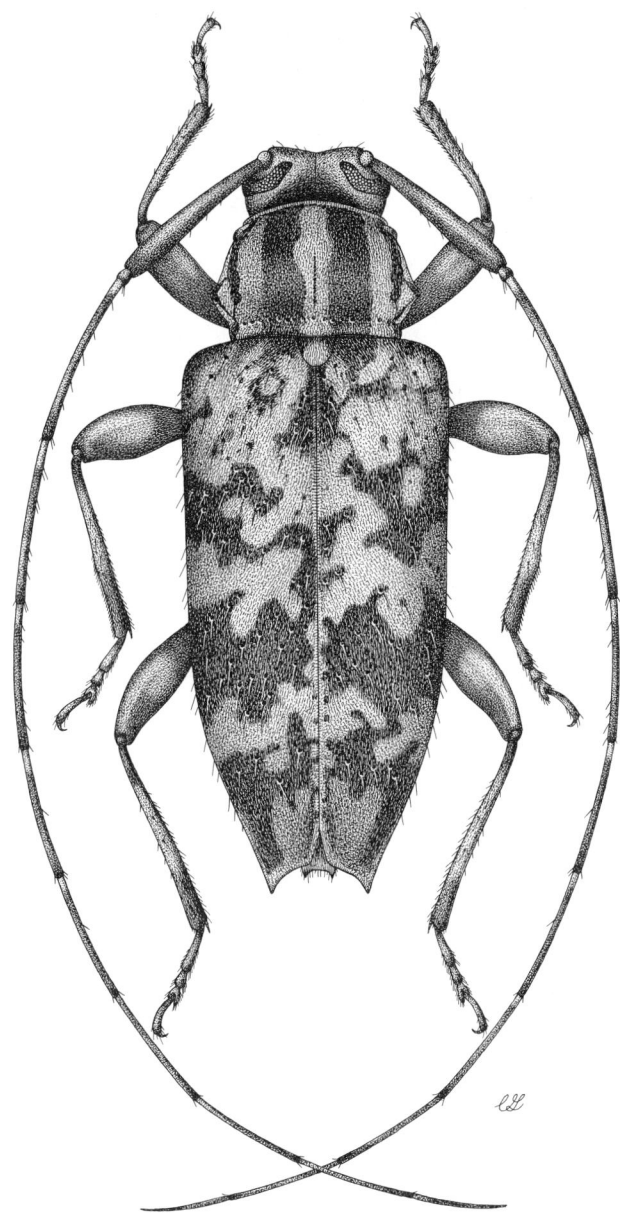

Figure 1. *Alcidion umbraticum* (Jacquelin du Val), male.

Male. Form moderate-sized; integument reddish brown; pubescence yellowish, appressed, occurring in dense patches, long, suberect setae interspersed over elytra. Head with front convex, quadrate, densely micropunctate, densely clothed with fine, appressed pubescence; genae slightly shorter than lower eye lobes; antennae about 1-1/2 times longer than body, basal segments white annulate at apices, erect setae more numerous on basal segments. Pronotum broader than long, lateral tubercles small, placed before basal impression; disk with a narrow, glabrous median line; punctures minute, dense, basal margin with a row of coarse puncture; pubescence forming maculae, a longitudinal one at extreme outer margin, a basally joined pair at sides, and a median macula divided by median line, base with a few long, erect hairs; prosternum finely, densely, pale pubescent; meso- and metasternum with patches of yellowish pubescence at sides. Elytra slightly less than twice as long as broad; base vaguely gibbose; pubescence fine, short, denser yellowish appressed pubescence arranged into reticulating maculae, long erect setae each arising from a puncture; apices shallowly emarginate, outer angles dentate. Legs robust, finely pubescent. Abdomen finely, densely pubescent, sides with patches of yellowish appressed pubescence; last sternite rounded at apex. Length, 6.5-11 mm.

Female. Form similar. Antennae shorter. Length, 6.5-12 mm.

Type locality. Of *umbraticum*, Cuba; *pulchra*, South Miami, Florida.

Range. Southern Florida, Cuba.

Flight period. May - June.

Host plants. Not known.

This species, endemic to Cuba, is apparently now established in Florida but not commonly collected.

The yellowish maculae of the body and erect setae of the elytra are typical of *A. umbraticum*.

Genus *Glaucotes* Casey

Glaucotes Casey, 1913, Memoirs on the Coleoptera, 4: 305; Bradley, 1930, Man. Genera Beetles, p. 246; Dillon, 1956, Ann. Entomol. Soc. Amer., 49: 155; Arnett, 1962, Beetles U.S., 103: 871.

Form moderate-sized, robust, slightly depressed. Head with front short, convex, broader than long; eyes with lower lobes slightly transverse, upper lobes separated by about twice their width; genae convergent, longer than lower eye lobes; antennal tubercles widely separated; antennae a little longer than body in males, as long as body or slightly shorter in females, scape slender, gradually enlarging to apex, extending almost to pronotal lateral tubercles, third segment longer than scape, fourth shorter than scape, remaining segments gradually decreasing in length, basal segments with a number of long, suberect setae beneath. Pronotum broader than long, sides obtusely tuberculate behind middle; disk with a flattened median callus and two behind apical margin, two vague calluses present at sides near lateral tubercles; basal sulcus broad, extending onto sides; prosternum narrow, excavated, intercoxal process arcuate, angulate at sides, medially grooved, at least one-half as broad as coxal cavities; metasternum with episternum very narrow, subparallel. Elytra about twice as long as broad, tapering at apical one-fourth; disk vaguely bigibbose near base, with numerous, short, erect hairs; apices rounded. Legs robust; femora strongly clavate, with short suberect hairs; tarsi short, third segment cleft to base. Abdomen normally segmented.

Type species. Leptostylus yuccivorus Fall (by original designation).

This genus is characterized by the robust body form, relatively short antennae, angled sides of the prosternal process and the erect hairs on the elytra and femora. A single species is known.

Glaucotes yuccivorus (Fall)
(Figure 2)

Leptostylus yuccivorous Fall, 1907, J. N.Y. Entomol. Soc., 15: 84.
Glaucotes yuccivorus: Casey, 1913, Memoirs on the Coleoptera, 4: 305; Fattig, 1947, Emory Univ. Mus. Bull., 5: 33; Dillon, 1956, Ann. Entomol. Soc. Amer., 49: 156, fig. 6; Linsley, Knull, and Statham, 1961, Amer. Mus. Nov., 2050: 27; Lewis, 1979, Pan-Pac. Entomol., 55: 24; Turnbow and Franklin, 1980, J. Ga. Entomol. Soc., 15: 347.

Male. Form moderate-sized, robust; integument dark reddish brown to piceous, antennal segments from third paler basally; pubescence dense, short, appressed, grayish, elytra with short, erect setae. Head with front minutely punctate with a few larger punctures interspersed, pubescence moderately dense, gray and dark variegated; vertex densely gray pubescent, often with a median black spot and small black spots interspersed; antennae gray and dark pubescent, scape with small scattered black spots and apex dark, segments from third dark annulate at apices. Pronotum densely gray pubescent, anterior and median calluses usually glabrous, apical margin with a black spot on each side, basal margin with arcuately linear black spots on each side; deep rather fine punctures interspersed around calluses; prosternum densely gray pubescent; meso- and metasternum densely pubescent, minutely punctate, metasternum with larger punctures interspersed. Elytra densely grayish pubescent, each elytron with five more densely pubescent, longitudinal vittae, basal gibbosities usually with a small dark spot; punctures coarse, dense, contiguous, usually obscured by pubescence; erect setae numerous, short; apices rounded. Legs with femora densely gray and black pubescent, short, suberect hairs numerous; tibiae with suberect hairs rather dense. Abdomen minutely, densely punctate with larger punctures interspersed, pubescence dense; last sternite broadly rounded to subtruncate at apex. Length, 12-15 mm.

Female. Form similar. Antennae about as long as body. Abdomen with last sternite narrowing, shallowly emarginate at apex. Length, 12-15 mm.

Type locality. San Bernardino Ranch, Douglas, Arizona.
Range. Northwestern to southern Arizona.
Flight period. June to September.
Host plants. Yucca.

This species is easily recognizable by the broad form, short antennae, dark integument, and the rather vague pubescent vittae of the elytra.

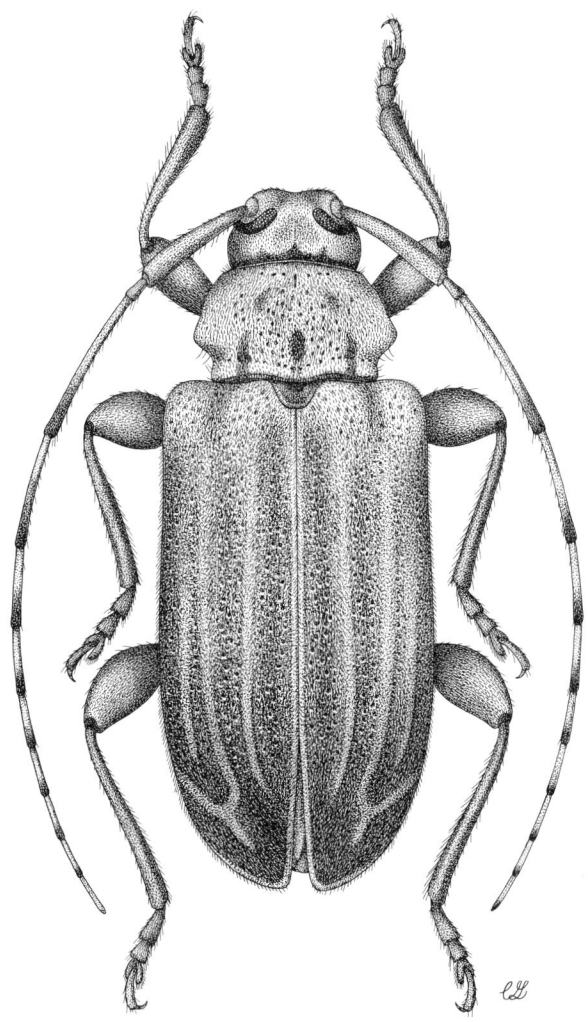

Figure 2. *Glaucotes yuccivorus* (Fall), male.

Genus *Urographis* Horn

Graphisurus LeConte 1852 (not Kirby, 1837), J. Acad. Nat. Sci. Philadelphia, 2: 174; Lacordaire, 1872, Genera des coléoptères, 9(2): 786 (part); LeConte, 1873, Smithson. Misc. Coll., 11(265): 339; Provancher, 1877, Pet. Fauna Entomol. Can., 1: 629; Horn, 1880, Trans. Amer. Entomol. Soc., 8: 128; LeConte and Horn, 1883, Smithson. Misc. Coll., 507: 324; Gahan, 1888, Trans. Amer. Entomol. Soc., 15: 299; Leng and Hamilton, 1896, Trans. Amer. Entomol. Soc., 23: 130; Wickham, 1897, Can. Entomol., 29: 203; Wickham, 1898, Can. Entomol., 30: 38; Blatchley, 1910, Coleoptera in Indiana, p. 1078; Craighead, 1923, Can. Dept. Agr. Bull., (n.s.) 27: 120; Bradley, 1930, Man. Genera Beetles, p. 246; Chagnon, 1933-40, Coleop. Prov. Quebec, p. 274; Doane et al., 1936, For. Ins., p. 189; Dillon, 1956, Ann. Entomol. Soc. Amer., 49: 163; Dillon and Dillon, 1961, Man. Beetles East. North America, p. 639; Arnett, 1962, Beetles U.S., 103: 872; Chagnon and Robert, 1962, Prin. Coleop. Prov. Quebec, p. 174; Bayer and Shenefelt, 1969, Univ. Wisc. Res. Bull., 275: 27; Rice and Enns, 1981, Trans. Mo. Acad. Sci., 15: 101.

Urographis Horn, 1880, Trans. Amer. Entomol. Soc., 8: 128; LeConte and Horn, 1883, Smithson. Misc., Coll., 507: 324; Casey, 1913, Memoirs on the Coleoptera, 4: 331; Leng and Mutchler, 1927, Supp. Cat. Coleop. Amer., p. 43; Knull, 1946, Ohio Biol. Surv. Bull., 39: 256.

Form moderate to moderately large, subdepressed; elytra with long, erect hairs. Head with front feebly convex, quadrate or slightly longer than broad; median line extending onto neck; mandibles feebly arcuate; genae convergent, half to almost twice as long as lower eye lobes; eyes moderate, lower lobes longer than wide, upper lobes small, separated by more than their widths; antennal tubercles prominent, widely divergent; antennae eleven-segmented, 1-1/2 times as long as body or slightly shorter, basal segments with a few short, erect hairs beneath, first segment subequal to or slightly shorter than third, fourth shorter than third, slightly longer than first. Pronotum broader than long, sides with a large blunt tubercle slightly behind middle; basal sulcus extending to sides; disk convex, median line narrow; prosternum with intercoxal process narrow, about one-sixth as wide as coxal cavities; mesosternum with intercoxal process broad, declivous in front, half or more as broad as coxal cavities; metasternum with episternum narrow, slightly tapering posteriorly. Elytra about twice as long as broad; basal gibbosities vague; apices truncate to emarginate truncate. Legs robust; femora strongly clavate; tarsi slender, first segment longer than two following together, third segment cleft to base. Abdomen normally segmented, female with last segment greatly elongated.

Type species. *Cerambyx fasciatus* Degeer (by present designation).

The erect hairs of the elytra, relatively short antennae, and elongated last abdominal segment of females make this genus readily recognizable.

There has been confusion in the use of the names *Graphisurus* Kirby, 1837 and *Urographis* Horn, 1880. Kirby originally proposed *Graphisurus* as a subgenus of *Acanthocinus,* with *pusillus* the type species by monotypy. LeConte (1852) expanded the definition of *Graphisurus* by including *triangulifer* Haldeman and *fasciatus* Degeer. These two species are not cogeneric with *pusillus.* Horn in 1880 correctly interpreted the situation and proposed the name *Urographis* for *triangulifer* and *fasciatus.* Casey (1913) and a few other authors followed Horn until Dillon (1956) inexplicably reverted to *Graphisurus,* listing *Cerambyx fasciatus* Degeer as the type species by original designation. *C. fasciatus* was not mentioned by Kirby.

Three species are known in our fauna.

KEY TO THE NORTH AMERICAN SPECIES OF *UROGRAPHIS*

1 Antennae with third segment longer than first; pronotum with disk sparsely punctate; elytra lacking a dark fascia around scutellum 2
   Antennae with third segment shorter than first; pronotum with disk impunctate; elytra with a dark fascia surrounding scutellum. Length, 12-17 mm. Ohio to Alabama and Texas .... *triangulifer*
2(1) Elytra with apices appearing truncate from above; scutellum uniformly dark pubescent; last abdominal tergite of females flattened dorsally. Length, 9-14 mm. Eastern United States .... *despectus*
   Elytra with apices emarginate at apex; scutellum pale pubescent medially; last abdominal tergite of females dorsally convex or vaguely keeled. Length, 9-15 mm. Eastern North America to Kansas and Texas .... *fasciatus*

Urographis despectus (LeConte), new combination

Aedilis despectus LeConte, 1850, Coleop. Lake Superior, Agassiz Cabot, p. 234.

Graphisurus pusillus Le Conte, 1852 (not Kirby, 1837), J. Acad. Nat. Sci. Philadelphia, 2: 175.

Graphisurus despectus: Dillon, 1956, Ann. Entomol. Soc. Amer., 49: 165; Gosling and Gosling, 1977, Gr. Lakes Entomol., 10: 25, fig. 142; Rice and Enns, 1981, Trans. Mo. Acad. Sci., 15: 101; Gosling, 1984, Gr. Lakes Entomol., 17: 71.

Urographis hebes Casey, 1913, Memoirs on the Coleoptera, 4: 333; Knull, 1946, Ohio Biol. Surv. Bull., 39: 257, pl. 24, fig. 104; Fattig, 1947, Emory Univ. Mus. Bull., 5: 38.

Graphisurus hebes: Aurivillius, 1923, Coleop., Cat., 74: 423; Craighead, 1923, Can. Dept. Agr. Bull., (n.s.) 27: 121; Leonard, 1928, Cornell Agr. Exp. Sta. Mem., 101: 453.

Male. Form moderate-sized, subparallel; integument dark reddish brown, appendages paler annulate; pubescence dense, appressed, grayish to dark brown. Head with front convex, quadrate, densely micropunctate, densely pubescent; eyes moderate, lower lobes about one-third longer than genae; upper lobes separated by about twice their width; antennae about one-third longer than body, segments broadly pale annulate at bases, third segment longer than first, fourth subequal to first. Pronotum broader than long, sides with blunt tubercles behind middle; base deeply impressed, apex narrowly impressed; disk convex, very sparsely punctate, middle with a narrow, linear callus; pubescence fine, dense, appressed grayish, dark maculae small, irregular; posternum densely grayish pubescent, intercoxal process about one-fifth as broad as coxal cavity; mesosternum with intercoxal process about two-thirds as broad as coxal cavity; metasternum finely, densely pubescent, long hairs obsolete. Elytra about twice as long as broad; basal gibbosities vague, dorsal impressions shallow; punctures at base moderate, well separated, denser at middle and becoming obsolete toward apex; pubescence dense, appressed, grayish, brownish pubescence arranged into small spots, each side behind humeri with a small dark fascia, disk behind middle with irregular, transverse fasciae, each almost attaining suture, sides near apex with a small dark spot; long, erect hairs numerous; apices truncate. Legs robust; femora dark biannulate dorsally; tibiae pale annulate at middle; tarsi with first segment pale pubescent. Abdomen finely, densely pubescent; last sternite deeply emarginate at apex; last tergite deeply emarginate at apex. Length, 9-13 mm.

Female. Form similar. Antennae extending about three segments beyond elytra. Abdomen with fifth segment elongate. Length (exclusive of ovipositor), 11-14 mm.

Type locality. Of *despectus*, Lake Superior; *hebes*, Keokuk, Iowa and New York.

Range. Eastern North America to Minnesota.

Flight period. May to July.

Host plants. Carya.

This species is very similar to *U. fasciatus*. *U. despectus* may be separated by the truncate elytral apices. Additionally, in *despectus*, the elytra are more grayish and lack some of the brownish spots near the basal one-third. In the females the last abdominal tergite is flattened.

Urographis fasciatus (Degeer), new combination

Cerambyx fasciatus Degeer, 1775, Mem. Ins., 5: 114, pl. 14, fig. 7; Goeze, 1777, Entomol. Beytr., 1: 475.

14   University of California Publications in Entomology

Figure 3. Known geographic range of *Urographis fasciatus* (Degeer).

Graphisurus fasciatus: LeConte, 1852, J. Acad. Nat. Sci. Philadelphia, 2: 175; Fitch, 1859, 5th Rept. Nox. Ben. Ins. N.Y., p. 794 (habits); Lacordaire, 1872, Genera des coléoptères, 9(2): 787, fn.; Provancher, 1877, Pet. Fauna Entomol. Can., 1: 630; Packard, 1881, U.S. Entomol. Comm. Bull., 7: 22, 108; Gahan, 1888, Trans. Amer. Entomol. Soc., 15: 299; Leng and Hamilton, 1896, Trans. Amer. Entomol. Soc., 23: 131; Wickham, 1898, Can. Entomol., 30: 39, fig. 3; Harrington, 1899, Ottawa Nat., 13: 67; Chagnon, 1905, Nat. Can., 32: 42; Blatchley, 1910, Coleoptera in Indiana,

p. 1079, fig. 466; Smith, 1910, N.J. St. Mus. Rept., 1909: 334; Felt, 1921, N.Y. St. Mus. Bull., 247-248: 88 (habits); Craighead, 1923, Can. Dept. Agr. Bull., (n.s.) 27: 121, pl. 16, fig. 6, pl. 7, figs. 3, 4, pl. 32, fig. 3; Chagnon, 1933-40, Coleop. Prov. Quebec, p. 274, pl. 18, fig. 10; Savely, 1939, Ecol. Mon., 9: 340; Hoffmann, 1942, USDA Misc. Pub., 466: 11; Loding, 1945, Geol. Surv. Ala. Mon., 5: 123; Dillon, 1956, Ann. Entomol. Soc. Amer., 49: 164; Dillon and Dillon, 1961, Man. Beetles East. North America, p. 639, pl. 64, no. 1; Chagnon and Robert, 1962, Prin. Coleop. Prov. Quebec, p. 274, pl. 18, fig. 10; Gardiner, 1966, Can. J. Zool., 44: 204, fig. 31; Gardiner, 1969, Can. Dept. Fish. For. Inst. Rept., 0-14: 87; Bayer and Shenefelt, 1969, Univ. Wisc. Res. Bull., 275: 28, fig. 36; Gardiner, 1970, Can. Entomol., 102: 113, figs. 1-4; Laliberte et al., 1977, Fabreries, 3: 93; Headstrom, 1977, Beetles of America, p. 377; Gosling and Gosling, 1977, Gr. Lakes Entomol., 10: 25, fig. 141; Rice and Enns, 1981, Trans. Mo. Acad. Sci., 15: 101; Waters and Hyche, 1984, Coleop. Bull., 38: 285; Gosling, 1984, Gr. Lakes Entomol., 17: 71; Gosling, 1986, Gr. Lakes Entomol., 19: 156; Kukor, Cowan, and Martin, 1988, Physiol. Zool., 61: 364.

Graphisurus fasciata: Leonard, 1928, Cornell Agr. Exp. Sta. Mem., 101: 453.
Acanthocinus fasciatus: White, 1855, Cat. Coleop. Ins. British Mus., 8: 370.
Urographis fasciata: Horn, 1880, Trans. Amer. Entomol. Soc., 8: 128; Casey, 1913, Memoirs on the Coleoptera, 4: 333; Blackman and Stage, 1924, N.Y. St. Coll. For. Tech. Pub., 17: 119; Mundinger, 1928, N.Y. St. Coll. For. Tech. Pub., 17: 320; Knull, 1946, Ohio Biol. Surv. Bull., 39: 256; Fattig, 1947, Emory Univ. Mus. Bull., 5: 37; Beal, 1952, Duke Univ. Sch. For. Bull., 14: 111 (habits).
Urographis fasciatus: Knobel, 1895, Beetles New England, p. 34, fig. 115; Beutenmuller, 1896, J. N.Y. Entomol. Soc., 4: 79.
Lamia mixta Fabricius, 1798, Entomol. Syst. Suppl., p. 144; Fabricius, 1801, Syst. Eleuth., 2: 290.
Cerambyx pensylvanicus Gmelin, in Linnaeus, 1790, Syst. Nat., ed. 13, 1(4): 290.
Urographis fasciata reducta Casey, 1913, Memoirs on the Coleoptera, 4: 333.
Graphisurus fasciatus var. *reducta*: Aurivillius, 1923, Coleop. Cat., 74: 423.

Male. Form moderate-sized, subparallel; integument pale to dark reddish brown, appendages pale annulate; pubescence dense, appressed, grayish to dark brown. Head with front convex, quadrate, densely micropunctate, densely pubescent; eyes moderate, lower lobes about one-third longer than genae, upper lobes separated by a little more than their width; antennae about one-third longer than body, segments pale annulate, third segment with a few, short, erect setae along inside margin, third segment longer than first, fourth shorter than first. Pronotum broader than long, sides with moderate, blunt tubercles slightly behind middle; base deeply, narrowly impressed, impression with a row of coarse punctures; apex very narrowly impressed; disk convex, very sparsely punctate, center with a thin linear callus behind middle; pubescence dense, appressed, yellowish brown with scattered dark spots and dark maculae at sides of middle; large punctures bearing long, erect hairs; prosternum densely pale pubescent, intercoxal process about one-fifth as broad as coxal cavity; mesosternum abruptly declivous anteriorly, intercoxal process about one-third as broad as coxal cavity; metasternum densely clothed with pale appressed pubescence, long, erect hairs rather sparse. Elytra a little more than twice as long as broad; basal gibbosities vague, dorsal impressions shallow; punctures at base moderately coarse, well separated, denser at middle and becoming obsolete toward apex; pubescence dense, appressed, grayish to pale brownish, dark fasciae and spots irregular, scattered, basal half with a vague fascia along margins and irregularly extending onto disk, fasciae behind middle transverse or slightly oblique, irregular, apices with dark spots near margins; long erect hairs numerous; apices shallowly emarginate. Legs robust; femora dark biannulate dorsally; tibiae narrowly pale at bases and broadly pale at middle; tarsi with first segment pale. Abdomen finely, densely appressed pubescent, short, suberect hairs rather sparse; last sternite deeply emarginate at apex; last tergite deeply notched at apex. Length, 9-15 mm.

Female. Form similar. Antennae extending about three segments beyond elytra. Legs less robust. Abdomen with last segment elongate. Length (exclusive of ovipositor), 10-15 mm.

Type locality. Of *fasciatus*, Pennsylvania; *mixta*, North America; *pensylvanicus*, Pennsylvania; *reducta*, West Point, N.Y.

Range. Eastern North America to Kansas and Texas (Figure 3).

Flight period. April to September.

Host plants. Acer, Betula, Carpinus, Carya, Castanea, Fagus, Juglans, Liquidambar, Magnolia, Malus, Pyrus, Quercus, Tilia, Toxicodendron, Ulmus.

This rather common species is easily recognized by the irregular maculae of the elytra and the emarginate elytra apices. The black markings of the elytra vary considerably in size and extent, but the vague antemedian macula is usually present.

Urographis triangulifer (Haldeman), new combination

Acanthoderes triangulifer Haldeman, 1847, Trans. Amer. Philos. Soc., (2)10: 45.

Graphisurus triangulifer: LeConte, 1852, J. Acad. Nat. Sci. Philadelphia, 2: 174; Lacordaire, 1872, Genera des coléoptères, 9(2): 787, fn.; Provancher, 1877, Pet. Fauna Entomol. Can. 1: 629; Gahan, 1888, Trans. Amer. Entomol. Soc., 15: 299; Packard, 1890, Ins. Inj. Trees, p. 610, fig. 200, a,b; Leng and Hamilton, 1896, Trans. Amer. Entomol. Soc., 23: 130; Wickham, 1898, Can. Entomol., 30: 39; Blatchley, 1910, Coleoptera in Indiana, p. 1078, fig. 465; Craighead, 1923, Can. Dept. Agr. Bull., (n.s.) 27: 121; Dillon, 1956, Ann. Entomol. Soc. Amer., 49: 163; Rice and Enns, 1981,

Trans. Mo. Acad. Sci., 15: 102; Hovore et al., 1987, Proc. Calif. Acad. Sci., 44: 316, fig. 17.
Urographis triangulifera: Horn, 1880, Trans. Amer. Entomol. Soc., 8: 128; Casey, 1913, Memoirs on the Coleoptera, 4: 334; Knull, 1946, Ohio Biol. Surv. Bull., 39: 257; Fattig, 1947, Emory Univ. Mus. Bull., 5: 38.
Urographis triangularis: Beutenmuller, 1896, J. N.Y. Entomol. Soc., 4: 79 (error).
Graphisurus triangulifera : Leonard, 1928, Cornell Agr. Exp. Sta. Mem., 101: 453.
Graphisurus trianguliferus: Kirk, 1970, S.C. Agr. Exp. Sta. Tech. Bull., 1038: 83.
Urographis texana Casey, 1924, Memoirs on the Coleoptera, 11: 292. New synonymy.
Graphisurus texanus: Chemsak and Linsley, 1982, Checklist of Ceram., p. 93.

Male. Form moderate-sized, robust; integument dark reddish brown to fuscous, appendages pale annulate, pronotum and elytra with dark fasciae; pubescence dense, appressed, grayish to dark brown, elytra with erect hairs. Head with front slightly longer than broad, densely micropunctate, densely pubescent; eyes moderate, lower lobes longer than broad, about one-third longer than genae, upper lobes separated by about their width; antennae extending two-three segments beyond body, segments from third broadly pale annulate at apices, third segment with a few very short, subdepressed setae beneath, third segment longer than first, fourth shorter than first. Pronotum broader than long, sides with moderate, blunt tubercles slightly behind middle; base deeply, narrowly impressed, impression with a row of coarse punctures; apex narrowly, shallowly impressed; disk convex, middle with a vague longitudinal callus; punctures almost absent on disk; pubescence dense, appressed, each side of middle with two rounded black maculae, sides grayish pubescent except for a black spot near apex and at least one small dark spot slightly behind middle; prosternum finely, densely pubescent, intercoxal process about one-fourth as wide as coxal cavity; mesosternum with intercoxal process abruptly declivous anteriorly, about half as broad as coxal cavity; metasternum densely pubescent, very sparsely punctate. Scutellum medially pale pubescent. Elytra less than twice as long as broad; basal gibbosities vague, dorsal impression shallow; punctures near base moderately coarse, moderately dense, coarser and denser at sides, punctures becoming obsolete behind middle; pubescence dense, appressed, base with a triangular-shaped black macula surrounding scutellum, sides with small maculae behind humeri, larger, irregular maculae behind middle rarely extending to suture, apex with small maculae at sides, remaining surface usually grayish pubescent often with small, irregular dark spots at middle, long erect hairs numerous; apices shallowly emarginate. Legs robust; femora usually biannulate; tibiae medially pale annulate; tarsi with first segment pale except narrowly at apices. Abdomen finely densely pubescent, first four sternites narrowly glabrous along posterior margins; last sternite about as long as two proceeding segments together, apex emarginate; last tergite deeply emarginate. Length, 12-17 mm.

Female. Form more robust. Antennae slightly shorter. Legs with femora less robust. Abdomen with last segment elongate. Length (exclusive of ovipositor), 14-17 mm.

Type locality. Of *triangulifer*, Alabama; *texana*, Comal Co., Texas.
Range. East-central United States to Kansas, Texas, and Alabama.
Flight period. May to September.
Host plants. Celtis, Acer.

The robust form and pattern of elytral maculae readily separate this species from other North American *Urographis*.

This species varies in the size and extent of the elytral maculae. The pronotum often has three small, pale spots on the disk.

Genus *Trichocanonura* Dillon

Trichocanonura Dillon, 1956, Ann. Entomol. Soc., Amer., 49: 229; Arnett, 1962, Beetles U.S., 103: 873.

Form moderate-sized, elongate. Head with front quadrate, convex, about as broad as long; vertex deeply impressed between eyes; mandibles arcuate, pointed at apices; genae shorter than lower eye lobes; eyes large, moderately coarsely faceted, deeply emarginate, upper lobes narrow, separated by about twice width of lobes; antennal tubercles prominent, broadly divergent; antennae slender, almost twice as long as body in males, shorter in females, segments lightly fringed beneath with erect setae, scape shorter than third segment, fourth equal to third, eleventh not elongate, segments biannulate with pale pubescence. Pronotum a little broader than long, sides broadly tuberculate behind middle, the short, broad spine directed backward; apex narrowly impressed, base broadly, deeply impressed, impression extending onto sides; disk with two calluses behind apex and a vague, longitudinal median callus; punctures moderately coarse, well separated; prosternum arcuate, intercoxal process narrow, less than one-fourth as broad as coxae; mesosternum with intercoxal process plane, about as wide as prosternal process; metasternum with episternum narrow, subparallel, tapering posteriorly. Elytra more than twice as long as broad; disk with tufted tubercles, costae distinct; long, erect hairs numerous; basal gibbosities very shallow; apices obliquely truncate. Legs robust; femora strongly clavate; front tibiae of males excavated beneath toward apex; middle tibiae with an external sinus; front tarsi of males broad, fringed laterally, hind tarsi longer than two following segments together. Abdomen normally segmented in males, last sternite elongate in females, ovipositor extruded.

Type species. Acanthocinus linearis Skinner (monobasic and by original designation).

This genus is distinctive by the erect setae of the elytra, narrow pro- and mesocoxal processes, relatively short, lightly fringed antennae, posteriorly directed apices of the lateral pronotal tubercles, and by the broad, fringed front tarsi of the males.

A single species is known.

Trichocanura linearis (Skinner)

Acanthocinus linearis Skinner, 1905, Entomol. News, 16: 290; Casey, 1913, Memoirs on the Coleoptera, 4: 341.
Acanthocinus (Acanthocinus) linearis: Aurivillius, 1923, Coleop. Cat., 74: 434.
Trichocanonura linearis: Dillon, 1956, Ann. Entomol. Soc. Amer., 49: 229.
Acanthocinus (Trichocanonura) linearis: Linsley, Knull, and Statham, 1961, Amer. Mus. Nov., 2050: 30.

Male. Form moderate-sized, tapering; integument reddish brown and piceous; pubescence dense, appressed, grayish, brownish, and black. Head with front micropunctate, pubescence mottled, dense, margins with long, erect setae; vertex micropunctate, mottled pubescent with a dark vitta behind eyes connecting those on pronotum; antennae slender, segments lightly fringed beneath, biannulate with pale pubescence, scape with small glabrous spots. Pronotum broader than long, apices of lateral tubercles directed back; punctures uniformly interspersed between calluses; pubescence dense, grayish and pale brownish at sides, middle broadly dark with three dark, longitudinal vittae, middle one often vague, sides beneath tubercles broadly dark vittate; prosternum densely pale pubescent; mesosternum dark at sides; metasternum densely pale pubescent with glabrous spots interspersed, sides dark. Elytra about 2-1/2 times as long as broad; basal gibbosities with a large dark tufted tubercle and several small ones; punctures coarse, well separated, becoming obsolete toward apex; each elytron tricostate, two discal costae uniting before apex; pubescence mottled grayish and brownish, usually with an angulate broad dark macula behind middle, sutural costae with dark tufts at middle, large tufts obliquely aligned on each costa behind middle, discal costae with a row of dark tufts forming a narrow V behind large tufts; erect black hairs fairly numerous; apices obliquely truncate. Legs with femora brown annulate at apices, gray pubescence mottled; tibiae dorsally triannulate; tarsi with first two segments pale at bases. Abdomen pale pubescent with numerous dark spots interspersed, sternites dark at sides; last sternite deeply emarginate at apex. Length, 11-16 mm.

Female. Form similar. Antennae a little longer than body. Abdomen with last sternite greatly elongated, apex narrow, deeply cleft, ovipositor extruded. Length (exclusive of ovipositor), 13-20 mm.

Type locality. Carr Canyon, Huachuca Mts., Arizona.
Range. Mountains of southern Arizona to southern Sierra Madre, Mexico.
Flight period. June to September.
Host plants. Quercus.

T. linearis may be recognized by the narrowly dark V-shaped elytral markings, coloration of the pronotum, shape of the lateral pronotal tubercles, and the very elongate last abdominal segment of the females.

Most variation is expressed in the paler or darker pubescence.

Genus *Eutrichillus* Bates

Eutrichillus Bates, 1885, Biol. Centr.-Amer., Coleoptera, 5: 397; Dillon, 1956, Ann. Entomol. Soc. Amer., 49: 221; Arnett, 1962, Beetles U.S., 103: 873.

Lepturgoides Schaeffer, 1905, Brooklyn Mus. Inst. Arts Sci. Bull., 1: 166; Bradley, 1930, Man. Genera Beetles, p. 246. (Type species: *Lepturgoides pini* Schaeffer, monobasic.)

Ceratographis Gahan, 1888, Trans. Amer. Entomol. Soc., 15: 300; Leng and Hamilton, 1896, Trans. Amer. Entomol. Soc., 23: 131; Wickham, 1897, Can. Entomol., 29: 203; Wickham, 1898, Can. Entomol., 30: 39; Blatchley, 1910, Coleoptera in Indiana, p. 1079; Bradley, 1930, Man. Genera Beetles, p. 246; Knull, 1946, Ohio Biol. Surv. Bull., 39: 255; Dillon, 1956, Ann. Entomol. Soc., Amer., 49: 224; Arnett, 1962, Beetles U.S., 103: 873; Bayer and Shenefelt, 1969, Univ. Wisc. Res. Bull. 275: 27. (Type species: *Liopus biguttatus* LeConte, by original designation.) New synonymy.

Leiopus: Lacordaire, 1872, Genera des coléoptères, 9(2): 775 (part).

Graphisurus Horn, 1880 (not Kirby 1837), Trans. Amer. Entomol. Soc., 8: 128.

Form moderate-sized, subdepressed; pubescence dense, appressed, elytra with long, erect setae. Head with front convex, quadrate to transverse; mandibles feebly arcuate; genae slightly convergent, shorter or longer than lower eye lobes; eyes moderate-sized, deeply emarginate, upper lobes small, separated by varying distances; antennal tubercles prominent, widely divergent at base; antennae eleven-segmented, slender, longer than body in both sexes, basal segments with some long, suberect setae beneath, scape longer or shorter than third segment, fourth subequal to or shorter than third. Pronotum broader than long, sides acutely tuberculate behind middle; disk with three, often vague, calluses, one median and two behind apical margin; base broadly, deeply impressed behind lateral spines; apex vaguely, narrowly impressed; prosternum narrow, intercoxal process slender, not laminiform; mesosternum with intercoxal process narrow, slightly broader than prosternal process; metasternum with episternum narrow, subparallel. Elytra about twice or slightly more as long as broad; epipleura partially vertical; disk with small basal crests and usually small tufted tubercles, costae usually distinct; pubescence short, appressed, erect setae of varying lengths numerous to rather sparse; apices variable. Legs short; femora clavate; tarsi long, slender, third segment cleft to base. Abdomen normally segmented, last segment of females moderately elongated.

Type species. Eleothinus comus Bates (monobasic).

This genus may be recognized by the erect setae of the elytra and the narrow pro- and mesosternal processes.

Four species are known from the United States. *E. comus* (Bates) from Mexico and Guatemala may be separated from these by the coloration, presence of more abundant elytral setae, and the reduced tufted tubercles of the elytra.

KEY TO THE NORTH AMERICAN SPECIES OF *EUTRICHILLUS*

1 Erect setae present only on elytra; scape of antennae with a few long setae beneath............................2
Erect setae present on elytra and legs; scape of antennae with numerous erect setae on dorsal surface. Length, 7-11 mm. Eastern United States.. *bigguttatus*

2(1) Genae longer than lower eye lobes; antennae with basal segments pale biannulate......................3
Genae shorter than lower eye lobes; antennae with basal segments singly annulate. Length, 8-12 mm. Arizona to Texas and Mexico................ *neomexicanus*

3(2) Eyes with upper lobes about one-fourth as wide as upper interocular space; elytra with distinct, angulate, lateral vittae behind middle. Length, 7-12 mm. California to Texas.................... *canescens*
Eyes with upper lobes about half as wide as interocular space; elytra with pubescence mottled, lacking distinct black vittae behind middle. Length, 6-10 mm. Colorado, Arizona, New Mexico, and Mexico.. *pini*

Eutrichillus biguttatus (LeConte), new combination

Liopus biguttatus LeConte, 1852, J. Acad. Nat. Sci. Philadelphia, 2:172.
Leiopus biguttatus: White, 1855, Cat. Coleop. Ins. British Museum, 8:388; Lacordaire, 1872, Genera des coléoptères, 9(2):776, fn.
Ceratographis biguttata: Gahan, 1888, Trans. Amer. Entomol. Soc., 15:300; Leng and Hamilton, 1896, Trans. Amer. Entomol. Soc., 23:131; Wickham, 1898, Can. Entomol., 30:39; Nicolay, 1919, Bull. Brooklyn Entomol. Soc., 14:71; Knull, 1946, Ohio Biol. Surv. Bull., 39:255, pl. 24, fig. 97; Fattig, 1947, Emory Univ. Mus. Bull., 5:37; Dillon, 1956, Ann. Amer. Entomol. Soc., 49:225, fig. 1.
Ceratographis biguttatus: Blatchley, 1910, Coleoptera in Indiana, p. 1079; Leonard, 1928, Cornell Agr. Exp. Sta. Mem., 101:453; Bayer and Shenefelt, 1969, Univ. Wisc. Res. Bull., 275:28, fig. 35.

Graphisurus pusilla Horn, 1880 (not Kirby 1837), Trans. Amer. Entomol. Soc., 8:129 (misidentified).
Leiopus crinicornis Casey, 1924, Memoirs on the Coleoptera, 11:291; Leng and Mutchler, 1927, Supp. Cat. Coleop. Amer., p. 43; Lewis, 1986, Pan-Pac. Entomol., 62:171 (synonymy).

Male. Form small to moderate-sized; integument pale to dark reddish brown; pubescence dense, appressed, brownish and gray. Head with front quadrate, finely, densely punctate, moderately densely pubescent, long, erect hairs numerous; genae shorter than lower eye lobes; vertex finely punctate, moderately pubescent; antennae about twice as long as body, segments from third narrowly white annulate at bases, scape with numerous curved, flying hairs on dorsal surface, basal segments with numerous setae beneath, third segment longer than scape, fourth subequal to third. Pronotum broader than long; disk with dorsal calluses vague or absent; pubescence grayish, dense, appressed; punctures fine, dense; prosternum finely pubescent; meso- and metasternum very finely, densely punctate, finely, densely pubescent. Elytra a little more than twice as long as broad; basal crests shallow, usually with narrow, longitudinal, low tufts; disk bicostate, costae uniting before apex, costae irregularly white and dark pubescent, often with a dark tuft at apical third; pubescence dense, fine, mottled brownish and gray, erect setae numerous, often curved; punctures moderately coarse, dense; apices rounded. Legs finely pubescent, erect, curved setae often numerous; femora usually with small denuded spots; tibiae narrowly white annulate at bases, broadly at middle; tarsi brown, vaguely pale annulate at bases of first and second segments. Abdomen very finely punctate, finely pubescent; last sternite emarginate at apex. Length, 7-10 mm.

Female. Form similar. Antennae about 1-1/2 times longer than body. Abdomen with last segment elongate; last sternite narrow at apex, notched medially. Length (exclusive of ovipositor), 6-11 mm.

Type locality. Of *biguttatus*, New York; *crinicornis*, Virginia.
Range. Eastern United States to Mississippi and Wisconsin.
Flight period. May to August.
Host plants. Pinus.

In addition to the coloration, the long setae of the head, antennal scape, and legs readily separate this species from other *Eutrichillus*.

Structurally, this species agrees well with other *Eutrichillus*. It differs primarily in the presence of erect setae on the head, antennal scape, and legs.

Eutrichillus neomexicanus (Champlain and Knull)
(Figure 4)

Lepturges neomexicanus Champlain and Knull, 1925, Ann. Entomol. Soc. Amer., 18:470.

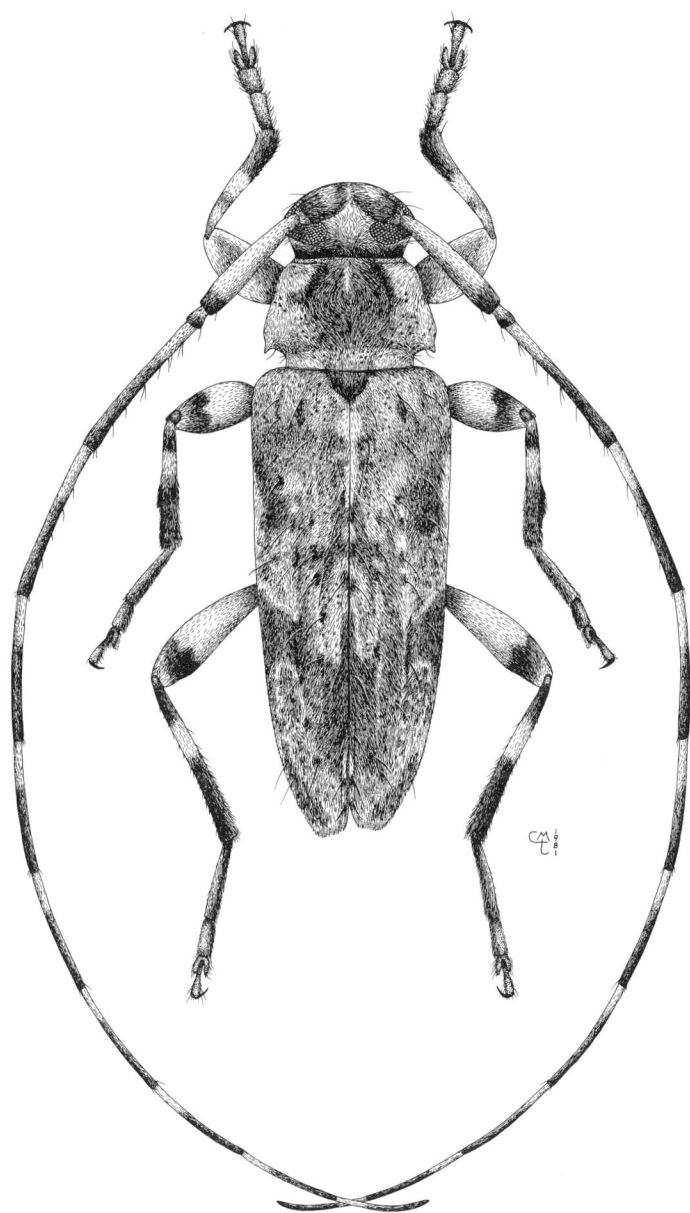

Figure 4. *Eutrichillus neomexicanus* (Champlain and Knull), male.

Eutrichillus neomexicanus: Dillon, 1956, Ann. Entomol. Soc. Amer., 49:224; Linsley, Knull, and Statham, 1961, Amer. Mus. Nov., 2050:29; Lewis, 1979, Pan-Pac. Entomol., 55:25; Cope, 1984, Coleop. Bull., 38:29.

Male. Form moderate-sized, subparallel; integument rufopiceous, parts of legs and antennae black; pubescence dense, appressed, mottled grayish and black, long, erect setae of elytra rather sparse. Head with front transverse, densely micropunctate, densely mottled gray and black pubescent, sides with several long, erect setae; genae shorter than lower eye lobes; vertex minutely punctate, dark pubescent; antennae about 1-1/2 times longer than body, segments pale annulate at bases, basal segments with several suberect setae beneath, scape shorter than third segment, fourth subequal to third. Pronotum broader than long; disk convex, vaguely tri-callused; pubescence dense, appressed, mottled dark and gray, apical calluses with an oblique, dark, short vitta; punctures moderately coarse, irregular; prosternum densely pubescent; meso- and metasternum densely minutely punctate, moderately densely clothed with fine, appressed, gray pubescense. Elytra slightly more than twice as long as broad; disk bicostate, costae uniting before apex; basal crests tufted, moderate-sized, small, dark tufts linearly arranged down suture and costae; pubescence grayish, mottled by dark tufts, sides behind humeri dark, each side behind middle with a lateral vitta extending partially onto disk, middle before apex dark; long erect setae rising from each tuft; punctures rather fine, dense, becoming sparser toward apex; apices obliquely truncate. Legs densely clothed with grayish appressed pubescence, femora apically dark annulate, tibiae pale annulate toward bases, tarsi dark; first segment of hind tarsi longer than following segments together. Abdomen very finely, densely punctate, finely pubescent; last sternite emarginate at apex. Length, 8-10 mm.

Female. Form similar. Abdomen with last segment moderately elongate; last sternite notched at apex. Length, 9-12 mm.

Type locality. Jemez Springs, New Mexico.
Range. Western Arizona to western Texas to Durango, Mexico.
Flight period. June to September.
Host plants. Pinus.

The short genae, color pattern of the elytra, and basally pale annulate segments of the antennae separate this species from other known *Eutrichillus*. Variation is expressed in the amount of black pubescence on the elytra and varying degrees of integumental infuscation.

Adults are usually taken at light.

Eutrichillus canescens Dillon

Eutrichillus canescens canescens Dillon, 1956, Ann. Entomol. Soc. Amer., 49:223.
Eutrichillus canescens: Lewis, 1979, Pan-Pac. Entomol., 55:25.

Eutrichillus canescens nelsoni Dillon, 1956, Ann. Entomol. Soc. Amer., 49:223; Linsley, Knull, and Statham, 1961, Amer. Mus. Nov., 2050:29. New synonymy.

Male. Form moderate-sized; integument dark reddish brown; pubescence dense, appressed, grayish, brown, and black. Head with front transverse, very finely, densely punctate, densely mottled grayish and brown pubescent, sides with a few long, erect setae; genae subequal in length to lower eye lobes; vertex minutely punctate, usually with two dark spots behind eyes; antennae about 1-1/2 times longer than body, segments to six or seven pale biannulate, remaining segments broadly pale basally, basal segments with several suberect setae beneath, scape shorter than third segment, fourth subequal to third. Pronotum broader than long; disk shallowly tri-callused; pubescence mottled gray and brownish, usually with a dark longitudinal vitta on each side of middle; punctures moderately coarse, scattered around median callus and along basal impression; prosternum finely, densely pubescent; meso- and metasternum finely, densely punctate, finely, densely pubescent. Elytra about twice as long as broad; disk bicostate, two additional costae present toward lateral margins beginning at about middle, costae uniting before apex; tufts of basal crests small, small black tufts linearly arranged down costae, tufts at apical one-third larger; pubescence mottled, mostly grayish, black vittae extending down sides behind humeri, expanding onto disk behind middle but not reaching suture, vittae angulate; setae uniformly long, arising out of tufts; punctures moderately coarse, dense, becoming sparse toward apex; apices obliquely truncate. Legs finely, densely pubescent, tibiae pale annulate medially; tarsi dark, first segment narrowly pale annulate at base. Abdomen minutely, densely punctate, finely, densely pubescent; last sternite emarginate at apex. Length, 7-11 mm.

Female. Form similar. Antennae a little shorter. Abdomen with last segment slightly elongate; last sternite narrowly emarginate at apex. Length, 7-12 mm.

Type locality. Of *canescens*, Mescalera Reservation, New Mexico; *nelsoni*, Prescott, Arizona.

Range. Southeastern California and southern Nevada to western Texas.

Flight period. June to September.

Host plants. Pinus monophylla.

This species may be recognized by the broader body form, length of the genae, and by the angled dark vittae and uniformly long setae of the elytra.

Eutrichillus pini (Schaeffer)

Leptosylus pini Schaeffer, 1905, Brooklyn Mus. Inst. Arts Sci. Bull., 1:165; Craighead, 1923, Can. Dept. Agr. Bull., (n.s.) 27:116.

Eutrichillus pini: Dillon, 1956, Ann. Entomol. Soc. Amer., 49:222; Linsley, Knull, and Statham, 1961, Amer. Mus. Nov., 2050:29.

Mle. Form small to moderate-sized; integument reddish brown to black; pubescence dense, appressed, gray, brownish, and black. Head with front quadrate, minutely, densely punctate, densely clothed with mottled gray and brownish pubescence, sides with a few long, erect setae; genae longer than lower eye lobes; vertex minutely punctate, usually with two dark spots behind eyes; antennae about 1-1/2 times longer than body, segments to six or seven pale biannulate, remaining segments pale basally, basal segments with several suberect setae beneath, third segment longer than scape, fourth subequal to third. Pronotum broader than long; disk convex, shallowly tricallused; pubescence mottled gray and black, often with two oblique, longitudinal, dark vittae; punctures moderately coarse, scattered around calluses; prosternum thinly pubescent; meso- and metasternum finely, densely punctate, rather thinly pubescent. Elytra slightly more than twice as long as broad; disk vaguely bicostate, costae uniting before apex; basal crests with moderate-sized black tufts, smaller tufts present down costae, inner costae with larger tufts behind middle, outer costae with larger tufts at about apical one-third; pubescence dense, appressed, usually with a gray patch behind basal crests and thinly gray down costae between tufts, middle usually with a vague broad dark transverse vitta and a narrow dark vitta before apex; moderately long, erect setae rising from each tuft, punctures moderately coarse, dense, becoming obsolete near apex; apices obliquely truncate, usually highly emarginate. Legs densely gray pubescent, femora narrowly dark annulate near apices, tibiae dark, broadly pale annulate medially, narrowly pale annulate at apices; tarsi dark, segments narrowly pale at bases. Abdomen finely, densely punctate, rather thinly, densely pubescent; last sternite broadly emarginate at apex. Length, 7-10 mm.

Female. Form similar. Abdomen with last segment moderately elongate; last sternite narrowly emarginate at apex. Length, 6-10 mm.

Type locality. Carr's Peak, Huachuca Mts., Arizona.

Range. Colorado, Arizona, and New Mexico to Durango, Mexico.

Flight period. June to August.

Host plants. Pinus (including *P. ponderosa*, and *P. edulis*).

This species is characterized by the long genae, color pattern of the elytra, and the erect setae of the elytra decreasing in length toward the apex.

Genus *Dectes* LeConte

Dectes LeConte, 1852, J. Acad. Nat. Sci. Philadelphia, 2: 144; Thomson, 1860, Class. ceramb., p. 128; Thomson, 1864, Syst. ceramb., p. 26; Lacordaire, 1872, Genera des coléoptères, 9(2): 724 (part); LeConte, 1873, Smithson. Misc. Coll., 11(265): 339; Horn, 1880, Trans. Amer. Entomol. Soc., 8: 119, 126; Bates, 1881, Biol. Centr.-Amer., Coleoptera, 5: 173; LeConte and Horn, 1883, Smithson. Misc. Coll., 507: 324; Bates, 1885, Biol. Centr.-Amer., Coleop., 5:408; Leng and Hamilton, 1896, Trans. Amer. Entomol. Soc., 23: 126; Blatchley, 1910, Coleoptera in Indiana, p. 1075; Casey, 1913,

Memoirs on the Coleoptera, 4: 341; Bradley, 1930, Man. Genera Beetles, p. 246; Knull, 1946, Ohio Biol. Surv. Bull., 39: 251; Dillon, 1956, Ann. Entomol. Soc. Amer., 49:352; Dillon and Dillon, 1961, Man. Beetles East. North America, p. 645; Arnett, 1962, Beetles U.S., 103: 873; Bayer and Shenefelt, 1968, Univ. Wisc. Res. Bull. 275: 27; Rice and Enns, 1981, Trans. Mo. Acad. Sci., 15: 99.

Dectes (Dectes): Aurivillius, 1923, Coleop. Cat., 74: 410.

Form small to moderate-sized, subcylindrical. Head with front slightly convergent, convex; median line extending length of front onto vertex; mandibles feebly arcuate; genae longer than or subequal to lower eye lobes, sides convergent; eyes moderate, finely faceted, upper lobes small, widely separated; antennal tubercles prominent, divergent; antennae slender, eleven-segmented, longer than body in both sexes, scape with several long, suberect hairs beneath, third segment subequal to or longer than scape, fourth shorter than third, remaining segments gradually decreasing in length. Pronotum broader than long, subcylindrical, sides acutely spined before basal impression; apex not transversely impressed, base shallowly, broadly impressed; disk convex, vaguely callused medially behind middle; prosternum narrow, intercoxal process laminiform, coxal cavities closed behind; mesosternum with intercoxal process narrow; metasternum with episternum narrow, subparallel. Elytra a little more than twice as long as broad, sides subparallel, tapering near apical fourth; pubescence dense, appressed longer, suberect setae numerous; apices variable. Legs short; femora gradually clavate; tibiae slightly arcuate, apical spurs short; tarsi rather robust, first segment of hind pair longer than two following segments together. Abdomen normally segmented.

Type species: Lamia spinosa Say (= *Dectes sayi* Dillon and Dillon) (monobasic).

This genus is distinctive by the cylindrical body form, very narrow prosternal process, cylindrical, laterally spined pronotum, and smooth elytra. The dense, appressed uniform pubescence with short, suberect setae of the elytra is also characteristic.

Two species are known from the United States.

KEY TO THE NORTH AMERICAN SPECIES OF *DECTES*

Antennae with second segment distinctly longer than broad, third segment longer than or equal to scape; genae longer than lower eye lobes. Length, 6-11 mm. Northeastern United States to Kentucky, west to Kansas and North Dakota .... *sayi*
Antennae with second segment as long as broad, third segment shorter than scape; genae subequal in length to lower eye lobes. Length, 5-11 mm.

Eastern United States to Montana and Arizona,
Mexico...*texanus*

Dectes sayi Dillon and Dillon

Dectes sayi Dillon and Dillon, 1953, Entomol. News, 64: 260 (new name for *spinosus* Say); Dillon, 1956, Ann. Entomol. Soc. Amer., 49: 352; Dillon and Dillon, 1961, Man. Beetles East. North America, p. 645, pl. 64, no. 10; Bayer and Shenefelt, 1969; Univ. Wisc. Res. Bull. 275: 28, fig. 35; Harris and Piper, 1970, Commonwealth Inst. Biol. Contr. Tech. Bull. 13: 128; Gosling and Gosling, 1976, Gr. Lakes Entomol., 10: 27, fig. 149; Headstrom, 1977, Beetles of America, p. 379; Piper, 1978, Coleop. Bull., 32: 299 (habits); Rice and Enns, 1981, Trans. Mo. Acad. Sci., 15: 99.

Lamia spinosa Say, 1826 (not Drury, 1773), J. Acad. Nat. Sci. Philadelphia, (1)5:271.

Leiopus? spinosus: Haldeman, 1847, Trans. Amer. Philos. Soc., (2)10: 50.

Dectes spinosa: LeConte, 1852, J. Acad. Nat. Sci. Philadelphia, 2: 144.

Dectes spinosus: Horn, 1880, Trans. Amer. Entomol. Soc., 8: 126; Bates, 1881, Biol. Centr.-Amer., Coleoptera, 5: 173; Leng and Hamilton, 1896, Trans. Amer. Entomol. Soc., 23: 126; Beutenmuller, 1896, J. N.Y. Entomol. Soc., 4: 79; Townsend, 1902, Trans. Texas Acad. Sci., 5: 78; Blatchley, 1910, Coleoptera in Indiana, p. 1075; Smith, 1910, N.J. St. Mus. Rept., 1909: 334; Casey, 1913, Memoirs on the Coleoptera, 4: 342; Garnett, 1918, Can. Entomol., 50: 282; Nicolay, 1919, Bull. Brooklyn Entomol. Soc., 14: 70; Craighead, 1923, Can. Dept. Agr. Bull., (n.s.) 27: 114, pl. 5, fig. 7, pl. 12, fig. 4, pl. 16, fig. 3 (larva); Balduf, 1923, Ohio Agr. Exp. Sta. Bull., 366: 171; Leonard, 1928, Cornell Agr. Exp. Sta. Mem., 101: 454; Kelly, 1931, Aust. J. Council Sci. Indust. Res., 4: 163, 171; Linsley and Martin, 1933, Entomol. News, 44: 182; Linsley, 1935, Pan-Pac. Entomol., 11: 74; Linsley, 1942, Proc. Calif. Acad. Sci., (4)24: 74; Knull, 1946, Ohio Biol. Surv. Bull., 39: 251, pl. 20, fig. 79; Fattig, 1947, Emory Univ. Mus. Bull., 5: 38; Papp, 1959, Bull. So. Calif. Acad. Sci., 58: 92; Anon., 1968, Coop. Econ. Ins. Rept., 18: 1113; Kirk, 1970, S.C. Agr. Exp. Sta. Bull., 1038: 83; Laliberte et al., 1977, Fabreries, 3: 92; Goeden, 1978, USDA Agr. Handbook 480: 384.

Dectes texanus Bates, 1881 (not LeConte, 1862), Biol. Centr.-Amer. Coleoptera, 5:173.

Dectes (Dectes) spinosus: Aurivillius, 1923, Coleop. Cat., 74: 411.

Dectes spirrosus: Beckham and Tippins, 1972, J. Econ. Entomol., 65: 865 (error); Hilgendorf and Goeden, 1981, Bull. Entomol. Soc. Amer., 27: 103 (error).

Male. Form small to moderate-sized, integument black, appendages and elytra often reddish black, pubescence dense, appressed, grayish, suberect setae of elytra dark. Head with front densely, minutely punctate, densely pubescent; genae longer than lower eye lobes; antennae extending about three segments beyond elytra, basal segments gray pubescent beneath, segments from fourth dark annulate at apices, third segment longer than scape, second segment longer than broad, fourth shorter than third. Pronotum a little broader than long, lateral spines small; disk convex, finely, densely punctate; pubescence dense, appressed, obscuring surface, sides behind spines with two or three long, erect hairs; prosternum densely pubescent; meso- and metasternum densely, minutely punctate, densely pubescent. Elytra a little more than twice as long as broad; punctures dense, larger than those on pronotum; pubescence dense, obscuring surface, suberect setae numerous; apices shallowly emarginate truncate. Legs densely gray pubescent; hind tibiae with short, black, suberect bristles over apical half. Abdomen densely pubescent; last sternite longer than preceding segments, shallowly emarginate at apex. Length, 6-10.5 mm.

Female. Form more robust. Antennae slightly shorter. Abdomen with last sternite truncate at apex with numerous long setae. Length, 7-11 mm.

Type locality. United States (not restricted by Say).

Range. Northeastern United States to Kentucky westward to Kansas and North Dakota.

Flight period. May to September.

Host plants. Ambrosia, Eupatorum, Glycine, Helianthus, Meibomia, Xanthium.

The elongate second antennal segment and longer genae readily separate this species from *texanus*.

Larvae bore in the stems of living plants.

Dectes texanus LeConte
(Figure 5)

Dectes texanus LeConte, 1862, Proc. Acad. Nat. Sci. Philadelphia, 1862: 39; Dillon and Dillon, 1961, Man. Beetles East. North America, p. 646, pl. 64, no. 11; Linsley, Knull, and Statham, 1961, Amer. Mus. Nov., 2050: 30; Hatchett et al., 1973, Ann. Entomol. Soc. Amer., 66: 519 (habits); Phillips et al., 1973, Texas Agr. Exp. Sta. MP-1116: 3, fig. 2; Hatchett et al., 1975, Ann. Entomol. Soc. Amer., 68: 209 (habits); Gosling and Gosling, 1976, Gr. Lakes Entomol., 10:27, fig. 150; Headstrom, 1977, Beetles of America, p. 380; Rogers, 1977, Envir. Entomol., 6: 833; Carter, 1978, Sunflower Sci. Tech. Agron., 19, p. 210; Hilgendorf and Goeden, 1981, Bull. Entomol. Soc. Amer., 27: 103; Genung and Green, 1983, Fla. Entomol., 66: 207.

Dectes texanus texanus: Dillon, 1956, Ann. Entomol. Soc. Amer., 49: 353; Gordon and Breuer, 1972, Coop. Econ. Ins. Rept., 22: 762; Patrick and White, 1972, J. Ga. Entomol. Soc., 7: 264; Patrick, 1973, J. Ga. Entomol. Soc., 8: 277 (habits); Stein and Tagestad, 1976, USDA For. Serv. Res. Pap., RM-171: 11; Rice and Enns, 1981, Trans. Mo. Acad. Sci., 15: 99.

Dectes (Dectes) spinosus var.? *texanus*: Aurivillius, 1923, Coleop. Cat., 74: 411.

Dectes spinosus var. *texanus*: Alexander, 1958, Proc. Okla. Acad. Sci., 38: 46.
Dectes alticola Casey, 1913, Memoirs on the Coleoptera, 4: 342. New synonymy.
Dectes (Dectes) spinosus var. *alticola*: Aurivillius, 1923, Coleop. Cat., 74: 411.
Dectes texanus alticola: Dillon, 1956, Ann. Entomol. Soc. Amer., 49: 354.
Dectes spinosus var. *alticola*: Alexander, 1958, Proc. Okla. Acad. Sci., 38: 46.
Dectes latitarsus Casey, 1913, Memoirs on the Coleoptera, 4: 342; Vogt, 1949, Pan-Pac. Entomol., 25: 182.
Dectes thoracicus Casey, 1913, Memoirs on the Coleoptera, 4: 342. New synonymy.

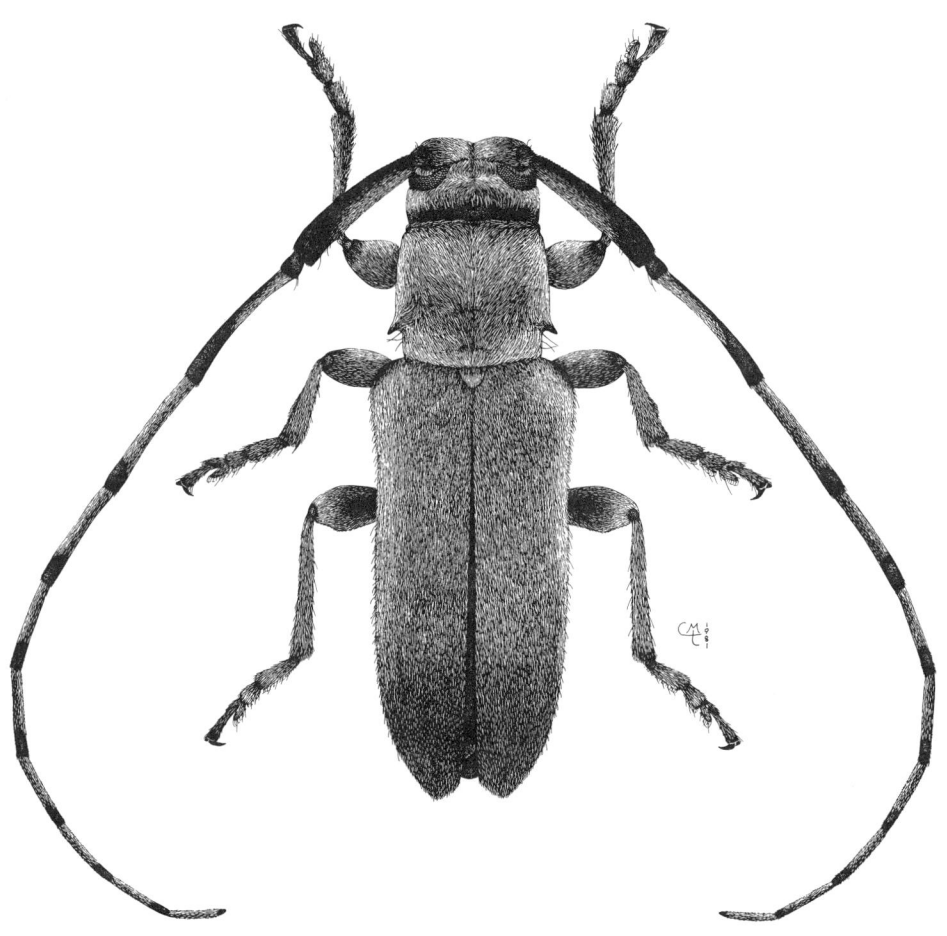

Figure 5. *Dectes texanus* LeConte, male.

Dectes texanus thoracicus: Dillon, 1956, Ann. Entomol. Soc. Amer., 49: 355.
Dectes brevis Casey, 1913, Memoirs on the Coleoptera, 4: 342.
Dectes brevisetosus Casey, 1913, Memoirs on the Coleoptera, 4: 343.
Dectes discolor Casey, 1913, Memoirs on the Coleoptera, 4: 343.
Dectes aridus Casey, 1913, Memoirs on the Coleoptera, 4: 343. New synonymy.
Dectes texanus aridus: Dillon, 1956, Ann. Entomol. Soc. Amer., 49: 355; Hovore et al., 1987, Proc. Calif. Acad. Sci., 44: 319.
Dectes texanus murinus Dillon, 1956, Ann. Entomol. Soc. Amer., 49: 355; Hovore, 1988, Wasmann J. Biol., 46: 24. New synonymy.

Male. Form small to moderate-sized; integument black to reddish; pubescence dense, appressed, grayish. Head with front densely, minutely punctate, densely pubescent; genae shorter than lower eye lobes; antennae extending about four segments beyond elytra, scape gray pubescent beneath, black above, segments usually dark annulate at apices, third segment subequal to or shorter than scape, second segment short, about as long as broad, fourth shorter than third. Pronotum broader than long, sides acutely tuberculate before basal impression; disk convex, finely, densely punctate, densely pubescent, two or three long, erect hairs present at sides behind tubercles; prosternum densely pubescent; meso- and metasternum finely, densely punctate, densely pubescent. Elytra a little more than twice as long as broad; punctures dense, larger than those on pronotum; pubescence dense, appressed, suberect setae numerous; humeri with black spots beneath; apices rounded to emarginate truncate. Legs densely gray pubescent, apices of femora and tibiae usually black. Abdomen finely, densely punctate, densely pubescent; last sternite emarginate at apex. Length, 5-10 mm.

Female. Form more robust. Antennae extending about three segments beyond elytra. Abdomen with last sternite narrowly truncate at apex. Length, 6-11 mm.

Type locality. Of *texanus*, Texas; *brevis*, Willets Point, Long Island; *latitarsus*, Brownsville, Texas; *alticola*, Colorado, Montana, New Mexico; *murinus*, Los Animas, Sierra Laguna, Baja California; *thoracicus*, Arizona; *brevisetosus*, Arizona; *discolor*, Arizona; *aridus*, Tepehuanes and Durango City, Mexico.

Range. Eastern United States to Montana and Arizona to central Mexico and Cape region of Baja California.

Flight period. June to September.

Host plants. Ambrosia, Baccharis, Baileya, Cucurbita, Gaillardia, Glycine, Gossypium, Haploppapus, Helenium, Helianthus, Heterotheca, Kallstroemia, Lepidium, Physalis, Solanum, Solidago, Sphaeralcea, Verbesina, Xanthium, Zalazonia.

The short second antennal segment is the best character to separate this species from *D. sayi*.

Dillon and Dillon (1956) recognized five subspecies of *texanus*. These were based on highly variable characters such as coloration and shape of the elytral apices. We have been unable to satisfactorily separate populations geographically on the basis of either of those characteristics. The shape of the elytral apices varies with body size, and reddish individuals occur throughout the range of the species. The utilization of different larval hosts in the same area may produce the observed variation.

The larvae girdle the stems of *Kallstroemia grandiflora*, and whole fields have been observed with the vertical stem drooping.

Genus *Valenus* Casey

Valenus Casey, 1891, Ann. N.Y. Acad. Sci., 6: 49; Casey, 1913, Memoirs on the Coleoptera, 4: 322; Bradley, 1930, Man. Genera Beetles, p. 246; Dillon, 1956, Ann. Entomol. Soc. Amer., 49: 345; Arnett, 1962, Beetles U.S., 103: 873.

Form moderate-sized, subdepressed. Head with front strongly convex, broader than long; eyes with lower lobes almost rounded, upper lobes broadly separated; genae convergent, longer than lower eye lobes; antennal tubercles widely separated; antennae extending about five segments beyond elytra, scape slender, extending to about basal sulcus of pronotum, third segment subequal to scape, fourth slightly shorter than third, remaining segments gradually decreasing in length, basal segments with a few very short, suberect setae beneath. Pronotum broader than long; sides broadly arcuate to acute spines behind middle, then abruptly constricted; disk with a feeble callus at middle and one on each side behind apical margin; basal sulcus punctate, extending onto sides; prosternum narrow, intercoxal process slightly arcuate, narrow, coxae not contiguous; mesosternum with intercoxal process plane, narrow, slightly broader than prosternal process; metasternum with episternum narrow, subparallel. Elytra about twice as long as broad, slightly broader behind middle, tapering at apical one-fourth; disk with long erect setae; apices subtruncate. Legs rather short, lacking erect hairs, femora clavate, tarsi short, third segment cleft to base. Abdomen normally segmented, last sternite of females about twice as long as fourth, narrowing apically.

Type species. Valenus inornatus Casey (monobasic).

This genus may be recognized by the erect hairs of the elytra, absence of hairs on the legs and antennae, strongly convex front of the head, and by the widely separated upper eye lobes. A single species is known.

Valenus inornatus Casey
(Figure 6)

Valenus inornatus Casey, 1891, Ann. N.Y. Acad. Sci., 6: 50; Leng and Hamilton, 1896, Trans. Amer. Entomol. Soc., 23: 127; Schaeffer, 1908, Brooklyn Mus. Inst. Arts Sci. Bull., 1: 346; Linsley, Knull, and Statham,

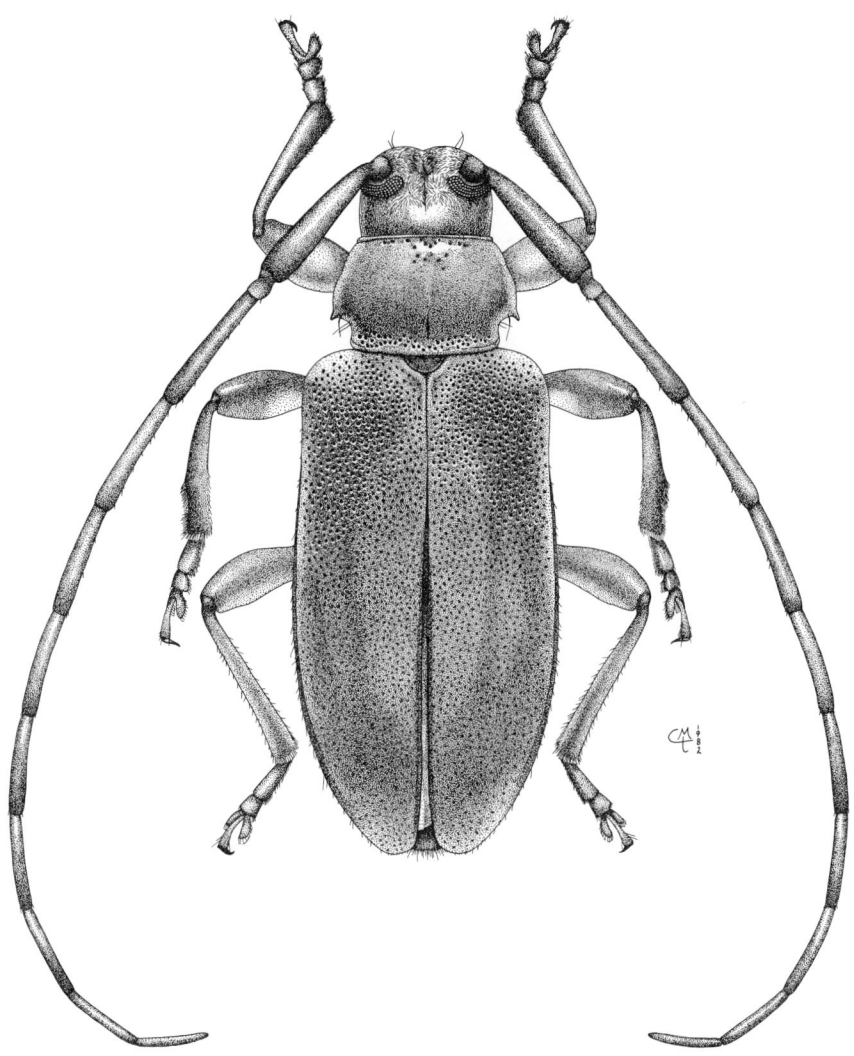

Figure 6. *Valenus inornatus* Casey, male.

1961, Amer. Mus. Nov., 2050: 30; Lewis, 1979, Pan-Pac. Entomol., 55: 25; Hovore et al., 1987, Proc. Calif. Acad. Sci., 44: 319; MacKay, Zak, and Hovore, 1987, Coleop. Bull., 41: 366.

Male. Form moderate-sized, subdepressed; integument reddish brown, legs and underside often paler; pubescence sparse, fine, appressed, grayish;

long, erect hairs numerous on elytra. Head minutely punctate, finely pubescent, several long erect setae present around eyes on lower margin of front; antennae moderately densely, finely clothed with pale, appressed pubescence. Pronotum moderately densely clothed with appressed pale pubescence, more densely at sides; punctures irregular, rather sparse, calluses impunctate; prosternum finely, moderately densely pubescent; meso- and metasternum densely, minutely punctate, finely, densely pubescent. Elytra moderately coarsely, separately punctate on basal half, punctures becoming sparser toward apex; pubescence very fine, short, disk subopaque appearing, long suberect hairs fairly numerous. Legs very finely punctate and pubescent. Abdomen minutely, densely punctate, finely pubescent; last sternite emarginate at apex. Length, 8-9 mm.

Female. Form similar. Legs with femora less strongly clavate. Abdomen with last sternite narrowing at apex. Length, 8.5-10 mm.

Type locality. El Paso, Texas.

Range. Southwestern Texas to western Arizona, northern Mexico.

Flight period. April to August.

Host plants. Agave, Yucca.

The unicolorous, reddish brown integument, fine pubescence, and erect hairs on the elytra make this species distinctive.

Adults are commonly taken at lights or beaten from their host plants.

Genus *Sternidius* LeConte

Sternidius LeConte, 1873, Smithson. Misc. Coll., 11(264): 234; LeConte, 1873, Smithson. Misc. Coll., 11(265):338.

Astyleiopus Dillon, 1956, Ann. Entomol. Soc. Amer., 49: 161; Dillon and Dillon, 1961, Man. Beetles East. North America, p. 638; Arnett, 1962, Beetles U.S., 103:872; Bayer and Shenefelt, 1968, Univ. Wisc. Res. Bull., 275: 26; Rice and Enns, 1981, Trans. Mo. Acad. Sci., 15: 98. (Type species: *Amniscus variegatus* Haldeman, by original designation.) New synonymy.

Leiopus: Lacordaire, 1872, Genera des coléoptères, 9(2): 775 (part).

Liopus: Leng and Hamilton, 1896, Trans. Amer. Entomol. Soc., 23: 121 (part); Blatchley, 1910, Coleoptera in Indiana, p. 1073 (part).

Form moderate to moderately large, elongate-ovate, robust. Head with front feebly convex, quadrate, median line extending onto neck; mandibles feebly arcuate; genae convergent, longer than lower eye lobes; eyes moderate, lower lobes longer than wide, upper lobes small, widely separated; antennal tubercles prominent, widely divergent; antennae eleven-segmented, about one-fifth longer than body, basal segments with very few short, erect hairs beneath, third segment slightly longer than first, fourth shorter than first. Pronotum broader than long, sides with a small, acute tubercle behind middle; basal sulcus not extending to sides; disk convex, tricallused, median callus linear; prosternum with intercoxal process narrow, less than one-sixth as wide as coxal cavities; mesosternum with intercoxal process half as broad as coxal cavities; metasternum with episternum narrow, tapering posteriorly. Elytra about twice as long as broad; basal gibbosities distinct, shallowly elevated; costae usually distinct; apices rounded. Legs robust; femora strongly clavate; tarsi slender, first segment of posterior pair longer than two following segments together. Abdomen normally segmented, female with fifth sternite twice as long as fourth.

Type species. Amniscus variegatus Haldeman (LeConte designation, 1873).

This genus resembles *Astylopsis,* but *Sternidius* may be separated by the placement of the lateral pronotal tubercles just before the basal sulcus which does not extend onto the sides.

The status of the generic name is discussed under the genus *Liopinus.*

A single species is known in our fauna.

Sternidius variegatus (Haldeman)

Amniscus variegatus Haldeman, 1847, Trans. Amer. Philos. Soc., (2)10: 47.

Liopus variegatus: LeConte, 1852, J. Acad. Nat. Sci. Philadelphia, 2: 172; Horn, 1880, Trans. Amer. Entomol. Soc., 8: 124; Chittenden, 1893, Proc. Entomol. Soc. Wash., 3: 100 (habits); Knobel, 1895, Beetles New England, p. 34, fig. 107; Leng and Hamilton, 1896, Trans. Amer. Entomol. Soc., 23: 122; Beutenmuller, 1896, J. N.Y. Entomol. Soc., 4: 79; Wickham, 1898, Can. Entomol., 30: 37; Smith, 1910, N.J. St. Mus. Rept., 1909: 333; Blatchley, 1910, Coleoptera in Indiana, p. 1073; Garman, 1916, Kentucky Agr. Exp. Sta. Bull., 200: 122; Rosewall, 1920, Can. Entomol., 52: 203; Craighead, 1923, Can. Dept. Agr. Bull., (n.s.) 27: 116, pl. 24, fig. 6.

Leiopus variegatus: White, 1855, Cat. Coleop. Ins. British Mus., 8: 388; Lacordaire, 1872, Genera des coléoptères, 9(2): 776, fn.; Champlain and Knull, 1925, Entomol. News, 36: 140; Leonard, 1928, Cornell Agr. Exp. Sta. Mem., 101:452; Knull, 1930, Entomol. News, 41: 102; Chagnon, 1933-40, Coleop. Prov. Quebec, p. 237; Hoffmann, 1942, USDA Misc. Pub., 466: 11; Loding, 1945, Geol. Surv. Ala. Mon., 11: 122; Knull, 1946, Ohio Biol. Surv. Bull., 39:248; Fattig, 1947, Emory Univ. Mus. Bull., 5: 35; Steyskal, 1951, Coleop. Bull., 5: 76; Chagnon and Robert, 1962, Prin. Coleop. Prov. Quebec, p. 173; Furth, 1985, Conn. Acad. Arts Sci., 46: 192.

Leptosylus variegatus: Lacordaire, 1872, Genera des coléoptères, 9(2): 772, fn.

Sternidius variegatus: LeConte, 1873, Smithson. Misc. Coll., 11(264): 234; LeConte, 1873, Smithson. Misc. Coll., 11(265): 338.

Astyleiopus variegatus: Dillon, 1956, Ann. Entomol. Soc. Amer., 49: 162, fig. 7; Dillon and Dillon, 1961, Man. Beetles East. North America, p. 638, pl. 63, no. 10; Beyer and Shenefelt, 1969, Univ. Wisc. Res. Bull., 275: 27, fig. 35; Gardiner, 1969, Can. Dept. Fish. Forest Int. Rept., 0-14: 91 (larvae); Stein and Tagestad, 1976, USDA For. Serv. Res. Pap., RM-171: 4; Gosling and Gosling, 1976, Gr. Lakes Entomol., 10: 27, fig. 148; Hovore and

Giesbert, 1976, Coleop. Bull., 30: 358; Headstrom, 1977, Beetles of America, p. 377; Laliberte et al., 1977, Fabreries, 3: 91; Rice and Enns, 1981, Trans. Mo. Acad. Sci., 15: 102; Gosling, 1984, Gr. Lakes Entomol., 17: 71; Hovore et al., 1987, Proc. Calif. Acad. Sci., 44: 318.

Amniscus variegatus var. *trifasciatus* Haldeman, 1847, Trans. Amer. Philos. Soc., (2)10: 47.

Liopus variegatus var. *trifasciatus*: Leng and Hamilton, 1896, Trans. Amer. Entomol.. Soc., 23: 122.

Leiopus variegatus ab. *trifasciatus*: Aurivillius, 1923, Coleop. Cat., 74: 409.

Amniscus variegatus var. *obscurus* Haldeman, 1847, Trans. Amer. Philos. Soc., (2)10: 47.

Liopus variegatus var. *obscurus*: Leng and Hamilton, 1896, Trans. Amer. Entomol. Soc., 23: 122.

Leiopus variegatus ab. *obscurus*: Aurivillius, 1923, Coleop. Cat., 74: 409.

Male. Form moderate-sized, robust; integument reddish brown to piceous, appendages paler annulate; pubescence dense, short, appressed, variegated grayish, brownish, and black. Head with front densely micropunctate, densely pubescent; genae longer than lower eye lobes; antennae extending about four segments beyond elytra, segments broadly dark annulate at apical half, third segment slightly longer than first, fourth shorter or subequal to first. Pronotum broader than long; lateral tubercles small, acute, placed before basal impression; basal impression broad, shallow, not extending onto sides, disk with three dark calluses, median callus linear; pubescence dense, pale brownish, often mottled with darker pubescence, occasionally with a narrow dark, longitudinal vitta on each side; punctures around calluses moderately coarse, dense; prosternum with intercoxal process narrow; mesosternum with intercoxal process one-third to half as broad as coxal cavities; metasternum finely, uniformly pubescent. Elytra about twice as long as broad; punctures coarse, dense, becoming finer and sparser toward apex; pubescence dense, variegated, small dark maculae variably interspersed, two vaguely rounded dark maculae present behind middle, usually not attaining suture, two oblong whitish maculae present behind dark postmedian pair, irregular narrow whitish maculae often extending along costae; apices rounded. Legs with femora dark annulate at middle; tibiae dark biannulate; tarsi with first segment basally pale pubescent. Abdomen densely micropunctate, finely, densely pubescent; last sternite emarginate at apex. Length, 7-11 mm.

Female. Form similar. Antennae slightly shorter. Abdomen with fifth sternite twice as long as fourth. Length, 7-11 mm.

Type locality. Of *variegatus*, none (North America); *obscurus*, none (North America); *trifasciatus*, Alabama.

Range. Eastern North America to Arizona and North Dakota.

Flight period. June to September.

Host plants. Acer, Aesculus, Ampelopsis, Caragana, Castanea, Celastris, Celtis, Citrus, Gaylussacia, Gleditsia, Juglans, Morus, Parthenocissus, Populus, Rhus, Robinia, Salix, Toxicodendron, Ulmus.

Parasites. Pimpla irritator (F.).

The very distinctive ovoid whitish maculae at the apical one-third of the elytra, broadly dark annulate antennae, and characters of the pronotum make this species easily recognizable.

The coloration and pubescence vary in the amount of pale and dark spots, and the postmedian dark maculae may be reduced.

Genus *Acanthocinus* Megerle

Acanthocinus Megerle, 1821, in Dejean, Cat. Coleop., p. 106; White, 1855, Cat. Coleop. Ins. British Mus., 8:368; Thomson, 1860, Class. ceramb., p. 13; Thomson, 1864, Syst. ceramb., p. 28; Lacordaire, 1872, Genera des coléoptères, 9(2):790; LeConte, 1873, Smithson. Misc. Coll., 11(265):339; Horn, 1880, Trans. Amer. Entomol. Soc., 8:130; LeConte and Horn, 1883, Smithson. Misc. Coll., 507:324; Gahan, 1888, Trans. Amer. Entomol. Soc., 15:300; Leng and Hamilton, 1896, Trans. Amer. Entomol. Soc., 23:131; Wickham, 1897, Can. Entomol., 29:203; Wickham, 1898, Can. Entomol., 30:38; Craighead, 1923, Can. Dept. Agr. Bull., (n.s.) 27:119; Leng and Mutchler, 1927, Suppl. Cat. Coleop. Amer., p. 43; Bradley, 1930, Man. Genera Beetles, p. 246; Chagnon, 1933-40, Coleop. Prov. Quebec, p. 274; Knull, 1946, Ohio Biol. Surv. Bull., 39:257; Chagnon and Robert, 1962, Prin. Coleop. Prov. Quebec, p. 274; Baker, 1972, USDA Misc. Pub., 1175:195; Marinoni, 1977, Dusenia, 10:39; Rice and Enns, 1981, Trans. Mo. Acad. Sci., 15:97.

Acanthocinus (Graphisurus) Kirby, 1837, in Richardson, Fauna Bor. Amer., 4:16; Bethune, 1872, Can. Entomol., 4:55. (Type species: *Acanthocinus (Graphisurus) pusillus* Kirby, monobasic.) New synonymy.

Astynomus Haldeman, 1847 (not Stephens, 1839), Trans. Amer. Philos. Soc., (2)10:46.

Aedilis LeConte, 1852 (not Audinet-Serville, 1835), J. Acad. Nat. Sci. Philadelphia, 2:173.

Graphisurus (Graphisurus) Casey, 1913, Memoirs on the Coleoptera, 4:334; Leng, 1920, Cat. Coleop. Amer., p. 283.

Acanthocinus (Acanthocinus): Aurivillius, 1923, Coleop. Cat., 74:434

Graphisurus (Acanthocinus): Keen, 1929, Calif. Div. For. Bull., 7: 62.

Graphisurus: Doane et al., 1936, For. Ins., p. 189.

Graphisurus (Canonura) Casey, 1913, Memoirs on the Coleoptera, 4: 335. (Type species: *Aedilis spectabilis* LeConte, Dillon designation, 1956.) New synonymy.

Acanthocinus (Canonura): Aurivillius, 1923, Coleop. Cat., 74: 434.

Canonura: Dillon, 1956, Ann. Entomol. Soc. Amer., 49: 225; Arnett, 1962, Beetles U.S., 103: 873; Hatch, 1971, Univ. Wash. Pub. Biol., 16: 151.

Graphisurus (Tylocerina) Casey, 1913, Memoirs on the Coleoptera, 4: 335. (Type species: *Cerambyx nodosus* Fabricius, Dillon designation, 1956.) New synonymy.

Acanthocinus (Tylocerina): Aurivillius, 1923, Coleop. Cat., 74: 434.
Tylocerina: Dillon, 1956, Ann. Entomol. Soc. Amer., 49: 230; Arnett, 1962, Beetles U.S., 103: 873; Drooz, 1985, USDA For. Serv. Misc. Pub., 1426: 304.
Neacanthocinus Dillon, 1956, Ann. Entomol. Soc. Amer., 49: 231; Arnett, 1962, Beetles U.S., 103: 873; Bayer and Shenefelt, 1969, Univ. Wisc. Res. Bull., 275: 27; Hatch, 1971, Univ. Wash. Pub. Biol., 16: 151; Drooz, 1985, USDA For. Serv. Misc. Pub., 1426: 304. (Type species: *Cerambyx obsoletus* Olivier, by original designation.) New synonymy.

Form moderate-sized to large, somewhat depressed. Head with front rather short, quadrate, slightly convex; clypeus narrowly margined apically; eyes with lower lobes oblong, usually longer than genae, upper lobes widely separated above; genae subparallel; antennal tubercles prominent, divergent, bases contiguous; antennae slender, 2-1/2 to four times longer than body in males, 1-1/2 to twice as long in females, segments four and/or five usually apically produced in males, segments three to at least five densely fimbriate beneath, scape usually extending to lateral tubercles of pronotum, third segment longer than first, fourth subequal to third. Pronotum broader than long, sides moderately to strongly tuberculate behind middle, tubercles acutely spined at basal one-third; basal sulcus extending onto sides; disk bicallused behind apical margin; prosternum narrow, shallowly excavated, intercoxal process narrow, less than one-fourth as broad as coxal cavities; mesosternum with intercoxal process about one-third as broad as coxal cavities, at least twice as broad as procoxal process; metasternum with episternum subparallel, abruptly tapering posteriorly. Elytra more than twice as long as broad, slightly tapering posteriorly to subparallel; disk vaguely to strongly tricostate, base feebly bigibbose; pubescence depressed, dense, often condensed into spots or vittae along costae, erect hairs absent; apices rounded to subtruncate. Legs robust; femora clavate, lacking erect hairs; tarsi with first segment of hind pair longer than two following segments together, third segment cleft to base. Abdomen normally segmented in males, last segment elongate in females, terminated by a long, visible ovipositor.

Type species. Cerambyx aedilis Linnaeus.

Members of this genus are easily recognized by the long antennae, fimbriated basal antennal segments, lack of erect hairs on the elytra, and by the elongate ovipositor of the females. The North American species fit the generic definition of *Acanthocinus* as represented by the European type species, *A. aedilis*. We consider the characters utilized by Dillon (1956) to separate the group into several genera to be specific, not generic.

According to Craighead (1923), the North American species are all pine feeders (as are the Palearctic ones). The entire larval period is passed in the bark and, with the exception of *A. obliquus*, the larvae never penetrate the sapwood. The eggs are laid in deep pits gnawed by the females or in scolytid holes, and the larvae penetrate into the deeper layers of bark. Pupation occurs near the surface in the dryer tissues. The life cycle normally requires one year.

Eight species are known from North America.

KEY TO THE NORTH AMERICAN SPECIES OF *ACANTHOCINUS*

1	Elytra simply punctate at base	2
	Elytra granulate punctate at base	6
2(1)	Elytra punctate over apical half, pubescent crests of basal gibbosities, if present, continuous with subsutural costae. Moderate-sized species	3
	Elytra impunctate over apical half, basal pubescent crests terminating behind gibbosities. Males with a large knob at apex of fourth segment of antennae. Large species. Length, 18-28 mm (exclusive of extruded ovipositor). Eastern United States to Texas	*nodosus*
3(2)	Tarsi with first two segments pale annulate. Elytra with apices rounded	4
	Tarsi totally dark, not pale annulate. Elytra obliquely subtruncate. Length, 7-15 mm. Eastern North America to Texas	*obsoletus*
4(3)	Elytra strongly costate, with rows or tufts of longer, condensed, black pubescence	5
	Elytra not or feebly costate, without black tufts of pubescence. Length, 7-13 mm. Alaska to northeastern North America	*pusillus*
5(4)	Elytra with outer costae extending to apical margins, joining discal costae before apex; black bands very strongly angulate posteriorly, costae usually continuously black pubescent	*angulosus*
	Elytra with outer costae not extending to apical margin, usually not joining with discal costae; black bands more transverse, moderately angulate at most, costae with pubescence interrupted. Length, 8-17 mm. British Columbia to Baja California, South Dakota to central Mexico	*obliquus*
6(1)	Body with underside entirely whitish pubescent. Pronotum with a transverse row of two or four pale spots behind apical margin	7
	Body with underside brownish pubescent at sides. Pronotum with a whitish interrupted macula around each lateral tubercle. Length, 18-26 mm. Rocky Mts. from Montana to southern Arizona	*spectabilis*

7(6) Elytra with four well defined, dark maculae, these mottled with brownish pubescence; genae subequal in length to lower eye lobes; pronotum with four pale brownish spots behind apical margin. Length, 13-24 mm. Pacific Coast from British Columbia to California .... *princeps*
Elytra with maculae ill defined, all black; genae shorter than lower eye lobes; pronotum with two pale brownish spots behind apical margin. Length, 16-22 mm. Montane Central Arizona .... *leechi*

Acanthocinus nodosus (Fabricius)

Cerambyx nodosus Fabricius, 1775, Syst. Entomol., p. 164; Fabricius, 1781, Species Ins., 1:209; Fabricius, 1787, Mant. Ins., 1:131; Olivier, 1790, Encycl. Meth. Ins., 5:291; Olivier, 1795, Entomol., 4(67):75, pl. 14, fig. 103; Fabricius, 1801, Syst. Eleuth., 2:289.

Astynomus nodosus: Castelnau, 1840, Hist. Nat. Anim. Art., 2:463; Haldeman, 1847, Trans. Amer. Philos. Soc., (2)10:46.

Aedilis nodosus: LeConte, 1852, J. Acad. Nat. Sci. Philadelphia, 2:174; Packard, 1881, U.S. Entomol. Comm. Bull., 7:159.

Acanthocinus nodosus : White, 1855, Cat. Coleop. Ins. British Mus., 8:368; Lacordaire, 1872, Genera des coléoptères, 9(2):791, fn.; LeConte, 1873, Smithson. Misc. Coll., 11(265):339; Horn, 1880, Trans. Amer. Entomol. Soc., 8:130; LeConte and Horn, 1883, Smithson. Misc. Coll., 507:324; Gahan, 1888, Trans. Amer. Entomol. Soc., 15:300; Beutenmuller, 1896, J. N.Y. Entomol. Soc., 4:79; Leng and Hamilton, 1896, Trans. Amer. Entomol. Soc., 23:132; Smith, 1910, N.J. St. Mus. Rept., 1909:334; Craighead, 1923, Can. Dept. Agr. Bull., (n.s.) 27:119; Leonard, 1928, Cornell Agr. Exp. Sta. Mem., 101:453; Knull, 1932, Entomol. News, 43:64 (habits); Savely, 1939, Ecol. Mon., 9:333; Loding, 1945, Geol. Surv. Ala. Mon., 11:123; Knull, 1946, Ohio Biol. Surv. Bull., 39:258, pl. 29, fig. 1; Fattig, 1947, Emory Univ. Mus. Bull., 5:38; Craighead, 1950, USDA Misc. Pub., 657:236; Beal, 1952, Duke Univ. Sch. For. Bull., 14:111 (habits); Baker, 1972, USDA Misc. Pub., 1175:196, fig. 67 (habits).

Graphisurus (Tylocerina) nodosus: Casey, 1913, Memoirs on the Coleoptera, 4:340.

Acanthocinus (Tylocerina) nodosus: Aurivillius, 1923, Coleop. Cat., 74:434.

Tylocerina nodosus: Dillon, 1956, Ann. Entomol. Soc. Amer., 49:230; Kirk, 1969, S.C. Agr. Exp. Sta. Tech. Bull., 1033:87; Finn et al., 1972, Ann. Entomol. Soc. Amer., 65:644.

Tylocerina nodosa: Drooz, 1985, USDA For. Serv. Misc. Pub., 1426:304, fig. 139.

Lamia bifidator Fabricius, 1801, Syst. Eleuth., 2:286.

Graphisurus (Tylocerina) laticollis Casey, 1913, Memoirs on the Coleoptera, 4:341.
Acanthocinus laticollis: Fattig, 1947, Emory Univ. Mus. Bull., 5:38.

Male. Form large, depressed, subparallel; integument reddish brown, pubescence short, dense, appressed, grayish and black. Head with front very finely, densely punctate, vertex with a row of small punctures behind eyes; pubescence fine, dense, brownish on front, gray on vertex; antennae about 2-1/2 times longer than body, segments three and four densely fimbriate beneath, fifth sparsely, fourth segment with a dense apical tuft, segments finely pale pubescent, segments to ninth dark annulate at apices, eleventh segment elongate. Pronotum broader than long, sides broadly tuberculate behind middle, apices of tubercles acute; disk with four vague calluses behind apex, median callus prominent; punctures sparse, irregular; pubescence fine, dense, grayish, apex and base with two short black vittae, sides below tubercles broadly dark vittae; prosternum thinly pale pubescent; meso- and metasternum densely pale pubescent. Elytra about twice as long as broad, sides narrowly explanate behind middle; basal gibbosities fairly prominent, bearing a linear, velvetly covered crest, subbasal impression deep; disk with three costae, first two coalescing near apex, outer pair ending before apex; basal punctures coarse, confluent, ending a little behind middle; pubescence fine, grayish, sides with broad black vittae extending to about middle, base with two short linear vittae, two short narrow vittae present at middle and two elongate black, broadly v-shaped vittae present behind middle, a few small dark spots present on costae; apices rounded. Legs robust; femora finely pale pubescent with numerous small glabrous spots interspersed; tibiae broadly dark at apices and bases; tarsi dark. Abdomen finely pale pubescent, dark brown at sides; last sternite emarginate at apex. Length, 18-27 mm.

Female. Form similar. Antennae about twice as long as body, fourth segment lacking apical tuft. Abdomen with last sternite elongate, deeply emarginate at apex, ovipositor strongly extruded. Length (exclusive of ovipositor), 20-26 mm.

Type locality. Of *nodosus*, Maryland; *bifidator*, America boreali; *laticollis*, Southern Pines, N.C.

Range. Eastern United States from Pennsylvania to Florida west to Texas.

Flight period. April to October.

Host plants. Pinus rigida.

This species is readily recognizable by the unique maculation of the elytra and the apical tuft of the fourth antennal segment.

Acanthocinus obsoletus (Olivier)

Cerambyx obsoletus Olivier, 1795, Entomol., 4(67):130, pl. 13, fig. 90.
Aedilis obsoletus: LeConte, 1852, J. Acad. Nat. Sci. Philadelphia, 2:174; Packard, 1881, U.S. Entomol. Comm. Bull., 7:159.
Graphisurus obsoletus: Lacordaire, 1872, Genera des coléoptères, 9(2):787, fn.

Acanthocinus obsoletus: LeConte, 1873, Smithson. Misc. Coll., 11(265):339; Horn, 1880, Trans. Amer. Entomol. Soc., 8:130; Kingsley, 1884, Riverside Nat. Hist., 2:327; Gahan, 1888, Trans. Amer. Entomol. Soc., 15:300; Leng and Hamilton, 1896, Trans. Amer. Entomol. Soc., 23:132; Beutenmuller, 1896, J. N.Y. Entomol. Soc., 4:79; Wickham, 1898, Can. Entomol., 30:38; Schaeffer, 1902, Entomol. News, 13:236; Felt, 1907, Ins. Affect. Trees, p. 662; Smith, 1910, N.J. St. Mus. Rept., 1909:334; Blatchley, 1910, Coleoptera in Indiana, p. 1079; Craighead, 1923, Can. Dept. Agr. Bull., (n.s.) 27:120; Leonard, 1928, Cornell Agr. Exp. Sta. Mem., 101:453; Chagnon, 1933-40, Coleop. Prov. Quebec, p. 274; Savely, 1939, Ecol. Mon., 9:334; Loding, 1945, Geol. Surv. Ala. Mon., 11:123; Knull, 1946, Ohio Biol. Surv. Bull., 39:258; Fattig, 1947, Emory Univ. Mus. Bull., 5:38; Craighead, 1950, USDA Misc. Pub., 657:236; Beal, 1952, Duke Univ. Sch. For. Bull., 14:111 (habits); Chagnon and Robert, 1962, Prin. Coleop. Prov. Quebec, p. 274; Baker, 1972, USDA Misc. Pub., 1175:196; Rice and Enns, 1981, Trans. Mo. Acad. Sci., 15:102.

Lamia obsoleta: LeConte and Horn, 1883, Smithson. Misc. Coll., 507:324.

Graphisurus (Graphisurus) obsoletus: Casey, 1913, Memoirs on the Coleoptera, 4:335.

Acanthocinus (Acanthocinus) obsoletus: Aurivillius, 1923, Coleop. Cat., 74:434.

Neacanthocinus obsoletus: Dillon, 1956, Ann. Entomol. Soc. Amer., 49:232, fig. 4; Bayer and Shenefelt, 1969, Univ. Wisc. Res. Bull., 275:29, fig. 37; Kirk, 1969, S.C. Agr. Exp. Sta. Tech. Bull., 1033:87; Wallace and Franklin, 1970, J. Ga. Entomol. Soc., 5:25 (parasite); Finn et al., 1972, Ann. Entomol. Soc. Amer., 65:644; Gosling and Gosling, 1976, Gr. Lakes Entomol., 10:25, fig. 140; Laliberte et al., 1977, Fabreries, 3:95; Waters and Hyche, 1984, Coleop. Bull., 38:285; Drooz, 1985, USDA For. Serv. Misc. Pub., 1426:304.

Graphisurus (Graphisurus) floridanus Casey, 1913, Memoirs on the Coleoptera, 4:336.

Graphisurus punctatus Casey, 1924, Memoirs on the Coleoptera, 11:293; Chemsak and Linsley, 1982, Checklist of Ceramb., p. 93. New synonymy.

Male. Form moderate-sized, sides tapering; integument reddish brown; pubescence fine, dense, appressed, gray and brownish. Head with front and vertex micropunctate only; pubescence very fine, pale, appressed, margins of front with a few dark, erect setae; genae shorter than lower eye lobes; antennae about 2-1/2 times longer than body, segments three to eight fringed beneath, fringe on outer segments sparser, basal segments barely enlarged apically, segments dark annulate at apices, eleventh segment elongate. Pronotum broader than long, sides acutely tuberculate behind middle; disk with two calluses behind apex and a linear median one; punctures rather fine, moderately dense; pubescence fine, dense, appressed, usually mottled grayish and brownish, post-apical calluses brownish; prosternum finely pale pubescent; meso- and metasternum densely clothed with pale, appressed pubescence, sides with glabrous punctures. Elytra a little more than twice as long as broad; basal gibbosities rounded, moderately elevated; punctures moderately coarse, deep, well separated, becoming very sparse at apex; each elytron medially costate; pubescence fine, short, pale with three or four irregular, transverse maculae, these more apparent to the naked eye; apices obliquely truncate. Legs robust; femora with glabrous spots; tibiae biannulate; tarsi dark. Abdomen finely pale pubescent, sides with glabrous spots; last sternite emarginate at apex. Length, 9-17 mm.

Female. Form similar. Antennae less than 1-1/2 times longer than body, segments not fringed beneath. Abdomen with last sternite elongate, deeply emarginate at apex, ovipositor strongly extruded. Length (exclusive of ovipositor), 11-17 mm.

Type locality. Of *obsoletus*, Carolina and Pennsylvania; *floridanus*, Florida; *punctatus*, Marquette, Lake Superior, Michigan.

Range. Eastern North America from Ontario to Florida west to Minnesota and Texas.

Flight period. January to November.

Host plants. Pinus (including *P. echinata, P. palustris, P. ponderosa, P. rigida, P. strobus, P. taeda, P. virginiana*).

Parasite. Eutheresia interrupta Curran (Tachinidae).

This species may be separated from *A. pusillus* by the dark pubescent tarsi.

The dark maculae of the elytra are variable. Usually four of these are visible to the naked eye, but under magnification the premedian band is very irregular and usually interrupted. This is also true of the apical band, which is often greatly reduced.

Acanthocinus pusillus Kirby

Acanthocinus (Graphisurus) pusillus Kirby, 1837, in Richardson, Fauna Bor.-Amer., 4:55; Bethune, 1872, Can. Entomol., 4:75.

Graphisurus pusillus: LeConte, 1852, J. Acad. Nat. Sci. Philadelphia, 2:175; Lacordaire, 1872, Genera des coléoptères, 9(2):787, fn.; Horn, 1880, Trans. Amer. Entomol. Soc., 8:129; Gahan, 1888, Trans. Amer. Entomol. Soc., 15:299; Mundinger, 1924, N.Y. St. Coll. For. Tech. Pub., 17:320.

Graphisurus (Graphisurus) pusillus: Casey, 1913, Memoirs on the Coleoptera, 4:336.

Acanthocinus pusillus: White, 1855, Cat. Coleop. Ins. British Mus., 8:369; Schaeffer, 1902, Entomol. News, 13:236; Smith, 1910, N.J. St. Mus. Rept., 1909:334; Leonard, 1928, Cornell Agr. Exp. Sta. Mem., 101:453; Knull, 1946, Ohio Biol. Surv. Bull., 39:258; Fattig, 1947, Emory Univ. Mus. Bull., 5:38; Gardiner, 1955, Can. Entomol., 87:219, 4 figs. (larva); Gardiner, 1957, Can. Entomol., 89:250, fig. 6; Baker, 1972, USDA Misc. Pub., 1175:196.

Acanthocinus (Acanthocinus) pusillus: Aurivillius, 1923, Coleop. Cat., 74:434.

Neacanthocinus pusillus: Dillon, 1956, Ann. Entomol. Soc. Amer., 49:233; Gardiner, 1969, Can. Dept. Fish. For. Int. Rept., 0-14:89 (pupa); Bayer and Shenefelt, 1969, Univ. Wisc. Res. Bull., 275:29, fig. 37; Gosling and Gosling, 1976, Gr. Lakes Entomol., 10:25, fig. 140; Drooz, 1985, USDA For. Serv. Misc. Pub., 1426:304; Gosling, 1986, Gr. Lakes Entomol., 19:156.

Male. Form small to moderate-sized, subparallel; integument reddish brown to piceous; pubescence dense, short, appressed, grayish and brownish. Head with front and vertex densely, minutely punctate, vertex with a few scattered large punctures; pubescence fine, dense, grayish and brown, margins of front with a few long, erect setae; genae slightly shorter than lower eye lobes; antennae about twice as long as body, segments three to five almost vaguely fimbriate, fringe very short and sparse, segments three to five barely expanded at apices, segments narrowly pale annulate at apices, eleventh segment about as long as tenth. Pronotum broader than long, sides strongly tuberculate behind middle, spines short; disk with two low calluses behind apical margin, median callus vague; punctures moderately coarse, scattered; pubescence fine, denser at sides, sides grayish, middle pale brownish, usually with four pale brownish spots behind apex; prosternum finely, densely pubescent; meso- and metasternum densely pubescent with glabrous spots at sides. Elytra a little more than twice as long as broad; basal gibbosities low, elongate; punctures moderately coarse, dense, separated, becoming obsolete toward apex; each elytron with a low median costa and vague, short inner one; pubescence mottled grayish and brown, usually with four transverse vague, brown maculae, postmedian pair distinct, oblique; apices broadly rounded. Legs robust; femora dark annulate at apices; tibiae bi-annulate; tarsi with first two segments pale pubescent at apices. Abdomen finely, densely pubescent, pubescence interrupted by glabrous spots; last sternite rather shallowly emarginate at apex. Length, 7-13 mm.

Female. Form similar. Antennae less than 1-1/2 times as long as body, segments not fimbriate beneath. Abdomen with last sternite elongate, notched at apex, ovipositor strongly extruded. Length (exclusive of ovipositor), 8-12 mm.

Type locality. New York to Cumberland House.
Range. Alaska to northeastern North America.
Flight period. May to July.
Host plants. Pinus, Picea, Tsuga, Abies.

The rounded elytral apices and pale annulate tarsi separate this species from *A. obsoletus*, and the rather vague elytral costae from *A. obliquus*.

Acanthocinus angulosus (Casey)

Graphisurus (Graphisurus) angulosus Casey, 1913, Memoirs on the Coleoptera, 4:338.

Acanthocinus (Neacanthocinus) angulosus: Linsley, Knull, and Statham, 1961, Amer. Mus. Nov., 2050:30.

Male. Form moderate-sized, slightly tapering; integument reddish brown, vertex of head and abdomen piceous; pubescence short, dense, appressed, grayish and black. Head with front finely, irregularly punctate, vertex moderately coarsely, irregularly punctate; pubescence dense, grayish, and interrupted on genae, front dark pubescent with an angulate pale vitta at bases of antennal tubercles, pubescence pale at apices and behind antennal tubercles, front margined with long erect, black setae; genae longer than lower eye lobe; antennae about three times longer than body, segments three to six fringed beneath, expanded at apices, segments to eighth dark annulate at apices, scape dark beneath, eleventh segment elongate. Pronotum broader than long, sides tuberculate behind middle, spines short, acute; disk with two prominent calluses behind apical margin, outside calluses vague, median callus linear; punctures moderately coarse, deep, irregular; pubescence dense, appressed, grayish and irregular at sides, darker medially, two apical calluses dark pubescent at outside, pale pubescence giving impression of two divergent vittae, extending from apical calluses to base, sides behind lateral tubercles with a few long, erect setae; prosternum densely pale pubescent; meso- and metasternum densely pale pubescent, pubescence interrupted by glabrous spots. Elytra a little more than twice as long as broad; basal gibbosities slightly crested by dark pubescent costae; punctures rather fine, dense, becoming finer toward apex; each elytron prominently tricostate, costae uniting before apex, outer pair extending to apical margin; pubescence short, dense, grayish, costae velvety black pubescent, transverse dark maculae rather than vague, strongly angulate, pre-median one usually interrupted, elongate v-shaped near suture, extending back beyond middle, postmedian macula narrow, strongly, narrowly v-shaped behind, extending to near apex; apices narrowly rounded. Legs robust; femora with grayish pubescent interrupted by numerous glabrous spots, apices with a narrow, dark angulate vitta; tibiae biannulate with dark pubescence; tarsi with first two segments white pubescent at apices. Abdomen densely pubescent, pubescence strongly interrupted by glabrous spots; last sternite deeply emarginate at apex. Length, 11-16 mm.

Female. Form similar, more robust. Antennae about 1-1/2 times longer than body, all segments dark annulate, segments three to five with a short fringe. Abdomen with last sternite elongate, deeply notched at apex, ovipositor strongly extruded. Length (exclusive of ovipositor), 13-17 mm.

Type locality. New Mexico.
Range. Southern Arizona to central New Mexico.
Flight period. July and August.
Host plants. Pinus chihuahuana.

This species has been confused with *A. obliquus*. The two are sympatric, although the known distribution of *angulosus* is much more restricted. The maculate pattern of the elytra of *angulosus* is quite different from that of *obliquus*. The transverse dark maculae are much more angulate posteriorly and form a narrow, elongate v on each side. Also, the costae are clothed with uninterrupted bands of black, velvety pubescence. In *A. obliquus*, only the two inner costae join near the apex and the outer pair do not extend to the apical margins.

Thus far, *A. angulosus* is known only from the Chiricahua Mts., Santa Catalina Mts., Apache National Forest in Arizona, Manzano Mts. of New Mexico, Cloudcroft, N.M. and the north rim of the Grand Canyon.

Acanthocinus obliquus (LeConte)
(Figure 7)

Aedilis obliquus LeConte, 1862, Proc. Acad. Nat. Sci. Philadelphia, 14:39.
Acanthocinus obliquus: LeConte, 1872, Smithson. Misc. Coll., 11(265):339; Horn, 1880, Trans. Amer. Entomol. Soc., 8:130; LeConte and Horn, 1883, Smithson. Misc. Coll., 507:324; Gahan, 1888, Trans. Amer. Entomol. Soc., 15:300; Leng and Hamilton, 1896, Trans. Amer. Entomol. Soc., 23:132; Garnett, 1918, Can. Entomol., 50:282; Craighead, 1923, Can. Dept. Agr. Bull., (n.s.) 27:120; Keen, 1938, USDA Misc. Pub., 273:137; Keen, 1952, USDA Misc. Pub., 27:174.
Graphisurus (Graphisurus) obliquus: Casey, 1913, Memoirs on the Coleoptera, 4:337.
Acanthocinus (Acanthocinus) obliquus: Aurivillius, 1923, Coleop. Cat., 74:434.
Graphisurus obliquus: Hardy, 1926, Rept. Prov. Mus., 1925:10; Keen, 1929, Calif. Div. For. Bull., 7:64; Doane et al., 1936, For. Ins., p. 189; Hardy, 1945, Victoria Nat., 2:90; Edwards, 1957, Pan-Pac. Entomol., 33:52; Essig, 1958, Ins. West. North America, p. 460.
Neacanthocinus obliquus obliquus: Dillon, 1956, Ann. Entomol. Soc. Amer., 49:234, fig. 2; Tyson, 1966, Pan-Pac. Entomol., 42:205; Hatch, 1971, Univ. Wash. Pub. Biol., 16:151; Kirk and Balsbaugh, 1975, S.D. Agr. Exp. Sta. Tech. Bull., 42:100; Lewis, 1979, Pan-Pac. Entomol., 55:25.
Acanthocinus (Neacanthocinus) obliquus obliquus: Linsley, Knull, and Statham, 1961, Amer. Mus. Nov., 2050:30.
Neacanthocinus obliquus: Horning and Barr, 1970, Univ. Idaho Coll. Agr. Misc. Ser., 8:38.
Graphisurus (Graphisurus) acomanus Casey, 1913, Memoirs on the Coleoptera, 4:337.
Graphisurus (Graphisurus) obliquus sedulus Casey, 1913, Memoirs on the Coleoptera, 4:337.
Acanthocinus (Acanthocinus) obliquus var. *sedulus*: Aurivillius, 1923, Coleop. Cat., 74:434.
Graphisurus (Graphisurus) pacificus Casey, 1913, Memoirs on the Coleoptera, 4:338. New synonymy.
Acanthocinus (Acanthocinus) pacificus: Aurivillius, 1923, Coleop. Cat., 74:434.

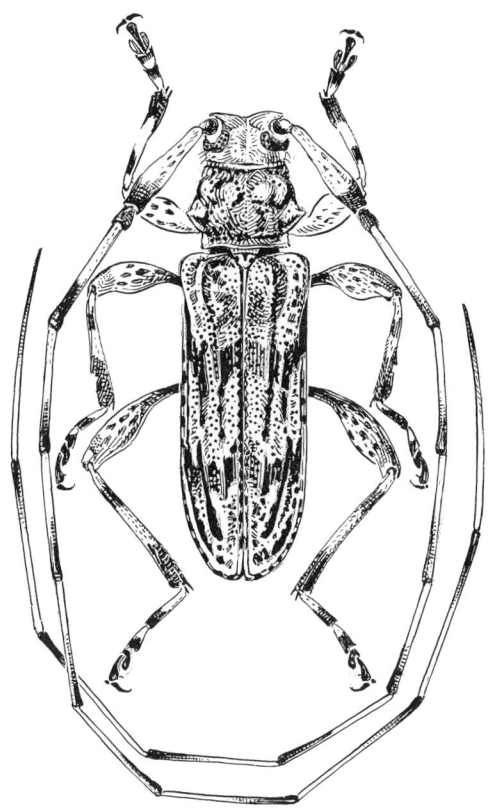

Figure 7. *Acanthocinus obliquus* (LeConte), male.

Neacanthocinus obliquus pacificus: Dillon, 1956, Ann. Entomol. Soc., Amer., 49:235, fig. 3; Gilmour, 1965, Cat. Lam. du Monde, 8:559; Tyson, 1966, Pan-Pac. Entomol., 42:205; Tyson, 1970, Pan-Pac. Entomol., 46:298; Hatch, 1971, Univ. Wash. Pub. Biol., 16:151.

Graphisurus obtusus Casey, 1924, Memoirs on the Coleoptera, 11:292. New synonymy.

Graphisurus (Graphisurus) obliquus chihuahuae Casey, 1913, Memoirs on the Coleoptera, 4:337. New synonymy.

Acanthocinus (Acanthocinus) obliquus var. *chihuahuae*: Aurivillius, 1923, Coleop. Cat., 74:434; Duffy, 1960, Mon. Neotrop. Timber Beetles, p. 247.

Neacanthocinus obliquus v. (?ssp.) *chihuahuae*: Gilmour, 1965, Cat. Lam. du Monde, 8:559.

Neacanthocinus obliquus chihuahuae: Chemsak and Linsley, 1975, Checklist of Beetles, Longhorn Beetles, 1(6):179.

Male. Form moderate-sized, slightly tapering; integument reddish brown to piceous; pubescence dense, short, appressed, grayish and black. Head micropunctate, front and vertex with a few scattered, large punctures; pubescence fine, appressed, mostly grayish, usually dark on antennal tubercles, margins of front with a few long, erect setae; genae subequal in length to lower eye lobes; antennae 2-1/2 to three times longer than body, segments three to five fringed beneath, slightly produced at apices, eleventh segment elongate. Pronotum broader than long, sides tuberculate behind middle, spines short, acute; disk with four calluses behind apical impression, inner pair larger, middle with a small callus; punctures rather fine, well separated; pubescence dense, appressed, pale and black, two inner calluses usually dark pubescent, median area behind calluses usually darker, sides behind lateral tubercles with a few long, erect setae; prosternum moderately densely pale pubescent; meso- and metasternum densely pale pubescent, sides with glabrous spots. Elytra a little more than twice as long as broad; basal gibbosities feeble; punctures fine, dense, becoming finer and sparser toward apex; each elytron tricostate, inner pair joining before apex, outer pair not uniting and usually not extending to apex; pubescence short, appressed, grayish, two dark transverse, slightly oblique maculae present, one before middle and one behind, costae with interrupted black spots, these often coalescing to form short lines; apices broadly rounded. Legs robust; femora pale pubescent with numerous glabrous spots; tibiae biannulate with dark bands; tarsi with first two segments pale at bases. Abdomen densely pale pubescent with numerous glabrous spots; last sternite deeply emarginate at apex. Length, 8-15 mm.

Female. Form similar. Antennae about twice as long as body, third segment with a few suberect setae beneath, segments not fimbriate. Abdomen with last sternite elongate, deeply notched at apex, ovipositor extruded. Length (exclusive of ovipositor), 10-17 mm.

Type locality. Of *obliquus*, Kansas; *sedulus*, Fort Wingate, New Mexico; *acomanus*, New Mexico; *obtusus*, Priest Lake, Idaho; *pacificus*, Siskiyou Co., California; *chihuahuae*, Chihuahua, Mexico.

Range. British Columbia to northern Baja California east to South Dakota and south to New Mexico and central Mexico (Figure 8).

Flight period. April to September.

Host plants. Pinus spp. (including *P. ponderosa, P. scopulorum, P. murrayana, P. chihuahuana, P. contorta, P. flexilis, P. jeffreyi, P. edulis, P. leiophylla, P. sabiana, P. strobiformis, P. coulteri), Picea*.

The prominent costae and slightly oblique dark maculae of the elytra make this species distinctive. It may be separated from *A. angulosus* by the irregular, dark pubescent lines of the costae, only two costae joining before the apex, and by the outer costae not extending to the apex of the elytra. The black elytral maculae are also less angulate in *obliquus*.

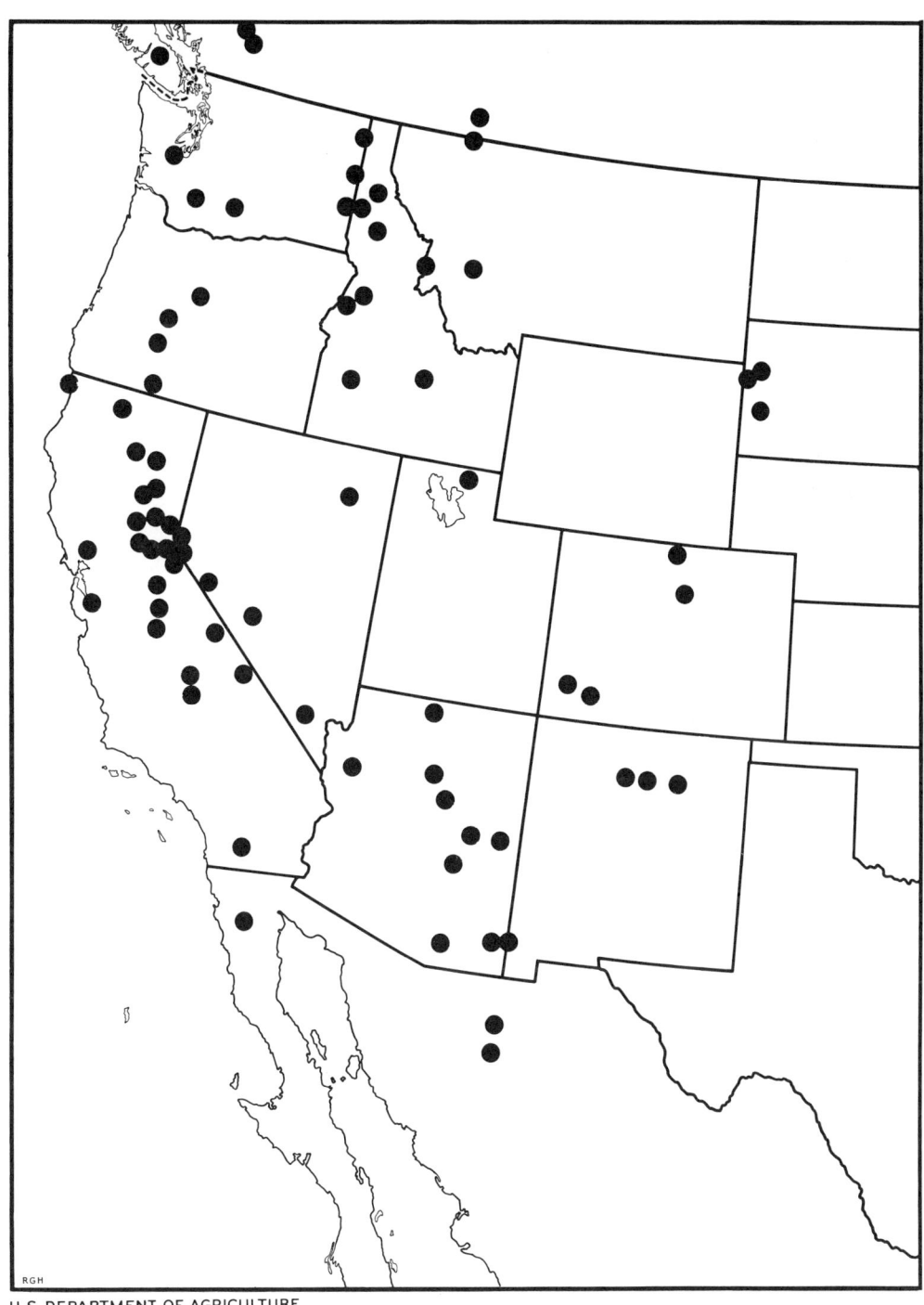

Figure 8. Known geographic range of *Acanthocinus obliquus* (LeConte).

Dillon (1956) recognized two subspecies, *A. obliquus obliquus* from the Rocky Mountain states and *A. obliquus pacificus* from the West Coast. In the large series of material now available for study, we can find no basis for this distinction. Differences such as reduction of the costal pubescence and more grayish overall appearance are present throughout the range of the species. The only area where this characteristic seems to prevail is in Craters of the Moon National Monument. The population there is fairly uniform in the more grayish appearance. Otherwise there appear to be no recognizable geographical differences even among populations ranging from British Columbia to Chihuahua, Mexico.

Acanthocinus spectabilis (LeConte)
(Figure 9)

Aedilis spectabilis LeConte, 1854, Proc. Acad. Nat. Sci. Philadelphia, 7:82.

Acanthocinus spectabilis: Horn, 1880, Trans. Amer. Entomol. Soc., 8:130; LeConte and Horn, 1883, Smithson. Misc. Coll., 507:324; Gahan, 1888, Trans. Amer. Entomol. Soc., 15:300; Leng and Hamilton, 1896, Trans. Amer. Entomol. Soc., 23:132 (part); Garnett, 1918, Can. Entomol., 50:282; Craighead, 1923, Can. Dept. Agr. Bull., (n.s.) 27:120, pl. 13, fig. 4 (part); Keen, 1938, USDA Misc. Pub., 273:136, fig. 67; Keen, 1952, USDA Misc. Pub., 273:174, fig. 81.

Acanthocinus (Canonura) spectabilis: Casey, 1913, Memoirs on the Coleoptera, 4:339; Aurivillius, 1923, Coleop. Cat., 74:434; Linsley, Knull, and Statham, 1961, Amer. Mus. Nov., 2050:30; Lewis, 1979, Pan-Pac. Entomol., 55:25.

Graphisurus spectabilis: Keen, 1929, Calif. Div. For. Bull., 7:62, figs. 7a, 31; Edwards, 1957, Pan-Pac. Entomol., 33:52; Essig, 1958, Ins. West. North America, p. 460.

Canonura spectabilis: Dillon, 1956, Ann. Entomol. Soc. Amer., 49:226, fig. 6; Tyson, 1966, Pan-Pac. Entomol., 42:203; Tyson, 1967, Pan-Pac. Entomol., 43:85 (habits); Kirk and Balsbaugh, 1975, S.D. Agr. Exp. Sta. Tech. Bull., 42:100.

Male. Form moderate-sized to large, sides tapering; integument reddish brown, antennae paler; first five segments brown annulate at apices; pubescence very short, appressed, grayish, pale and dark brownish. Head with front micropunctate, vertex with sparse, small punctures behind eyes; pubescence dense, appressed, brownish, narrowly grayish along margins; antennae almost four times longer than body, segments three to five densely fimbriate beneath, fifth segment with a large tuft at apex, segments pale pubescent, first five brown annulate at apices, eleventh segment elongate. Pronotum broader than long, sides acutely tuberculate; disk with a row of four calluses behind apex, median pair more prominent; punctures moderately coarse, rather dense; pubescence dense, appressed, brownish, each

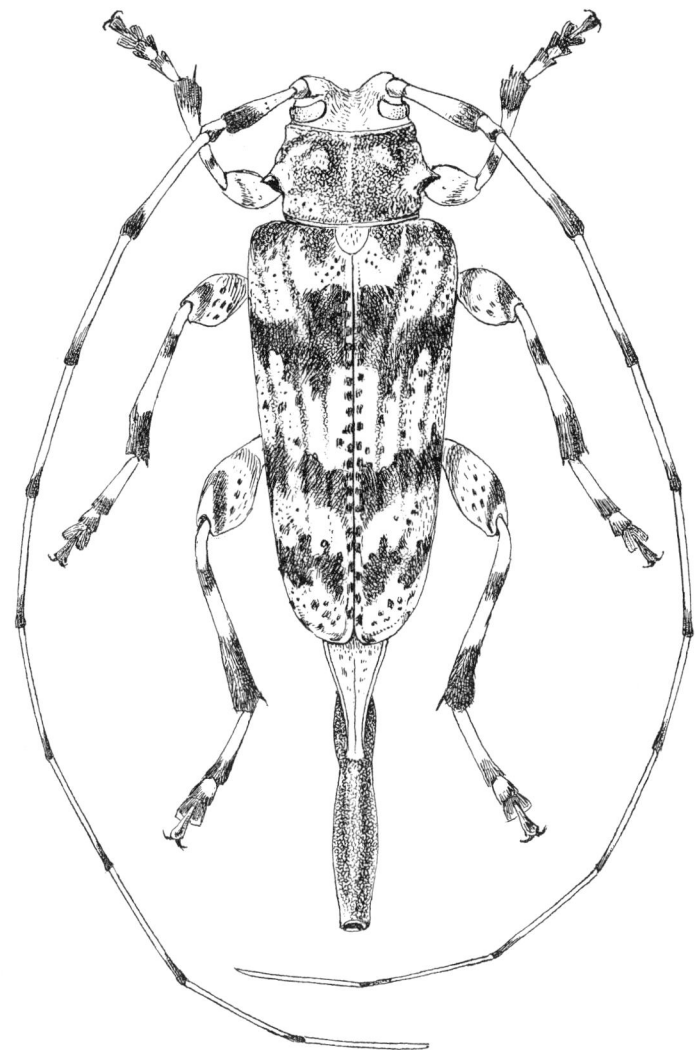

Figure 9. *Acanthocinus spectabilis* (LeConte), female.

side with a vaguely rounded grayish macula enclosing three large brownish spots, four post-apical calluses grayish pubescent, whitish maculae extending onto sides across anterior edges of lateral tubercles, middle with a vague longitudinal whitish stripe, sides with several long, erect setae; prosternum densely pale pubescent, pale brownish at sides; meso- and metasternum densely pale pubescent at middle, sides broadly pale brownish, pubescence at sides interrupted by glabrous punctures. Elytra less than 2-1/2 times as long as broad; each elytra vaguely costate at middle; base finely, densely asperate, punctures fine, rather sparse, becoming obsolete toward apex; four dark bands present, a narrow basal one, a slightly oblique, strongly bilobed one before middle, a broad irregularly margined one behind middle, and an angulate band before apex, three postbasal narrowly white margined, postmedian band suffused with pale brownish near suture, two dark brown spots present behind scutellum and remaining pale surface sparsely interspersed with small dark spots, grayish pubescence very fine; apices subtruncate to rounded. Legs robust; femora triannulate with brownish pubescence; tibiae biannulate; tarsi dark, first segment pale pubescent at apex. Abdomen densely pale pubescent, brownish at sides; last sternite deeply emarginate at apex. Length, 18-26 mm.

Female. Form similar. Antennae about twice as long as body, fifth segment not tufted at apex. Abdomen with last sternite elongate, deeply emarginate, ovipositor strongly extruded. Length (exclusive of ovipositor), 21-26 mm.

Type locality. Fort Union, New Mexico.

Range. Montana and South Dakota south to Arizona and New Mexico and northern Mexico.

Flight period. June to September.

Host plants. Pinus spp. (including *P. ponderosa, P. scopulorum, P. chihuahuana*).

This handsome species may be readily recognized by the brownish elytral bands and the pale brownish pubescence along the sides of the venter. The absence of large glabrous black spots on the antennal scape further serves to separate it from *A. princeps*.

Acanthocinus princeps (Walker)

Eutrypanus princeps Walker, 1866, in Lord, Nat. in British Columbia, 2:331.
Graphisurus (Canonura) princeps: Casey, 1913, Memoirs on the Coleoptera, 4:339.
Acanthocinus (Canonura) princeps: Aurivillius, 1923, Coleop. Cat., 74:434.
Canonura princeps: Dillon, 1956, Ann. Entomol. Soc. Amer., 49:227; Gilmour, 1965, Cat. Lam. du Monde, 8:559; Hatch, 1971, Univ. Wash. Pub. Biol., 16:151.
Acanthocinus spectabilis (part): LeConte, 1873, Smithson. Misc. Coll., 11(265):339; Horn, 1880, Trans. Amer. Entomol. Soc., 8:130; Leng and Hamilton, 1896, Trans. Amer. Entomol. Soc., 23:132; Craighead, 1923, Can. Dept. Agr. Bull., (n.s.) 27:120.
Graphisurus spectabilis (part): Doane et al., 1936, For. Ins., p. 189.
Graphisurus (Canonura) vinctus Casey, 1913, Memoirs on the Coleoptera, 4:340.

Male. Form moderate-sized to large, sides tapering; integument piceous, antennae pale, first seven segments dark at apices, parts of underside and apices of femora often paler; pubescence dense, short, appresed, gray, brownish and black. Head with front sparsely punctate, vertex with deep, scattered punctures; pubescence dense, appressed, mottled gray and brownish, interrupted by punctures, margins of front with a few long, erect, black hairs; genae subequal in length to lower eye lobes; antennae extending beyond elytra by more than six segments, pale appressed pubescence of basal half of scape interrupted by punctures, segments three to five densely fimbriate beneath, fifth segment with a large dark tuft at apex, segments finely, densely clothed with white, appressed pubescence, fourth segment slightly longer than third, fifth longer than fourth, eleventh segment very long. Pronotum broader than long, sides acutely tuberculate; disk with two rounded calluses behind apex on each side of middle and a vague median one; punctures deep, well separated, denser near middle; pubescence dense, appressed, pale brownish on calluses, dark at middle and gray at sides, pubescence interrupted by punctures; prosternum finely, densely pale pubescent; meso- and metasternum densely pale pubescent, interrupted by large glabrous spots at sides. Elytra a little more than twice as long as broad; each elytron bicostate over middle three-fifths; base rather densely asperate, asperites sparser behind scutellum; punctures small, moderately dense before middle, becoming obsolete toward apex; four dark transverse bands present, a narrow basal one, a medially broadened one before middle, an irregularly margined band behind middle, and a similar one before apex, dark bands mottled with brownish, gray portions with black pubescent spots interspersed throughout; apices rounded. Legs robust; femora pale pubescent with dark bands near apices, glabrous spots interrupting pubescence; tibiae narrowly dark annulate near base and broadly at apex; tarsi broadly white annulate at bases of first two segments. Abdomen densely pale pubescent with glabrous spots interspersed; apex of last sternite deeply emarginate. Length, 13-24 mm.

Female. Form similar, more robust. Antennae about twice as long as body, fifth segment lacking apical tuft, eleventh segment not elongate. Abdomen with last sternite elongated, deeply emarginate at apex, ovipositor strongly extruded. Length (exclusive of ovipositor), 14-23 mm.

Type locality. of *princeps*, British Columbia; *vinctus*, Siskiyou Co., California.

Range. Pacific Coast from British Columbia to California.

Flight period. May to August.

Host plants. Pinus spp. (including *P. ponderosa, P. sabiniana, P. jeffreyi*).

The four irregular, blackish transverse bands and the four pale brownish maculae of the pronotum distinguish this species from the others in the genus.

Acanthocinus leechi (Dillon)
(Figure 10)

Canonura leechi Dillon, 1956, Ann. Entomol. Soc. Amer., 49:228, fig. 5.

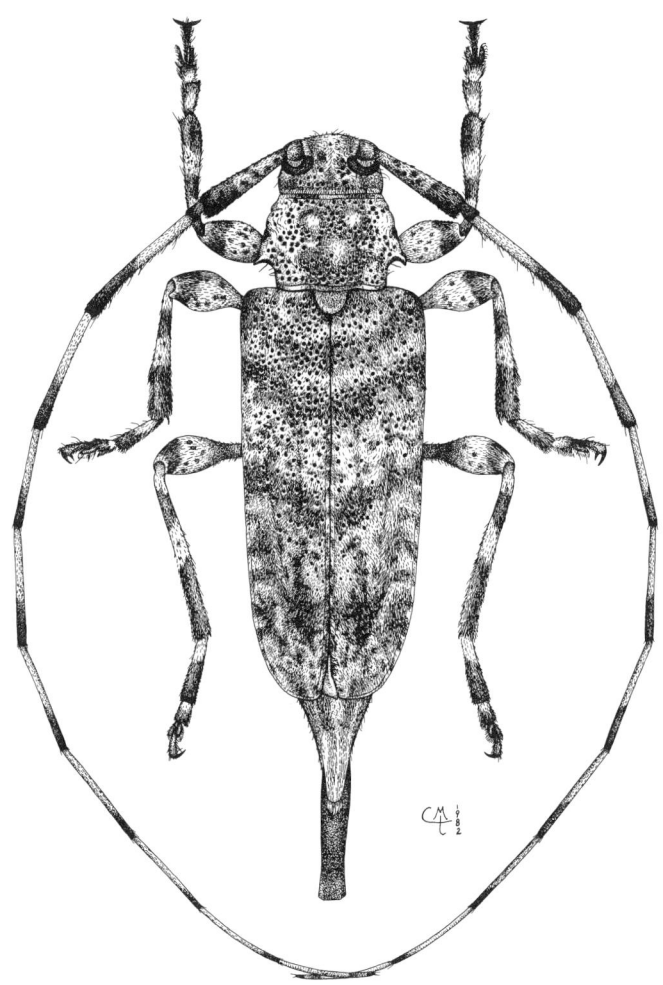

Figure 10. *Acanthocinus leechi* (Dillon), female.

Acanthocinus (Canonura) leechi: Lewis, 1976, Pan-Pac. Entomol., 52:203; Lewis, 1979, Pan-Pac. Entomol., 55:25.

Male. Form rather large, sides tapering; integument piceous, parts of antennae and apices of femora paler; pubescence dense, short, appressed, grayish, black, and pale brownish. Head with front minutely punctate with a few larger punctures interspersed, several large punctures present along eye margins, vertex coarsely, densely punctate; pubescence mottled grayish, front with dark patches, margins of front with long erect setae; genae shorter than lower eye lobes; antennae more than three times longer than body, segments dark annulate at apices, vaguely from eighth, segments three to five densely fimbriate beneath, expanded at apices, fifth segment with a dense tuft at apex, segments three to five subequal in length, eleventh a little longer than tenth. Pronotum broader than long, sides acutely tuberculate behind middle; disk with four calluses behind apical margin, two inner larger, median callus vague, a callus present at sides near posterior edges of lateral tubercles; punctures deep, dense; pubescence fine, appressed, mottled grayish at sides, dark at middle, inner apical calluses very pale brownish pubescent, several long, erect setae present on posterior half of lateral tubercles; prosternum mottled grayish pubescent; meso- and metasternum densely grayish pubescent, pubescence strongly interrupted by glabrous spots. Elytra a little more than twice as long as broad; base narrowly asperate; punctures behind base moderately coarse, well separated, becoming obsolete behind middle; each elytron vaguely bicostate; pubescence mottled black and grayish, transverse bands except postmedian vague, black pubescence usually dominant; apices broadly rounded. Legs robust; femora with gray pubescence interrupted by glabrous spots, with a large dark spot dorsally at apices; tibiae biannulate; tarsi with first two segments pale pubescent at apices. Abdomen densely pale pubescent with numerous glabrous spots; last sternite emarginate at apex. Length, 18-22 mm.

Female. Form similar. Antennae less than twice as long as body, segments three and four weakly fimbriate beneath. Abdomen with last sternite elongate, deeply notched at apex, ovipositor strongly extruded. Length (exclusive of ovipositor), 16-20 mm.

Type locality. Jerome, Arizona.
Range. Montane Central Arizona.
Flight period. July to September.
Host plants. Pinus monophylla.

This species is closely related to *A. princeps. A. leechi* may be separated by the indistinct, all-black maculae of the elytra, the shorter genae, and the proportions of the antennal segments. The absence of brownish pubescence on the underside readily distinguishes it from *A. spectabilis.*

Genus *Sternidocinus* Dillon

Sternidocinus Dillon, 1956, Ann. Entomol. Soc. Amer., 49:166; Arnett, 1962,
    Beetles U.S., 103:872.

Form moderate-sized, subcylindrical. Head with front quadrate, rather broad, convex in males, plane in females; mandibles feebly arcuate, apices acute; genae longer than lower eye lobes, sides divergent in females, convergent in males; eyes moderate, upper lobes widely separated; antennal tubercles moderate, divergent; antennae slender, eleven-segmented, about one-third longer than body in males, a little longer than body in females, scape and third segment with several suberect hairs beneath, scape subequal in length to segments three and four, remaining segments gradually decreasing in length. Pronotum broader than long, sides acutely tuberculate behind middle; apical impression vague, shallow, basal impression broad, deep, extending onto sides; disk with vague, shallow calluses at sides of middle; prosternum with intercoxal process narrow, about one-sixth as broad as coxal cavity; mesosternal process plane, about two-fifths as broad as coxal cavity; metasternum with episternum narrow, tapering posteriorly. Elytra a little more than 2-1/2 times as long as broad; each elytron tricostate, with another short costa behind middle; basal gibbosities shallow; apices rounded to subtruncate. Legs robust; femora strongly clavate; tibiae with short apical spurs, middle pair with an external sinus; tarsi with first segment of hind pair longer than two following segments together. Abdomen with last sternite longer than preceding segments.

Type species. Liopus barbarus Van Dyke (monobasic and by original designation).

The sexual differences of the head, relatively short antennae, proportions of the antennal segments, and the short tibial spurs separate this genus from other North American acanthocines.

A single species is known.

Sternidocinus barbarus (Van Dyke)
(Figure 11)

Liopus barbarus Van Dyke, 1920, Bull. Brooklyn Entomol. Soc., 15:45.
Leropus barbarus: Doane et al., 1936, For. Ins., p. 189 (error).
Sternidocinus barbarus: Dillon, 1956, Ann. Entomol. Soc. Amer., 49:166;
    Hovore and Giesbert, 1976, Coleop. Bull., 30:358 (habits).

Male. Form moderate-sized, subparallel; integument pale reddish brown, underside partially infuscated; pubescence dense, appressed, grayish and brownish. Head with front micropunctate, densely gray and brownish pubescent, genae with several long, erect setae; vertex minutely punctate, densely pubescent; antennae extending about four segments beyond elytra, segments dorsally brown annulate at apices, segments four and five with excavated poriferous areas at apices. Pronotum broader than long; disk with two shallow calluses and a vague elongate, median one; punctures irregular, sparse; pubescence dense, grayish with pale brown sparsely interspersed, a few erect setae present behind lateral tubercles; prosternum densely gray pubescent; meso- and metasternum densely pubescent, erect hairs absent. Elytra a little more than twice as long as broad; punctures moderately dense, becoming obsolete toward apex; pubescence dense, appressed, grayish, base pale brownish, small brownish spots usually interspersed along suture and costae, an angulate brownish macula present behind middle and usually

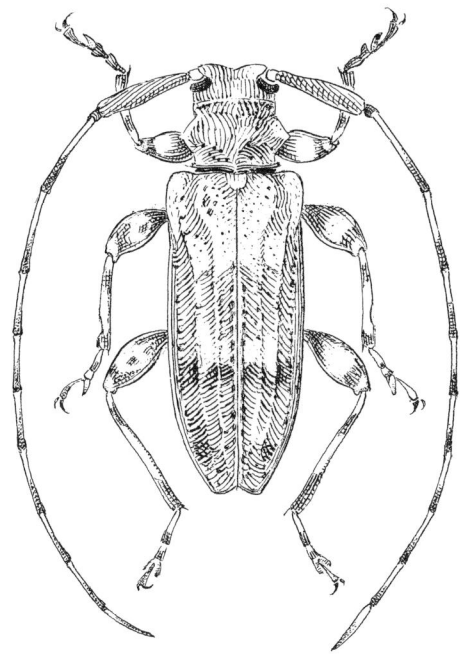

Figure 11. *Sternidocinus barbarus* (Van Dyke), male.

larger brownish spots present just before apex; apices subtruncate. Legs densely gray pubescent; femora dorsally brownish annulate at apices; tibiae dorsally brownish biannulate; tarsi gray pubescent. Abdomen densely grayish pubescent; last sternite shallowly emarginate at apex. Length, 9-12 mm.

Female. Form more robust. Antennae a little longer than body. Head with front plane. Abdomen with last sternite narrow at apex, lightly fringed. Length, 11-15 mm.

Type locality. Carpenteria, Santa Barbara County, California.

Range. Santa Barbara to Los Angeles County and Santa Cruz Island, California.

Flight period. April to August.

Host plants. Quercus agrifolia.

The short antennae, coloration, and sexual dimorphism of the head make this species easily recognizable.

According to Hovore and Giesbert (1976), larvae utilize dead, thick bark and mine the cambium layer and inner bark surface. The galleries are packed with fine, shredded frass. The pupal cell is constructed inside the bark, similar to that of *Acanthocinus.*

Genus *Astylopsis* Casey

Astylopsis Casey, 1913, Memoirs on the Coleoptera, 4:309; Aurivillius, 1923, Coleop. Cat., 74:398; Bradley, 1930, Man. Gen. Beetles, p. 246; Chagnon, 1933-40, Coleop. Prov. Quebec, p. 272; Knull, 1946, Ohio Biol. Surv. Bull., 39:245; Chagnon and Robert, 1962, Prin. Coleop. Prov. Quebec, p. 272.

Amniscus Haldeman, 1847 (not Dejean, 1837), Trans. Amer. Philos. Soc., (2)10:46; White, 1855, Cat. Coleop. Ins. British Mus., 8:390; Lacordaire, 1872, Genera des coléoptères, 9(2):761; Dillon, 1956, Ann. Entomol. Soc. Amer., 49:151; Dillon and Dillon, 1961, Man. Beetles East. North America, p. 638; Arnett, 1962, Beetles U.S., 103:872; Bayer and Shenefelt, 1969, Univ. Wisc. Res. Bull., 275:26, 27; Rice and Enns, 1981, Trans. Mo. Acad. Sci., 15:98. (Type species: *Lamia macula* Say, Dillon designation, 1956.)

Leptostylus: Leng and Hamilton, 1896, Trans. Amer. Entomol. Soc., 23:116 (part).

Form moderately small, subdepressed. Head with front slightly to moderately convex, about as long as broad; mandibles moderately arcuate; genae slightly convergent, about as long as lower eye lobes; eyes moderate, deeply emarginate, upper lobes small, widely separated above; antennal tubercles prominent, widely divergent; antennae slender, eleven-segmented, slightly longer than body in females, a little longer in males, first segment extending a little beyond middle of pronotum, third segment longer than first, fourth slightly shorter than third. Pronotom broader than long, sides acutely to obtusely tuberculate behind middle; disk with five obtuse calluses, median not larger than others, surface between calluses usually densely punctate; base and apex broadly impressed transversely; prosternum with intercoxal process arcuate, one-third to half as broad as coxal cavities; mesosternum with intercoxal process abruptly declivous anteriorly, about four-fifths as broad as coxal cavities, broadening anteriorly; metasternum rather short, episternum narrow. Elytra about 1.7 to two times as long as broad, sides tapering at apical one-third; basal gibbosities shallow; disk costate, often with tufted tubercles; apices narrowing, rounded to obliquely truncate. Legs with femora pedunculate; tibiae slender; tarsi slender, rather long, first segment shorter than or subequal to two following segments together, third segment cleft to base. Abdomen normally segmented, last sternite of females at least twice as long as fourth, narrowing apically.

Type species. Lamia macula Say (by original designation).

The distinctly, usually densely punctate pronotum, the five moderate calluses of the pronotum, the relatively short antennae, and broad mesosternal process serve to distinguish this genus.

There has been much confusion in the use of the generic name *Amniscus*. Most workers have failed to credit Dejean (1835) with the authorship, and there has been no designation of a type species until Monne and Giesbert (1992) who designated *Lamia praemorsus* Fabricius as the type of *Amniscus* Dejean. Gilmour (1963) included this species in his genus *Leptostyloides*.

Haldeman (1847) applied the name *Amniscus* to a group of North American species from which Dillon (1956) designated *Lamia macula* Say as the type. However, since this name is a primary homonym, the valid name for the North American components of the genus becomes *Astylopsis* Casey (1913), which also has *Lamia macula* Say as the type.

Amniscus Bates (1863) (not Dejean, 1835) is replaced by *Alcathous* Thomson (1864). These are isogenotypic with *Alcidion polyrhaphoides* White, the type by original designation.

Five species of *Astylopsis* are presently known in North America.

KEY TO THE NORTH AMERICAN SPECIES OF *ASTYLOPSIS*

1	Elytra lacking black vittae at sides	2
	Elytra with black vittae at sides extending from humeri almost to apex, apical one-third brownish. Length, 7-13 mm. Eastern United States from Maryland to Florida westward to Arkansas	*arcuata*
2(1)	Elytra with brownish spots or tubercles, lacking black tufted tubercles	3
	Elytra with black tufted tubercles especially on basal gibbosities, disk coarsely, densely punctate. Pubescence mottled grayish, brownish and black. Length, 6-10.5 mm. Eastern North America west to Texas	*sexguttata*
3(2)	Elytra without a distinct whitish, transverse macula behind middle. Pronotum not contrastingly pubescent	4
	Elytra with a transverse whitish fascia behind middle. Pronotum whitish pubescent at sides. Length, 6-10 mm. Eastern North America to North Carolina westward to Kansas	*macula*
4(3)	Elytra with a narrow dark fascia behind middle extending obliquely back from suture. Pronotum separately punctate. Larger species. Length, 10-15 mm. Southeastern United States from North Carolina to Florida west to Texas	*perplexa*

Elytra with numerous brownish spots, dark area at apical one-third, if present, extending obliquely upward from suture. Pronotum confluently punctate. Smaller species. Length, 6-10 mm. Eastern United States to Georgia, west to Minnesota...*collaris*

Astylopsis arcuata (LeConte), new combination
(Figure 12)

Leptostylus arcuatus LeConte, 1878, Proc. Amer. Philos. Soc., 17:414; Horn, 1880, Trans. Amer. Entomol. Soc., 8:121; Leng and Hamilton, 1896, Trans. Amer. Entomol. Soc., 23:118; Knull, 1942, Entomol. News, 53:227; Fattig, 1947, Emory Univ. Mus. Bull., 5:34.
Amniscus arcuatus: Dillon, 1956, Ann. Entomol. Soc. Amer., 49:154; Turnbow and Franklin, 1980, J. Ga. Entomol. Soc., 15:344.
Amniscus arcuatus arcuatus: Dillon, 1956, Ann. Entomol. Soc. Amer., 49:154; Kirk, 1969, S.C. Agr. Exp. Sta. Tech. Bull., 1033:86.
Leptostylus floridanus Champlain and Knull, 1922, Entomol. News, 33:148.
Leptostylus knulli Fisher, 1925, Proc. Entomol. Soc. Wash., 27:103; Knull, 1937, Entomol. News, 48:42; Fattig, 1947, Emory Univ. Mus. Bull., 5:34. New synonymy.
Amniscus arcuatus knulli: Dillon, 1956, Ann. Entomol. Soc. Amer., 49:154; Wray, 1967, Insects N.C., 3rd supp., p. 47; Kirk, 1969, S.C. Agr. Exp. Sta. Tech. Bull., 1033:86; Perry, 1974, Coleop. Bull., 28:216.

Male. Form moderate-sized; integument pale reddish brown, head, antennae, legs, and underside partly black, elytra with black vittae; pubescence dense, short, appressed, grayish, brownish and black. Head with front slightly convex, densely micropunctate, densely clothed with mottled brownish, appressed pubescence; genae longer than lower eye lobes; upper eye lobes small, separated by more than diameter of antennal scape; antennae a little longer than body, segments dark annulate at apices, basal segments dark spotted between annulae, last segment dark, pubescence fine, grayish and dark brown or black, third segment longer than first, fourth subequal to first. Pronotum with sides distinctly tuberculate behind middle, tubercles usually acute; disk with calluses low; apical impression shallow; basal impression broad, deeper; punctures fine, dense; pubescence dense, appressed, grayish and brownish, usually with a longitudinal dark vitta on each side, sides beneath lateral tubercles with a dark vitta extending from tubercles to margin; prosternum with intercoxal process slightly more than half as broad as coxal cavities; meso- and metasternum finely pubescent, spotted, mesosternal process about as broad as coxal cavities. Elytra slightly less than twice as long as broad, sides slightly emarginate before middle; basal gibbosities broad; costae distinct, epipleura almost vertical; tufted tubercles irregularly placed on costae, those on basal gibbosities and at apical

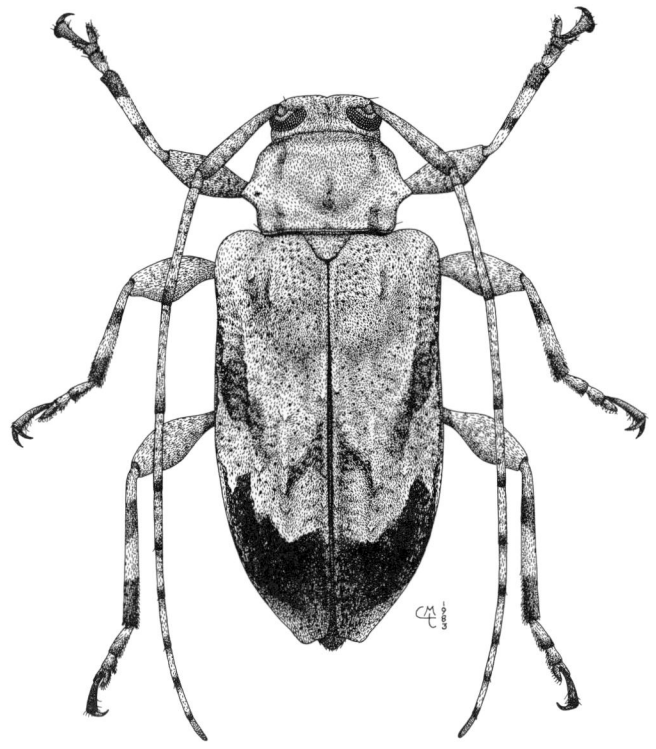

Figure 12. *Astylopsis arcuata* (LeConte), male.

one-third more prominent; punctures dense, finer toward apex; pubescence dense, appressed, mostly grayish over basal two-thirds, brownish on apical one-third, sides with black vittae extending from under humeri to about apical one-fifth, vittae obliquely extending into disk before middle and arcuately extending across to suture at apical one-third, suture with an inverted Y-shaped black vitta behind middle, tubercles on basal gibbosities tufted with black; apices narrowly, shallowly emarginate truncate. Legs short; femora finely pale pubescent; tibiae black biannulate; tarsi black. Abdomen finely pubescent; last sternite emarginate at apex. Length, 7-13 mm.

Female. Form similar. Antennae about as long as body. Abdomen with fifth sternite twice as long as fourth, apex narrowly rounded. Length, 7-12 mm.

Type locality. of *arcuata*, Tampa, Florida; *floridanus*, Miami, Florida; *knulli*, Dorchester Co. (near Lloyds), Maryland.
Range. Maryland to Florida west to Arkansas and Louisiana.
Flight period. March to December.
Host plants. Pinus.

The black lateral vittae of the elytra make this species distinctive.

As in many eastern United States Lamiinae, specimens from southern Florida tend to have paler pubescence (named *knulli* by Fisher). Northern individuals are more brownish and usually do not have as much grayish pubescence. There may be justification for recognizing these populations as subspecies, but defining the geographical range of each cannot now be done with available material.

Astylopsis sexguttata (Say)

Lamia 6-guttata Say, 1826, J. Acad. Nat. Sci. Philadelphia, 5:269; LeConte, 1859, Compl. Writings T. Say, 2: preface.

Leptostylus sexguttatus: Leng and Hamilton, 1896, Trans. Amer. Entomol. Soc., 23:119; Wickham, 1897, Can. Entomol., 29:207; Harrington, 1899, Ottawa Nat., 13:67; Smith, 1910, N.J. St. Mus. Rept., 1909:333; Blatchley, 1910, Coleop. in Indiana, p. 1072, fig. 460; Blackman and Stage, 1918, N.Y. St. Coll. For. Tech. Pub., 10:47, pl. 6, fig. 18 (habits); Craighead, 1923, Can. Dept. Agr. Bull., (n.s.) 27:116.

Amniscus sexguttatus: White, 1855, Cat. Coleop. Ins. British Mus., 8:390; Dillon, 1956, Ann. Entomol. Soc. Amer., 49:153; Dillon and Dillon, 1961, Man. Beetles East. North America, p. 638, pl. 64, no. 3; Kirk, 1970, S.C. Agr. Exp. Sta. Tech. Bull., 1038:82; Headstrom, 1977, Beetles of America, p. 377.

Amniscus sexguttata: Gardiner, 1957, Can. Entomol., 89:251; Bayer and Shenefelt, 1969, Univ. Wisc. Res. Bull., 275:27, fig. 35; Perry, 1975, Coleop. Bull., 29:59; Gosling and Gosling, 1976, Gr. Lakes Entomol., 10:25, fig. 144; Rice and Enns, 1981, Trans. Mo. Acad. Sci., 15:98; Gosling, 1984, Gr. Lakes Entomol., 17:70.

Lamia guttata: LeConte, 1859, Comp. Writings T. Say, 2:328 (error).

Astylopsis guttata: Casey, 1913, Memoirs on the Coleoptera, 4:309; Blackman and Stage, 1924, N.Y. St. Coll. For. Tech. Pub., 17:115; Chagnon, 1933-40, Coleop. Prov. Quebec, p. 272; Procter, 1946, Biol. Surv. Mt. Desert, 7:183; Chagnon and Robert, 1962, Prin. Coleop. Prov. Quebec, p. 272.

Astylopsis sexguttata: Aurivillius, 1923, Coleop. Cat., 74:399; Knull, 1946, Ohio Biol. Surv. Bull., 39:246; Fattig, 1947, Emory Univ. Mus. Bull., 5:35.

Amniscus commixtus Haldeman, 1847, Trans. Amer. Philos. Soc., (2)10:47; White, 1855, Cat. Coleop. Ins. British Mus., 8:393.

Leptostylus commixtus: LeConte, 1852, J. Acad. Nat. Sci. Philadelphia, 2:169; Fitch, 1859, 4th Rept. Nox. Ben. Ins. N.Y., p. 712; LeConte, 1873,

Smithson. Misc. Coll., 11(264):233; Horn, 1880, Trans. Amer. Entomol. Soc., 8:121; Packard, 1881, U.S. Entomol. Comm. Bull. 7:157; Packard, 1890, Ins. Inj. For. Trees, p. 697; fig. 233; Beutenmuller, 1896, J. N.Y. Entomol. Soc., 4:79.

Male. Form relatively small; integument reddish brown, partially infuscated on head, antennae, pronotum, elytra, legs, and underside; pubescence fine, dense, appressed, brownish, black, and white, elytra with black tufts. Head with front slightly convex, densely micropunctate, moderately densely clothed with appressed grayish or brownish mottled pubescence; genae longer than lower eye lobes; upper eye lobes small, separated by more than diameter of antennal scape; antennae a little longer than body, segments mostly dark brownish, segments four to ten narrowly pale annulate basally, third segment longer than first, fourth subequal to first. Pronotum with sides subacutely tuberculate behind middle; disk with calluses not prominent, anterior pair larger; apical impression usually shallow; basal impression broader; punctures dense, fine; pubescence fine, appressed, pale brownish, calluses usually dark, each side of middle with at least a vague, longitudinal dark vitta; prosternum with intercoxal process slightly less than half as broad as coxal cavities; mesosternum with intercoxal process almost as broad as coxal cavities; metasternum with pubescence interrupted by spots. Elytra a little less than twice as long as broad, side slightly emarginate before middle; basal gibbosities moderate, prominently dark-tufted; costae distinct, dark-tufted tubercles irregularly placed down costae, more numerous behind middle; punctures coarse, dense, becoming finer toward apex; pubescence fine, dense, mottled brownish and whitish, apical one-fourth with vague, pale sutural spots and dark spots immediately behind; apices lightly emarginate-truncate. Legs stout; femora pale basally; tibiae dark biannulate; tarsi dark. Abdomen finely pubescent; fifth sternite about as long as fourth, narrowly emarginate at apex. Length, 6-9 mm.

Female. Form similar, more robust. Antennae slightly longer than body. Abdomen with fifth sternite twice as long as fourth; apex narrowly rounded. Length, 7-10.5 mm.

Type locality. of *sexguttata*, United States; *commixtus*, Washington City.

Range. Eastern North America from Ontario to Florida west to Texas and Minnesota.

Flight period. April to September.

Host plants. *Larix, Picea, Pinus echinata*.

The black-tufted tubercles of the elytra and the two conspicuous postmedian pale, sutural elytral spots distinguish this species.

The color varies in the amount of brownish and grayish pubescence suffused over the surface. The degree of infuscation of the pronotum, legs, antennae, and underside also varies considerably.

Astylopsis macula (Say)

Lamia macula Say, 1826, J. Acad. Nat. Sci. Philadelphia, 5:268; LeConte, 1859, Compl. Writings T. Say, 2:327.

Leptostylus macula: LeConte, 1852, J. Acad. Nat. Sci. Philadelphia, 2:169; Fitch, 1857, 3rd Rept. Nox. Ben. Ins. N.Y., p. 462 (habits); Lacordaire, 1872, Genera des coléoptères, 9(2):772, fn.; LeConte, 1873, Smithson. Misc. Coll., 11(264):233; Provancher, 1877, Pet. Fauna Entomol. Can., 1:628; Horn, 1880, Trans. Amer. Entomol. Soc., 8:120; Packard, 1881, U.S. Entomol. Comm. Bull., 7:85; Packard, 1890, Ins. Inj. For. Trees, p. 337; Chittenden, 1894, Proc. Entomol. Soc. Wash., 3:100; Leng and Hamilton, 1896, Trans. Amer. Entomol. Soc., 23:121; Beutenmuller, 1896, J. N.Y. Entomol. Soc., 4:79; Wickham, 1897, Can. Entomol., 29:207; Lugger, 1899, Minn. Agr. Exp. Sta. Bull., 66:208; Harrington, 1899, Ottawa Nat., 13:67; Blatchley, 1910, Coleoptera in Indiana, p. 1072; Smith, 1910, N.J. St. Mus. Rept., 1909:333; Craighead, 1923, Can. Dept. Agr. Bull., (n.s.) 27:115, pl. 16, fig. 7, pl. 31, fig. 3.

Amniscus macula: Haldeman, 1847, Trans. Amer. Philos. Soc., (2)10:48; White, 1855, Cat. Coleop. Ins. British Mus., 8:391; Dillon, 1956, Ann. Entomol. Soc. Amer., 49:152; Gardiner, 1966, Can. J. Zool., 44:204, figs. 25, 52; Gardiner, 1969, Can. Dept. Fish. For. Int. Rept., 0-14:83; Gosling and Gosling, 1976, Gr. Lakes Entomol., 10:25, fig. 143; Chamberland, 1976, Fabreries, 2:89; Laliberte et al., 1977, Fabreries, 3:89; Rice and Enns, 1981, Trans. Mo. Acad. Sci., 15:98; Gosling, 1984, Gr. Lakes Entomol., 17:70.

Amniscus maculus: Dillon and Dillon, 1961, Man. Beetles East. North America, p. 638, pl. 64, no. 2; Headstrom, 1977, Beetles of America, p. 377.

Astylopsis macula: Casey, 1913, Memoirs on the Coleoptera, 4:309; Aurivillius, 1923, Coleop. Cat., 74:398, Leonard, 1928, Cornell Agr. Exp. Sta. Mem., 101:451; Knull, 1930, Entomol. News, 41:102; Park, 1931, Ecology, 12:192; Barrett, 1932, Univ. Calif. Pub. Entomol., 5:289; Chagnon, 1933-40, Coleop. Prov. Quebec, p. 272; Knull, 1934, Entomol. News, 45:211; Hoffmann, 1942, USDA Misc. Pub., 466:11; Knull, 1946, Ohio Biol. Surv. Bull., 39:245, pl. 22, fig. 88; Fattig, 1947, Emory Univ. Mus. Bull., 5:34; Chagnon and Robert, 1962, Prin. Coleop. Prov. Quebec, p. 272; Furth, 1985, Conn. Acad. Arts Sci., 46:192.

Astylopsis (Leptostylus) macula: Craighead, 1950, USDA Misc. Pub., 657:251.

Astylopsis maculata: Papp, 1955, Entomol. News, 66:219 (error).

Amniscus sticticus Haldeman, 1847, Trans. Amer. Philos. Soc., (2)10:48; Haldeman, 1847, Proc. Amer. Philos. Soc., 4:372.

Male. Form small to moderate-sized, rather broad, subdepressed; integument brownish to reddish brown, scutellum dark, antennae with segments dark annulate at apices; pubescence fine, appressed, whitish, pale and dark brown. Head with front convex, micropunctate, densely clothed with appressed grayish and brownish mottled pubescence; genae longer than lower eye lobes; upper eye lobes small, separated by more than diameter of

Figure 13. Known geographic range of *Astylopsis macula* (Say)

antennal scape; antennae slightly longer than body, segments dark annulated at apices, outer segments dark for about half their lengths, pubescence short, fine, grayish and brown, third segment longer than first, fourth shorter than third, subequal to first. Pronotum with sides vaguely tuberculate behind middle; disk with calluses moderately elevated; apical impression narrow; basal impression broad; punctures dense, deep, separated; pubescence short, appressed, white to grayish at sides, median portion grayish to pale brownish, each side with a brownish vitta extending down over outside calluses; prosternum with intercoxal process slightly more than half as broad as coxal cavities; meso- and metasternum finely grayish pubescent, pubescence interrupted by glabrous spots, mesosternal process about as broad as coxal cavities. Elytra about 1-1/2 times as long as broad, sides slightly broadening behind middle, then abruptly tapering toward apex; basal gibbosities shallow, impression behind moderate; disk shallowly costate; pubescence dense, fine, appressed, mottled pale brownish with darker brown spots irregularly placed down costae, an usually irregular, transverse, whitish macula present behind middle, usually broader at sides and narrowing toward suture, apical one-fourth with scattered whitish pubescence; basal punctures dense, coarse, punctures becoming finer and sparser toward apex; apices narrowly rounded to obliquely truncate. Legs short; femora finely clothed with grayish pubescence interrupted by small brownish spots; tibiae biannulate with dark brown pubescence; tarsi with two basal segments grayish pubescent, apices brownish. Abdomen finely pubescent; fifth sternite about half longer than fourth, apex narrow, slightly emarginate at middle. Length, 6-8 mm.

Female. Form similar, slightly more robust. Antennae about as long as body. Abdomen with last sternite twice as long as fourth, apex narrow, margin setose. Length, 6-9 mm.

Type locality. of *macula*, not indicated (United States); *sticticus*, none (United States).

Range. Eastern North America from Quebec to North Carolina west to Kansas and Minnesota (Figure 13).

Flight period. May to September.

Host plants. Acer, Aesculus, Carpinus, Carya, Castanea, Celastrus, Cornus, Fagus, Hamamelis, Juglans, Malus, Quercus, Rhus, Tilia, Toxicodendron, Ulmus.

This species may be recognized by the whitish fascia behind the middle of the elytra, the brownish, nontufted elytral spots, and the small, vague lateral tubercles of the pronotum.

Astylopsis perplexa (Haldeman), new combination

Amniscus perplexus Haldeman, 1847, Trans. Amer. Philos. Soc., (2)10:46; Casey, 1913, Memoirs on the Coleoptera, 4:309; Dillon, 1956, Ann. Entomol. Soc. Amer., 49:155, fig. 5; Rice et al., 1985, Coleop. Bull., 39:22;

Palmer, 1987, Proc. Entomol. Soc. Wash., 89:195 (habits); Palmer and Bennett, 1988, Proc. Entomol. Soc. Wash., 90:221.

Leptostylus perplexus: LeConte, 1852, J. Acad. Nat. Sci. Philadelphia, 2:169; White, 1855, Cat. Coleop. Ins. British Mus., 8:392; Lacordaire, 1872, Genera des coléoptères, 9(2):772, fn.; LeConte, 1873, Smithson. Misc. Coll., 11(264):233; Horn, 1880, Trans. Amer. Entomol. Soc., 8:120; Leng and Hamilton, 1896, Trans. Amer. Entomol. Soc., 23:120; Aurivillius, 1923, Coleop. Cat., 74:402; Fattig, 1947, Emory Univ. Mus. Bull., 10:34.

Male. Form moderate-sized; integument pale reddish brown to dark brown; pubescence fine, appressed, pale and dark brownish. Head with front convex, micropunctate, densely clothed with pale brownish pubescence interspersed with small, dark spots; genae shorter than lower eye lobes; upper lobes small, separated by about diameter of antennal scape; antennae a little longer than body, segments to fourth with numerous dark spots, third segment longer than first, fourth subequal to first. Pronotum with sides obtusely tuberculate behind middle; disk with calluses shallow, anterior pair more prominent; apical impression narrow, basal impression broader, more shallow; punctures sparse around calluses, denser along impressions; pubescence dense, appressed, mottled brownish; prosternum with intercoxal process less than half as broad as coxal cavities; mesosternum with intercoxal process more than half as broad as long; metasternum with pale brownish pubescence interrupted by numerous dark spots. Elytra about twice as long as broad, sides tapering behind middle; basal gibbosities shallow; costae distinct, small tubercles not tufted; punctures coarse, contiguous, becoming finer behind middle; pubescence fine, mottled pale brownish, apical one-third darker, forming an oblique chevron directed back from suture, pale brown pubescent tubercles numerous along costae; apices narrow, subtruncate to rounded. Legs stout; femora strongly clavate, pubescence pale brown with numerous dark brown spots; tibiae uniformly pubescent; tarsi dark or pale pubescent. Abdomen finely pubescent with small, dark spots interspersed; last sternite rounded at apex. Length, 10-14 mm.

Female. Form similar. Antennae about as long as body. Abdomen with last sternite twice as long as fourth, apex narrow, shallowly emarginate. Length, 11-15 mm.

Type locality. Not indicated (North America).

Range. South Carolina to Florida westward to Texas.

Flight period. May to September (reared specimens emerged in May and October).

Host plants. Baccharis halmifolia.

A. *perplexa* differs from others in the genus by the larger, more elongate form, more convex front of the head, and somewhat narrower prosternal process. Specimens from the eastern part of the range usually have paler pubescence than those from Texas and Louisiana, as well as pale pubescent tarsi.

This species is currently being studied as a biological control agent for *Baccharis* in Texas by W. A. Palmer, who kindly provided specimens for study.

Astylopsis collaris (Haldeman)

Amniscus collaris Haldeman, 1847, Trans. Amer. Philos. Soc., (2)10:46; White, 1855, Cat. Coleop. Ins. British Mus., 8:393; Casey, 1913, Memoirs on the Coleoptera, 4:309; Dillon, 1956, Ann. Entomol. Soc. Amer., 49:151; Gosling and Gosling, 1976, Gr. Lakes Entomol., 10:25, fig. 145; Laliberte et al., 1977, Fabreries, 3:89; Turnbow and Franklin, 1980, J. Ga. Entomol. Soc., 15:339; Gosling, 1984, Gr. Lakes Entomol., 17:70; Morris, 1987, J. Entomol. Sci., 22:140.

Leptostylus collaris: LeConte, 1852, J. Acad. Nat. Sci. Philadelphia, 2:169; Lacordaire, 1872, Genera des coléoptères, 9(2):772, fn.; LeConte, 1873, Smithson. Misc. Coll., 11(264):233; Horn, 1880, Trans. Amer. Entomol. Soc., 8:120; Chittenden, 1894, Proc. Entomol. Soc. Wash., 3:100; Leng and Hamilton, 1896, Trans. Amer. Entomol. Soc., 23:120; Wickham, 1897, Can. Entomol., 29:207; Blatchley, 1910, Coleoptera in Indiana, p. 1072; Smith, 1910, N.J. St. Mus. Rept., 1909:333; Craighead, 1923, Can. Dept. Agr. Bull., (n.s.) 27:115; Knull, 1934, Entomol. News, 45:211; Fattig, 1947, Emory Univ. Mus. Bull., 5:34.

Astylopsis collaris: Knull, 1946, Ohio Biol. Surv. Bull., 39:246.

Amniscus interruptus Haldeman, 1847, Trans. Amer. Philos. Soc., (2)10:48; White, 1855, Cat. Coleop. Ins. British Mus., 8:393.

Leptostylus interruptus: LeConte, 1852, J. Acad. Nat. Sci. Philadelphia, 2:170; Lacordaire, 1872, Genera des coléoptères 9(2):772, fn.

Leptostylus collaris var. *interruptus*: Leng and Hamilton, 1896, Trans. Amer. Entomol. Soc., 23:120; Aurivillius, 1923, Coleop. Cat., 74:400; Leonard, 1928, Cornell Agr. Exp. Sta. Mem., 101:451.

Male. Form moderate-sized; integument pale to dark reddish brown, antennae and legs dark annulate, underside partially infuscated; pubescence fine, appressed, grayish, light and dark brownish. Head with front slightly convex, densely micropunctate, moderately densely clothed with appressed grayish to pale brownish pubescence; genae longer than lower eye lobes; upper eye lobes small, separated by slightly more than diameter of antennal scape; antennae a little longer than body, segments dark annulate at apices, third segment longer than first, fourth subequal to first. Pronotum with sides obtusely tuberculate behind middle; disk with calluses not prominent, anterior pair larger; apical and basal impressions rather narrow, moderately deep; punctures dense, contiguous, moderate-sized; pubescence fine, appressed, grayish, outer pair of calluses with a dark brown vitta usually extending to base; prosternum with intercoxal process slightly more than half as broad as coxal cavities; mesosternum with intercoxal process about as broad as coxal cavities; metasternum with pubescence vaguely interrupted by spots. Elytra about twice as long as broad, sides slightly emarginate before middle; basal gibbosities shallow; costae distinct, not tufted; punctures coarse, dense, becoming finer toward apex; pubescence fine, dense, mottled brownish and gray, darker brown spots interspersed down costae and suture, disk behind middle often with a vague, irregular, dark brown transverse vitta preceded by two pale spots; apices rounded. Legs stout; femora grayish pubescent with brownish spots; tibiae dark biannulate; tarsi dorsally pale pubescent. Abdomen finely grayish pubescent; last sternite rounded at apex. Length, 6-10 mm.

Female. Form similar. Antennae slightly shorter. Legs with femora less strongly clavate. Abdomen with fifth sternite twice as long as fourth, apex narrow. Length, 7-11 mm.

Type locality. of *collaris*, not indicated (North America); *interruptus*, Pennsylvania.

Range. Northeastern United States south to Georgia and west to Minnesota.

Flight period. May to August.

Host plants. Castanea, Quercus.

A. collaris is rather similar to *A. macula*, but the latter may be distinguished by the distinct postmedian pale fascia of the elytra and the small, vague lateral tubercles of the pronotum. It is relatively rare in collections, and the specimens available for study show very little variation in the pubescent pattern.

Genus *Hyperplatys* Haldeman

Hyperplatys Haldeman, 1847, Trans. Amer. Philos. Soc., (2)10:49; Thomson, 1864, Syst. ceramb., p. 26; Lacordaire, 1872, Genera des coléoptères, 9(2):776; LeConte, 1873, Smithson. Misc. Coll., 11(265):338; Horn, 1880, Trans. Amer. Entomol. Soc., 8:119, 127; LeConte and Horn, 1883, Smithson. Misc. Coll., 507:324; Casey, 1891, Ann. N.Y. Acad. Sci., 6:50; Leng and Hamilton, 1896, Trans. Amer. Entomol. Soc., 23:129; Wickham, 1897, Can. Entomol., 29:203; Wickham, 1898, Can. Entomol., 30:38; Blatchley, 1910, Coleoptera in Indiana, p. 1077; Casey, 1913, Memoirs on the Coleoptera, 4:323; Bradley, 1930, Man. Genera Beetles, p. 246; Chagnon, 1933-40, Coleop. Prov. Quebec, p. 273; Knull, 1946, Ohio Biol. Surv. Bull., 39:254; Dillon, 1956, Ann. Entomol. Soc. America, 49:346; Dillon and Dillon, 1961, Man. Beetles East. North America, p. 644; Arnett, 1962, Beetles U.S., 103:872, 873; Chagnon and Robert, 1962, Prin. Coleop. Prov. Quebec, p. 273; Bayer and Shenefelt, 1969, Univ. Wisc. Res. Bull., 275:27; Hatch, 1971, Univ. Wash. Pub. Biol., 16:152; Rice and Enns, 1981, Trans. Mo. Acad. Sci., 15:101.

Liopus (part): LeConte, 1852, J. Acad. Nat. Sci. Philadelphia, 2:170.

Form small, subdepressed; pubescence very fine, appressed. Head with front convex, broader than long; mandibles feebly arcuate; genae convergent, subequal to or half as long as lower eye lobes; eyes small, upper lobes widely separated; antennal tubercles moderate, widely divergent; antennae slender, eleven-segmented, 1-1/2 to two times as long as body, third segment subequal to, shorter or longer than first, fourth subequal to third or first. Pronotum broader than long; lateral tubercles acute, placed before extreme base, directed posteriorly; disk vaguely convex, basal impression broad, not extending beneath lateral tubercles; apex very narrowly impressed; prosternum with intercoxal process one-fourth to half as broad as coxal cavities; mesosternal process subequal to or slightly broader than prosternal process; metasternum with episternum narrow, parallel-sided. Elytra a little over twice as long as broad; base not or vaguely gibbose, basal depressions shallow; epipleura vertical, usually with a longitudinal carina extending from humeri to near apex; disk occasionally with a costa on each elytron; pubescence dense, appressed, usually with small, rounded, dark spots, erect hairs absent; apices obliquely emarginate, angles usually dentate. Legs moderately long; femora strongly clavate in males, feebly in females; tarsi slender, first segment of hind pair longer than two following together, last segment cleft almost to base. Abdomen normally segmented.

Type species. Hyperplatys maculata Haldeman (Thomson designation, 1864).

This genus may be distinguished from related forms by the small, subdepressed form, the abruptly vertical epipleura, usual presence of a lateral carina on the elytra, and also by usually having small dark, round spots on the elytra.

Four species occur in the North American fauna.

KEY TO THE NORTH AMERICA SPECIES OF *HYPERPLATYS*

1   Integument reddish brown to black, pronotum and elytra with dark spots................................2
            Integument black except for bases of femora, spots lacking. Length, 5.5-8 mm. Southeastern United States .... *femoralis*

2(1)   Elytra without a distinct blackish vitta along sides behind humeri, disk lacking entire whitish pubescent lines 3
            Elytra with a distinct blackish vitta along sides behind humeri, disk usually with entire whitish lines. Length, 4.5-7.5 mm. Eastern North America to Texas .... *maculata*

3(2)   Elytra each with a distinct costa on disk extending from near humerus to apical one-third; lateral carinae of elytra sharply defined. Length, 5-8 mm. California, Arizona .... *californica*
            Elytra lacking discal costae, usually with only small spots; lateral carinae moderately elevated, whitish lines always absent on disk. Length, 4.5-9.5 mm. Eastern North America to Montana and Arizona ...... *aspersa*

Hyperplatys femoralis Haldeman

Hyperplatys femoralis Haldeman, 1847, Trans. Amer. Philos. Soc., (2)10:49; Lacordaire, 1872, Genera des coléoptères, 9(2):776, fn.; Horn, 1880, Trans. Amer. Entomol. Soc., 8:127; Casey, 1891, Ann. N.Y. Acad. Sci., 6:50; Leng and Hamilton, 1896, Trans. Amer. Entomol. Soc., 23:129; Casey, 1913, Memoirs on the Coleoptera, 4:330; Fattig, 1947, Emory Univ. Mus. Bull., 5:37; Dillon, 1956, Ann. Entomol. Soc. Amer., 49:351.
Liopus femoralis: LeConte, 1852, J. Acad. Nat. Sci. Philadelphia, 2:171.
Leiopus femoralis: White, 1855, Cat. Coleop. Ins. British Mus., 8:380.

Male. Form small, subdepressed; integument black, basal halves of femora reddish; pubescence very fine, appressed. Head with front convex, almost twice as broad as long, micropunctate, very finely pubescent; genae about half as long as lower eye lobes; upper eye lobes small, widely separated; antennae about 1-1/2 times longer than body; third segment slightly shorter than first, fourth subequal to first. Pronotum broader than long; punctures fine, dense; pubescence very fine, appressed; disk often with two dark spots at apical half; prosternum with intercoxal process about one-fifth as broad as coxal cavities; mesosternal process subequal to prosternal; mesosternum densely micropunctate, densely, finely pubescent. Elytra about 2-1/4 times longer than broad; lateral carinae prominent; basal depression very shallow; punctures moderately coarse, dense, becoming obsolete at apex; pubescence very fine, short, appressed; apices emarginate, outer angles strongly dentate. Legs with femora strongly clavate; pubescence very fine. Abdomen densely micropunctate; pubescence fine, appressed, grayish; last sternite rounded at apex. Length, 5.5-6.5 mm.

Female. Form more robust. Femora moderately clavate. Abdomen with last sternite subtruncate. Length, 6.5-8 mm.

Type locality. None (North America).
Range. Southeastern United States.
Flight period. June.
Host plants. Not known.

H. femoralis is readily recognizable by the wholly dark integument with only the basal halves of the femora reddish. It is rare in collections. We have seen only three specimens, one each from North Carolina, Florida and Alabama.

Hyperplatys maculata Haldeman

Hyperplatys maculata Haldeman, 1847, Trans. Amer. Philos. Soc., (2)10:49; Lacordaire, 1872, Genera des coléoptères, 9(2):776, fn.; Casey, 1891, Ann. N.Y. Acad. Sci., 6:50; Casey, 1913, Memoirs on the Coleoptera, 4:324; Leonard, 1928, Cornell Agr. Exp. Sta. Mem. 101:453; Knull, 1932, Entomol. News, 43:64; Chagnon, 1933-40, Coleop. Prov. Quebec, p. 273; Knull, 1946, Ohio Biol. Surv. Bull., 39:255; Fattig, 1947, Emory Univ.

Mus. Bull., 5:37; Dillon, 1956, Ann. Entomol. Soc. Amer., 49:347; Gardiner, 1961, Can. Entomol., 93:1011, figs. 1-4 (larva); Chagnon and Robert, 1962, Prin. Coleop. Prov. Quebec, p. 273; Gardiner, 1966, Can. J. Zool., 44:204; Bayer and Shenefelt, 1969, Univ. Wisc. Res. Bull., 275:28, fig. 36; Stein and Tagestad, 1976, USDA For. Serv. Res. Pap., RM-171:17; Gosling and Gosling, 1977, Gr. Lakes Entomol., 10:30, fig. 157; Laliberte et al., 1977, Fabreries, 3:94; Rice and Enns, 1981, Trans. Mo. Acad. Sci., 15:101; Waters and Hyche, 1984, Coleop. Bull., 38:284; Gosling, 1984, Gr. Lakes Entomol., 17:72; Gosling, 1896, Gr. Lakes Entomol., 19:156.

Hyperplatys maculatus: Horn, 1880, Trans. Amer. Entomol. Soc., 8:127; Hubbard, 1885, Ins. Affect. Orange, p. 174, pl. 14, fig. 3; Beutenmuller, 1896, J. N.Y. Entomol. Soc., 4:79; Leng and Hamilton, 1896, Trans. Amer. Entomol. Soc., 23:129; Wickham, 1898, Can. Entomol., 30:38; Chagnon, 1905, Nat. Can., 32:42; Blatchley, 1910, Coleoptera in Indiana, p. 1078; Craighead, 1921, J. Agr. Res., 22:217; Dillon and Dillon, 1961, Man. Beetles East. North America, p. 644, pl. 64, no. 13; Headstrom, 1977, Beetles of America, p. 379.

Liopus maculatus: LeConte, 1852, J. Acad. Nat. Sci. Philadelphia, 2:170.

Leiopus maculatus: White, 1855, Cal. Coleop. Ins. British Mus., 8:387.

Hyperplatys maculata var. *nigrellus* Haldeman, 1847, Trans. Amer. Philos. Soc., (2)10:49.

Hyperplatys nigrellus: Leng and Hamilton, 1896, Trans. Amer. Entomol. Soc., 23:129.

Hyperplatys nigrellus: Casey, 1891, Ann. N.Y. Acad. Sci., 6:50; Casey, 1913, Memoirs on the Coleoptera, 4:327; Leonard, 1928, Cornell Agr. Exp. Sta. Mem., 101:453; Fattig, 1947, Emory Univ. Mus. Bull., 5:37.

Hyperplatys aspersus var. *nigrellus*: Smith, 1910, N.J. St. Mus. Rept., 1909:334.

Hyperplatys lentiginosa Casey, 1913, Memoirs on the Coleoptera, 4:325.

Hyperplatys lentiginosus: Aurivillius, 1923, Coleop. Cat., 74:417.

Hyperplatys robustula Casey, 1913, Memoirs on the Coleoptera, 4:326.

Hyperplatys robustulus: Aurivillius, 1923, Coleop. Cat., 74:418.

Hyperplatys amnicola Casey, 1913, Memoirs on the Coleoptera, 4:326.

Hyperplatys delicata Casey, 1913, Memoirs on the Coleoptera, 4:327.

Hyperplatys delicatus: Aurivillius, 1923, Coleop. Cat., 74:417.

Hyperplatys cryptica Casey, 1913, Memoirs on the Coleoptera, 4:327.

Hyperplatys crypticus: Aurivillius, 1923, Coleop. Cat., 74:417.

Hyperplatys frigida Casey, 1913, Memoirs on the Coleoptera, 4:328.

Hyperplatys frigidus: Aurivillius, 1923, Coleop. Cat., 74:417.

Hyperplatys binocularis Casey, 1913, Memoirs on the Coleoptera, 4:388.

Hyperplatys variolata Casey, 1913, Memoirs on the Coleoptera, 4:328.

Hyperplatys variolatus: Aurivillius, 1923, Coleop. Cat., 74:418.

Male. Form small, subdepressed; integument pale to dark reddish brown, antennae and basal halves of femora paler, pronotum and elytra with dark spots; pubescence fine, dense, appressed, whitish to dark brown. Head with front convex, almost twice as broad as long, densely micropunctate, finely, densely pubescent; genae about half as long as lower eye lobes; upper eye lobes small, widely separated; antennae almost twice as long as body, third segment longer than first, fourth subequal to third. Pronotum broader than long; disk finely, separately punctate, middle often with a longitudinal, linear, glabrous callus; apical half with two rounded, dark spots; pubescence dense, appressed, grayish; prosternum with intercoxal process about one-fourth as wide as coxal cavities; mesosternal process slightly wider than prosternal; metasternum densely micropunctate, densely pubescent. Elytra a little over twice as long as broad; lateral carinae prominent; basal depression moderate; punctures moderately coarse, dense, becoming obsolete toward apex; pubescence fine, pale brownish, usually with longitudinal whitish lines interspersed, dark spots small, scattered, apical one-third usually with a pair of larger brown spots near middle, sides with dark vittae from humeri at least to middle; apices obliquely emarginate, angles dentate. Legs robust; femora strongly clavate; pubescence fine. Abdomen densely micropunctate, finely, densely pubescent; last sternite broadly rounded at apex. Length, 4.5-6 mm.

Female. Form similar. Femora feebly clavate. Abdomen with last sternite subtruncate. Length, 5-7.5 mm.

Type locality. Of *maculata*, Pennsylvania; *nigrellus*, none; *lentiginosa*, Watch Hill, Rhode Island; *robustula*, Indiana; *amnicola*, Keokuk, Iowa; *delicata*, Boston Neck, Rhode Island; *cryptica*, Indiana and New York; *frigida*, Monmouth, Maine; *binocularis*, New York; *variolata*, District of Columbia and New York.

Range. Eastern North America to Texas and North Dakota.

Flight period. May to October.

Host plants. Acer, Amelanchier, Carpinus, Carya, Castanea, Citrus, Cornus, Juglans, Liriodendron, Malus, Populus, Quercus, Rhus, Ribes, Robinia, Salix, Tilia, Ulmus.

As indicated by the synonymy, this is a variable species. It usually may be distinguished by the presence of dark lateral vittae, large postmedian dark maculae, and longitudinal whitish lines on the elytra. The elytral spots and white lines are occasionally reduced or absent, and specimens have been seen with expanded postmedian maculae.

There is no indication of geographic variation in material available for study.

Hyperplatys californica Casey
(Figure 14)

Hyperplatys californica Casey, 1891, Ann. N.Y. Acad. Sci., 6:51; Casey, 1913, Memoirs on the Coleoptera, 4:330; Barrett, 1932, Univ. Calif. Pub. Entomol., 5:290; Dillon, 1956, Ann. Entomol. Soc. Amer., 49:349; Essig, 1958, Ins. West. North America, fig. 370; Cope, 1984, Coleop. Bull., 38:30 (habits).

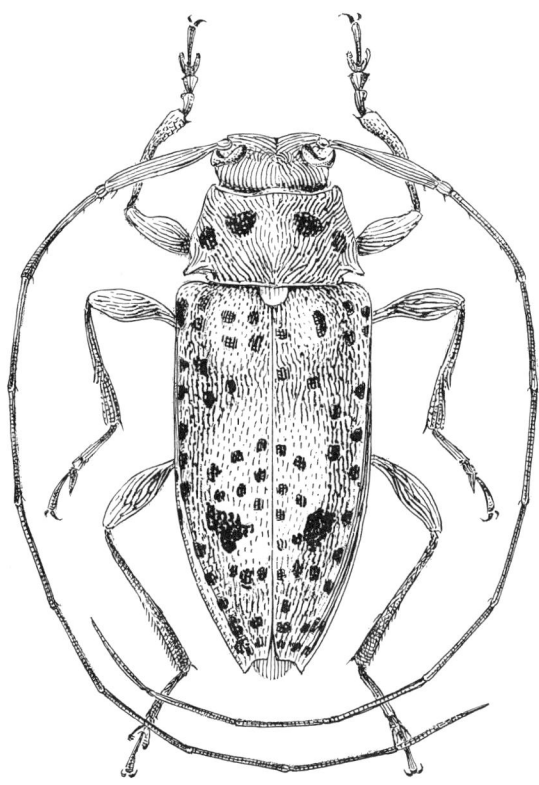

Figure 14. *Hyperplatys californica* Casey, male.

Hyperplatys californicus: Leng and Hamilton, 1896, Trans. Amer. Entomol. Soc., 23:129; Garnett, 1918, Can. Entomol., 50:282.
Hyperplatys asperatus californicus: Doane et al., 1936, For. Ins., p. 189.

Male. Form small, subdepressed; integument pale to dark brownish, antennae and basal halves of femora paler. Pronotum and elytra with brown spots; pubescence dense, appressed, grayish. Head with front convex, about twice as broad as long, densely micropunctate, densely pubescent; genae about half as long as lower eye lobes; upper lobes small, widely separated; antennae 1-1/2 to two times as long as body, third segment slightly longer than first, fourth subequal to third, outer segments often black annulate at apices. Pronotum broader than long; disk irregularly punctate, middle with a vague longitudinal callus; pubescence dense, appressed, pale grayish, anterior half with two brownish spots and usually with a smaller brown spot on each side slightly posterior to discal pair; prosternum with intercoxal process one-fourth to one-third as broad as coxal cavities; mesosternal process slightly broader than prosternal; metasternum densely, rather thickly pubescent. Elytra a little more than twice as long as broad; lateral carinae prominent; each elytron with an oblique costa extending from humerus to apical one-third; basal gibbosities moderate, basal depressions fairly deep; punctures rather coarse, contiguous, becoming obsolete toward apex; pubescence dense, grayish, appressed, small brown spots interspersed, often with two large postmedian spots; apices obliquely emarginate, angles feebly dentate. Legs robust, densely pubescent; femora strongly clavate. Abdomen densely pubescent; last sternite broadly rounded at apex. Length, 5-7 mm.

Female. Form similar. Femora feebly clavate. Abdomen with last sternite shallowly emarginate. Length, 5.5-8 mm.

Type locality. Santa Cruz. Co., California.

Range. Central to southern California, Arizona.

Flight period. May to July.

Host plants. Aesculus, Juglans, Populus, Prunus.

The presence of elytral costae separates this species from *H. aspersa*.

Specimens from Santa Cruz County and vicinity tend to have reduced elytral spots and denser grayish pubescence. The more southern individuals usually possess larger postmedian elytral spots and have somewhat darker integument.

Hyperplatys aspersa (Say)

Lamia aspersa Say, 1823, J. Acad. Nat. Sci. Philadelphia, 3:330; Le Conte, 1859, Compl. Writings T. Say, 2:187.

Hyperplatys aspersa: Haldeman, 1847, Trans. Amer. Philos. Soc., (2)10:49; Lacordaire, 1872, Genera des coléoptères, 9(2):776, fn.; Casey, 1891, Ann. N.Y. Acad. Sci., 6:51; Casey, 1913, Memoirs on the Coleoptera, 4:329; Blackman and Stage, 1924, N.Y. St. Coll. For. Tech. Pub., 17:119; Leonard, 1928, Cornell Agr. Exp. Sta. Mem., 101:453; Barrett, 1932, Univ. Calif. Pub. Entomol., 5:290; Knull, 1932, Entomol. News, 43:64; Knowlton and Thatcher, 1936, Utah Acad. Sci. Arts Letters, 13:280; Hoffman, 1940, Bull. Brooklyn Entomol. Soc., 35:59; Procter, 1946, Biol. Surv. Mt. Desert, 7:153; Knull, 1946, Ohio Biol. Surv. Bull., 39:255, pl. 25, fig. 107; Fattig, 1947, Emory Univ. Mus. Bull., 5:37; Dillon, 1956, Ann. Entomol. Soc. Amer., 49:348; Gardiner, 1961, Can. Entomol., 93:1012, figs. 5-10 (larva); Gardiner, 1966, Can. J. Zool., 44:204, fig. 54; Bayer and Shenefelt, 1969, Univ. Wisc. Res. Bull., 275:28, fig. 36; Gosling and Gosling, 1977, Gr. Lakes Entomol., 10:30, fig. 158; Laliberte et al., 1977, Fabreries, 3:94; Rice and Enns, 1981, Trans. Mo. Acad. Sci., 15:101; Waters and Hyche, 1984, Coleop. Bull., 38:284; Gosling, 1984, Gr. Lakes Entomol., 17:72; Gosling, 1986, Gr. Lakes Entomol., 19:156.

Liopus aspersus: Leconte, 1852, J. Acad. Nat. Sci. Philadelphia, 2:171.

Leiopus aspersus: White, 1855, Cat. Coleop. Ins. British Mus., 8:387.

Hyperplatys aspersus: Horn, 1880, Trans. Amer. Entomol. Soc., 8:127; Packard, 1881, U.S. Entomol. Comm. Bull., 7:115; Leng and Hamilton,

1896, Trans. Amer. Entomol. Soc., 23:129; Beutenmuller, 1896, J. N.Y. Entomol. Soc., 4:79; Wickham, 1898, Can. Entomol., 30:38; Blatchley, 1910, Coleoptera in Indiana, p. 1078; Smith, 1910, N.J. St. Mus. Rept., 1909:334; Garnett, 1918, Can. Entomol., 50:282; Craighead, 1923, Can. Dept. Agr. Bull., (n.s.) 27:117; Chagnon, 1933-40, Coleop. Prov. Quebec, P. 273, figs. 5-10; Dillon and Dillon, 1961, Man. Beetles East. North America, p. 645, pl. 64, no. 12; Stein and Tagestad, 1976, USDA For. Serv. Res. Pap., Rm. 171:16, Headstrom, 1977, Beetles of America, p. 379, fig. 529.

Hyperplatys asperatus: Hoffmann, 1942, USDA Misc. Pub., 466:11 (error).

Hyperplatys aspera: Chagnon and Robert, 1962, Prin. Coleop. Prov. Quebec, p. 273; Hatch, 1971, Univ. Wash. Pub. Biol., 16:152.

Hyperplatys vigilans Casey, 1913, Memoirs on the Coleoptera, 4:329.

Hyperplatys montana Casey, 1913, Memoirs on the Coleoptera, 4:325; Dillon, 1956, Ann. Entomol. Soc. Amer., 49:350; Cope, 1984, Coleop. Bull., 38:30 (habits). New synonymy.

Hyperplatys laceyi Dillon, 1956, Ann. Entomol. Soc. Amer., 49:351. New synonymy.

Male. Form small, subdepressed; integument pale reddish brown to black, antennae and bases of femora testaceous to reddish, pronotum and elytra with rounded brownish spots; pubescence dense, short, appressed, grayish to black. Head with front convex, usually a little more than 1-1/2 times broader than long, densely micropunctate, densely pubescent; genae usually shorter than lower eye lobes; upper lobes small, widely separated; antennae usually almost twice as long as body, third segment subequal to first, fourth subequal to third, segments usually narrowly dark annulate at apices. Pronotum broader than long; disk finely, densely punctate, middle often with a vague longitudinal callus; pubescence dense, grayish, appressed, anterior half with two dark spots and two smaller spots near sides; prosternum with intercoxal process narrow, less than one-fifth as broad as coxal cavities; mesosternal process slightly broader than prosternal; metasternum densely pubescent. Elytra about 2.2 times as long as broad; lateral carinae usually prominent; basal gibbosities vague, basal depressions shallow; punctures moderately coarse, dense, becoming obsolete at apex; pubescence fine, dense, appressed, grayish, rounded brownish spots numerously interspersed, larger postmedian maculae occasionally present; apices obliquely emarginate, angles dentate. Legs robust; femora strongly clavate, densely pubescent; hind tarsi slender, first segment longer than two following together. Abdomen densely micropunctate, densely pubescent; last sternite broadly subtruncate at apex. Length, 4.5-8 mm.

Female. Form similar. Hind femora narrowly clavate. Abdomen with last sternite narrow subtruncate at apex. Length, 5-9.5 mm.

Type locality. Of *aspersa*, Mississippi and Missouri; *vigilans*, District of Columbia; *montana*, Boulder Co., Colorado; *laceyi*, Palmerlee, Cochise Co., Arizona.

Range. Eastern North America to Utah and Arizona.

Flight period. March to September.

Host plants. Acer, Alnus, Amelanchier, Carya, Castanea, Celastrus, Celtis, Cercocarpus, Cornus, Fraxinus, Juglans, Lindera, Liriodendron, Malus, Populus, Prunus, Rhus, Ribes, Robinia, Salix, Tilia, Ulmus.

Parasites. Xorides humeralis (Say) (Ichneumonidae); *Cenocoelius provancheri* (Rohwer) (Braconidae); *Meteorus tibialis* Muesbeck (Braconidae).

The absence of lateral dark vitta along the sides of the elytra separates this species from *H. maculata,* and the oblique, discal costae of the elytra distinguish *H. californica.*

H. aspersa is rather variable. The populations in eastern North America are more uniform, and the larger postmedian maculae of the elytra are occasionally present. Westward, the integument becomes darker and the postmedian maculae are present more often.

The Arizona population may represent a distinct subspecies, but we lack adequate series to make a decision.

Genus *Lepturges* Bates

Lepturges Bates, 1863, Ann. Mag. Nat. Hist., (3)12:367 (reprint p. 84); Lacordaire, 1872, Genera des coléoptères, 9(2):777; LeConte, 1873, Smithson. Misc. Coll., 11(264):338; Provancher, 1877, Pet. Fauna Entomol. Can., 1:628; Horn, 1880, Trans. Amer. Entomol. Soc., 8:119, 126; Bates, 1881, Biol. Centr.-Amer., Coleoptera, 5:166; LeConte and Horn, 1883, Smithson. Misc. Coll., 507:324; Leng and Hamilton, 1896, Trans. Amer. Entomol. Soc., 23:127 (part); Wickham, 1897, Can. Entomol., 29:203; Wickham, 1898, Can. Entomol., 30:37; Blatchley, 1910, Coleoptera in Indiana, p. 1075 (part); Casey, 1913, Memoirs on the Coleoptera, 4:317; Bradley, 1930, Man. Genera Beetles, p. 246; Chagnon, 1933-40, Coleop. Prov. Quebec, p. 273; Knull, 1946, Ohio Biol. Surv. Bull., 39:251, Dillon, 1956, Ann. Entomol. Soc. Amer., 49:339; Dillon and Dillon, 1961, Man. Beetles East. North America, p. 644; Arnett, 1962, Beetles U.S., 103:873; Chagnon and Robert, 1962, Prin. Coleop. Prov. Quebec, p. 273; Bayer and Shenefelt, 1969, Univ. Wisc. Res. Bull., 275:27; Rice and Enns, 1981, Trans. Mo. Acad. Sci., 15:100.

Maculurges Dillon, 1956, Ann. Entomol. Soc. Amer., 49:338; Arnett, 1962, Beetles U.S., 103:873; Rice and Enns, 1981, Trans. Mo. Acad. Sci., 15:98. (Type species: *Liopus regularis* LeConte, by original designation.) New synonymy.

Form small to moderate-sized, subdepressed; pubescence fine, appressed. Head with front convex, transverse; mandibles feebly arcuate; genae convergent, shorter than lower eye lobes; eyes moderate-sized (except *megalops*), deeply emarginate, upper lobes small, widely separated; antennal tubercles moderate, widely divergent; antennae slender, eleven-segmented, 1-1/2 to two times as long as body, often with several short setae beneath on basal segments, scape subequal to or longer than third segment, fourth usually longer than third, fifth subequal to fourth. Pronotum broader than long, sides tapering toward apex; lateral tubercles acute, placed at or near extreme base; disk not tuberculate, narrowly impressed longitudinally at middle; apex very narrowly impressed, base more broadly impressed across to lateral tubercles, sulcus not extending below tubercles; punctures fine, scattered, rather sparse; prosternum narrow, intercoxal process narrow, no more than one-sixth as broad as coxal cavities, broadly expanded behind, coxal cavities closed; mesosternum with intercoxal process about as broad as prosternal process; metasternum with episternum narrow, subparallel. Elytra at least twice or more longer than broad; base not gibbose; disk shallowly impressed at about basal one-third; costae, if present, vague; pubescence short, appressed, erect setae absent; apices rounded to slightly obliquely truncate. Legs rather elongate, femora clavate; posterior femora slender in both sexes; tarsi slender, elongate, first segment of hind pair longer than two following segments together, third segment cleft to base. Abdomen normally segmented, last segment of females moderately elongated.

Type species. Lepturges elegantulus Bates (by present designation).

This genus is one of the most easily recognizable in the tribe. The basal sulcus of the pronotum terminating at the lateral tubercles, the basally placed lateral tubercles, the narrow pro- and mesosternal intercoxal processes, and the proportions of the antennal segments make *Lepturges* very distinctive.

Dillon (1956) decided to leave the question of a type species for *Lepturges* open until the South American species could be revised. We are unaware of a subsequent type designation, in spite of the large volume of work on this tribe by recent authors such as Gilmour, Martins, and Monne. Since a large number of Neotropical Acanthocinini still remain to be described, we have designated a type species for *Lepturges* to insure stability for this generic name.

Numerous species of *Lepturges* occur from Mexico through South America. Nine species are known from North America.

KEY TO THE NORTH AMERICAN SPECIES OF *LEPTURGES*

1	Pronotum and elytra immaculate or with brownish maculae; antennae with first segment longer than third	2
	Pronotum and elytra with rounded black spots; antennae with first segment subequal to third. Length, 7-10.5 mm. East-central states to Indiana and Missouri	*regularis*
2(1)	Eyes with lower lobes at least four times longer than genae	3
	Eyes with lower lobes about twice as long as genae	4
3(2)	Integument black, elytra with vague longitudinal vittae. Length, 5.5-8 mm. Florida to Panama	*megalops*

	Integument dark reddish brown, elytra with a broad pale macula at middle on basal half. Length, 7-10.5 mm. Arizona to Costa Rica .... *infilatus*
4(2)	Body entirely testaceous to pale reddish brown, not dark maculate..5
	Body distinctly maculate with dark brown 6
5(4)	Elytra uniformly testaceous; lateral tubercles of pronotum placed before basal margin. Length, 5.5-9 mm. Arizona to northern Mexico ....*yucca*
	Elytra with pale pubescent vittae; lateral tubercles of pronotum placed at basal margin. Length, 6-8.5 mm. Southern Texas .... *vogti*
6(4)	Elytra with dark vittae on disk unclouded with grayish 7
	Elytra with median dark vittae clouded with grayish 8
7(6)	Elytra with a dark vitta along each side from humerus to near apical one-third. Length, 6-9 mm. Northeastern North America to Minnesota ...... *symmetricus*
	Elytra lacking dark lateral vittae. Length, 8-10.5 mm. East-central states to Mississippi................... *pictus*
8(6)	Elytra with apices broadly rounded; integument usually dark reddish brown; elytra with distinct median maculae. Length, 6-10 mm. Eastern United States to Kansas and Texas ....*confluens*
	Elytra with apices obliquely truncate; integument usually pale reddish brown; elytra with maculae indistinct. Length, 5.5-10 mm. Eastern United States to Kansas and Texas .... *angulatus*

Lepturges regularis (LeConte), new combination

Liopus regularis LeConte, 1862, Proc. Acad. Nat. Sci. Philadelphia, 1862:39.
Lepturges regularis: Horn, 1880, Trans. Amer. Entomol. Soc., 8:127; Leng and Hamilton, 1896, Trans. Amer. Entomol. Soc., 23:127,129; Blatchley, 1910, Coleoptera in Indiana, p. 1077, fig. 464; Casey, 1913, Memoirs on the Coleoptera, 4:319; Champlain and Knull, 1925, Entomol. News. 36:140; Knull, 1946, Ohio Biol. Surv. Bull., 39:253, pl. 25, fig. 110; Fattig, 1947, Emory Univ. Mus. Bull., 5:36.
Maculurges regularis: Dillon, 1956, Ann. Entomol. Soc. Amer., 49:339; Gosling and Gosling, 1977, Gr. Lakes Entomol., 10:28, fig. 153; Rice and Enns, 1981, Trans. Mo. Acad. Sci., 15:100.

Male. Form moderately small; integument reddish brown, antennae and femora except apices testaceous, pronotum and elytra with dark spots; pubescence fine, dense, short, appressed, grayish except on dark spots. Head with front convex, broader than long, densely micropunctate, pubescence fine, appressed, lower margin with several long, erect setae; genae short; vertex deeply impressed between antennal tubercles; eyes large, lower lobes about twice as long as genae; antennae almost twice as long as body, third segment subequal to first, fourth slightly longer than third, fifth subequal to fourth, short suberect hairs sparse beneath. Pronotum broader than long; disk usually with an elevated median, longitudinal line, each side with two dark spots at apical one-third and base one-half; punctures around dark spots and median line moderately coarse, well separated; pubescence fine, appressed, dense; lateral tubercles placed slightly before base; prosternum finely, densely pubescent; meso- and metasternum finely, densely pubescent. Elytra a little more than twice as long as broad; basal gibbosities and impressions vague; pubescence dense, short, appressed, grayish except for dark rounded maculae, each elytron with dark maculae at humeri, near base, at basal one-fourth near suture, and another subtriangular one toward margin and slightly behind basal one, a larger median macula just behind middle, and a median one at apical one-fourth; punctures rather coarse, dense, becoming finer and sparser toward apex; apices rounded. Legs robust, finely pubescent. Abdomen densely micropunctate, densely pubescent; last sternite rounded at apex. Length, 7-9.5 mm.

Female. Form more robust. Head with front less broad. Abdomen with last sternite elongate, emarginate at apex. Length, 8.5-10.5 mm.

Type locality. Ohio.
Range. East-central states to Indiana and Missouri.
Flight period. June-July.
Host plants. Aesculus, Carya.

This species is easily recognized by the black spots of the pronotum and elytra. The characters used by Dillon (1956) to generically separate this species from other *Lepturges* are inadequate to justify that action.

Lepturges megalops Hamilton

Lepturges megalops Leng and Hamilton, 1896, Trans. Amer. Entomol. Soc., 23:127; Blatchley, 1928, Can. Entomol., 60:71; Dillon, 1956, Ann. Entomol. Soc. Amer., 49:345.

Male. Form small; integument black, femora often basally paler; pubescence fine, grayish, appressed, usually forming vague longitudinal vittae on elytra. Head with front convex, densely micropunctate, pubescence fine, appressed, lower margin with several long, erect setae; genae very short; eyes large, lower lobes occupying most of side of head, upper lobes moderate, separated by less than twice width of lobes; vertex deeply impressed; antennae extending about 5-1/2 segments beyond body, segments from third with two short, suberect setae at apices, third segment longer than fourth, fifth subequal to fourth. Pronotum broader than long; disk feebly convex, surface densely micropunctate, basal impression with a row of coarse punctures; pubescence fine, appressed; lateral spines placed just before basal angles; prosternum finely pubescent; meso- and metasternum densely micropunctate, finely pubescent. Elytra a little more than twice as long as broad; basal gibbosities shallow, impression behind shallow; basal punctures coarse, contiguous, becoming obsolete at apex; pubescence dense, fine, usually with longitudinal ashy vittae on disk and along suture at apex; apices narrowly rounded. Legs with femora often reddish at bases. Abdomen densely micropunctate, densely clothed with appressed grayish pubescence; last sternite broadly, shallowly emarginate at apex, apical margin with numerous long setae. Length, 5.5-8 mm.

Female. Form similar. Antennae slightly shorter. Abdomen with last sternite narrowing at apex, margin moderately deeply notched, setose. Length, 6.5-8 mm.

Type locality. Biscayne, Florida.

Range. Florida to Panama.

Flight period. April to September.

The black integument and large eyes make this species very distinctive.

We have seen five specimens of this species (collected and made available by E. Giesbert), three from Florida, one from Mexico, and one from Panama. We can find no significant morphological differences between these specimens, and can only assume that *L. megalops* is a widespread Neotropical species which extends its range into Florida.

Lepturges infilatus Bates

Lepturges infilatus Bates, 1872, Trans. Entomol. Soc. London, 1872:217; Bates, 1881, Biol. Centr.-Amer., Coleoptera, 5:166, pl. 13, fig. 3; Horn, 1886, Trans. Amer. Entomol. Soc., 13: x; Hovore and Penrose, 1982, Southwest. Nat., 27:26; Hovore et al., 1987, Proc. Calif. Acad. Sci., 44:319, fig. 19.

Male. Form small to moderate-sized; integument pale to dark reddish brown; pubescence fine, dense, appressed, brownish and grayish. Head with front convex, densely micropunctate, pubescence dense, appressed, lower margin with several long, erect setae; genae about one-third as long as lower eye lobes; eyes moderate, upper lobes small, widely separated; vertex deeply impressed between eyes; antennae about twice as long as body, scape extending to lateral pronotal tubercles, third segment shorter than first, fourth subequal to third, fifth longer than fourth, pronotum broader than long; disk feebly impressed at sides of middle; punctures sparse, irregular, vague median callus impunctate; pubescence fine, appressed, dark longitudinal vittae vague, usually a pair present down sides of middle and a narrower pair along base of lateral tubercles; lateral tubercles placed a little before basal margin; prosternum thinly pubescent; meso- and metasternum densely micropunctate, densely pubescent. Elytra a little more than twice as long as broad; punctures moderately coarse, well separated at base, denser toward middle, obsolete toward apex; pubescence very fine, appressed, basal half usually with a broad pale macula extending from behind scutellum to middle and from suture, apical half with narrow, angulate maculae; apices obliquely truncate. Legs minutely punctate and pubescent. Abdomen densely micropunctate, finely, densely pubescent; last sternite rounded at apex. Length, 7-9 mm.

Female. Form similar. Antennae extending about four segments beyond elytra. Abdomen with fifth abdominal sternite slightly longer than fourth, apex narrowed, emarginate. Length, 7-10.5 mm.

Type locality. Chontales, Nicaragua.
Range. Arizona to Costa Rica.
Flight period. May to August.
Host plants. Leucaena, Morus.

The short genae and color pattern of the elytra make this species readily recognizable among the North American *Lepturges*.

Lepturges yucca Schaeffer

Lepturges yucca Schaeffer, 1905, Brooklyn Inst. Arts Sci. Mus. Bull., 1:167; Schaeffer, 1908, Brooklyn Inst. Arts Sci. Mus. Bull., 1:346; Dillon, 1956, Ann. Entomol. Soc. Amer., 49:344, fig. 7; Linsley, Knull, and Statham, 1961, Amer. Mus. Nov., 2050:30; Lewis, 1979, Pan-Pac. Entomol., 55:25.

Male. Form small; integument testaceous to pale brownish; pubescence fine, uniform, pale, depressed. Head with front strongly convex, micropunctate, pubescence very fine, moderately dense, appressed, lower margin with several long, erect setae; genae shorter than lower eye lobes; eyes moderate, upper lobes small, widely separated; vertex deeply impressed between eyes; antennae about 1-2/3 longer than body, scape extending to lateral pronotal tubercles, third segment subequal to fourth, fifth shorter than fourth, segments densely clothed with very fine, appressed pubescence, not obscuring surface. Pronotum broader than long; disk feebly convex, surface micropunctate with moderately coarse punctures scattered over middle and along basal depression, apical one-fourth lacking large punctures; pubescence fine, appressed, pale; lateral tubercles placed before basal margins; prosternum thinly pubescent; meso- and metasternum micropunctate, finely pubescent. Elytra a little more than twice as long as broad; each elytron vaguely costate down middle; punctures moderately coarse, dense, obsolete at apex; pubescence uniform, moderately dense, fine, appressed; apices rather broadly rounded. Legs micropunctate, very finely pubescent. Abdomen micropunctate, finely pubescent; last lightly emarginate at apex. Length, 5.5-7 mm.

Female. Form similar. Antennae about 1-1/2 times longer than body. Abdomen with last sternite longer than fourth, apex narrow, shallowly emarginate, several long, erect, dark setae present along posterior margin. Length, 7-9 mm.

Type locality. Parmalee, Cochise Co., Arizona.

Range. Southern Arizona to northern Mexico.
Flight period. July-August.
Host plants. Yucca.
This species is easily recognizable by the uniformly pale brownish integument and uniform fine, pale, appressed pubescence of the body.

Lepturges vogti Hovore and Tyson

Lepturges vogti Hovore and Tyson, 1983, Coleop. Bull., 37:349; Hovore et al., 1987, Proc. Calif. Acad. Sci., 44:319.

Male. Form small; integument uniformly pale reddish brown; pubescence fine, pale, appressed, forming vague patterns on pronotum and elytra. Head with front convex, micropunctate, pubescence fine, dense, appressed, lower margin with several long, erect setae; genae shorter than lower eye lobes; eyes moderate, upper lobes small, widely separated; vertex deeply impressed between bases of antennal tubercles; antennae extending 5-1/2 segments beyond body, third segment subequal to fourth, fifth subequal to fourth, segments finely, densely pubescent, segments three to about seven with a few, short, suberect setae. Pronotum broader than long; disk feebly convex, surface micropunctate, coarse punctures sparse, irregular, clustered around middle and along basal impression; pubescence fine, appressed, with vague longitudinal vittae at sides; lateral tubercles placed at basal margin; prosternum finely pubescent; meso- and metasternum micropunctate, finely pubescent. Elytra a little more than twice as long as broad; each elytron vaguely costate down middle; punctures moderately coarse, irregular, obsolete at apex; pubescence depressed, whitish, thicker patches forming indistinct, irregular, longitudinal vittae and spots at apex; apices narrowly rounded. Legs with femora often slightly infuscated; pubescence fine. Abdomen densely micropunctate, densely clothed with fine, appressed pubescence; last sternite broadly truncate to feebly emarginate. Length: 6-8 mm.

Female. Form similar. Antennae about five segments longer than body. Abdomen with fifth sternite emarginate at apex, thinly fringed with long, black setae. Length, 6-8.5 mm.

Type locality. 5 miles N San Diego, Duval Co., Texas.
Range. Southern Texas.
Flight period. May to October.
Host plants. Yucca truculeana.

The indistinct vittae of the elytra, basally placed pronotal tubercles, less punctate pronotal disk, and more narrowly rounded elytral apices separate this species from *L. yucca* Schaeffer.

According to Hovore and Tyson (1983), larvae of *L. vogti* feed in the pith layer of recently dead leaves of *Yucca truculeana*. Pupation occurs in a chamber within the pith, and adults emerge through elliptical holes in the leaf surface.

Adults may be beaten from the host plant and are also attracted to lights.

Lepturges symmetricus (Haldeman)

Leiopus symmetricus Haldeman, 1847, Trans. Amer. Philos. Soc., (2)10:50; White, 1855, Cat. Coleop. Ins. British Mus., 8:388.
Liopus symmetricus: LeConte, 1852, J. Acad. Nat. Sci. Philadelphia, 2:171.
Leiopus symetricus: Lacordaire, 1872, Genera des coléoptères, 9(2):776, fn. (error).
Lepturges symmetricus: Horn, 1880, Trans. Amer. Entomol. Soc., 8:126; Chittenden, 1894, Proc. Entomol. Soc. Wash., 3:101; Leng and Hamilton, 1896, Trans. Amer. Entomol. Soc., 23:127; Beutenmuller, 1896, J. N.Y. Entomol. Soc., 4:79; Wickham, 1898, Can. Entomol., 30:38; Townsend, 1902, Trans. Texas Acad. Sci., 5:78; Smith, 1910, N.J. St. Mus. Rept., 1909:334; Casey, 1913, Memoirs on the Coleoptera, 4:318; Craighead, 1923, Can. Dept. Agr. Bull., (n.s.) 27:118, pl. 16, fig. 1, pl. 7, fig. 5; Champlain and Knull, 1925, Entomol. News, 36:140; Leonard, 1928, Cornell Agr. Exp. Sta. Mem., 101:452; Fletcher, 1929, Can. Entomol., 61:259; Chagnon, 1933-40, Coleop. Prov. Quebec, p. 273; Knull, 1934, Entomol. News, 45:211; Hoffmann, 1942, USDA Misc. Pub., 466:11; Knull, 1946, Ohio Biol. Surv. Bull., 39:252; Fattig, 1947, Emory Univ. Mus. Bull., 5:36; Dillon, 1956, Ann. Entomol. Soc. Amer., 49:344; Chagnon and Robert, 1962, Prin. Coleop. Prov. Quebec, p. 273; Bayer and Shenefelt, 1969, Univ. Wisc. Res. Bull., 275:29, fig. 37; Gosling and Gosling, 1977, Gr. Lakes Entomol., 10:28, fig. 152; Laliberte et al., 1977, Fabreries, 3:94; Rice and Enns, 1981, Trans. Mo. Acad. Sci., 15;101.

Male. Form small to moderate-sized; integument pale to dark reddish brown; appendages paler, legs often partially infuscated, antennae unicolorous. Head with front convex, densely micropunctate, finely, densely pubescent, lower margin with a few long, erect hairs; genae less than half as long as lower eye lobes; eyes small, upper lobes separated by about twice their width; vertex deeply impressed; antennae extending about five segments beyond body, third segment subequal to fourth, fifth subequal to fourth, basal segments with two or three short, stout setae. Pronotum broader than long; disk with larger punctures scattered around middle, sides, and basal impression, not attaining apical margin; pubescence fine, grayish, usually with brownish maculae down sides of middle and at bases of lateral tubercles; lateral tubercles basal; prosternum finely pubescent; meso- and metasternum very finely punctate and pubescent. Elytra about twice as long as broad, tapering apically; basal impressions shallow; punctures moderately coarse, dense, becoming obsolete toward apex; pubescence fine, grayish, brownish maculae usually indistinct, basal spots small, lateral maculae extending from humeri down margins, usually to apex and onto disk behind middle, small spots present along suture behind middle and near apex, suture narrowly brownish for most of its length; apices slightly dehiscent, narrowing, obliquely rounded. Legs stout; femora usually partially infuscated, tibiae dark over about apical half, tarsi with first segment basally pale. Abdomen very finely, densely punctate, finely pubescent; last sternite truncate to rounded at apex. Length, 6-8 mm.

Female. Form similar. Abdomen with last sternite elongate, narrowing, emarginate at apex. Length, 7-9 mm.

Type locality. Not indicated (North America).

Range. Northeastern North America to Minnesota.

Flight period. May to August.

Host plants. Acer, Carya, Castanea, Celtis, Cornus, Ficus, Juglans, Morus, Tilia, Ulmus.

Parasites. Cenocoelis rubriceps Prov. (Braconidae).

The narrow brownish sutural bands of the elytra and lateral dark maculae separate *symmetricus* from other eastern *Lepturges*.

Lepturges pictus (LeConte)

Liopus pictus LeConte, 1852, J. Acad. Nat. Sci. Philadelphia, 2:172.

Leiopus pictus: White, 1855, Cat. Coleop. Ins. British Mus., 8:388; Lacordaire, 1872, Genera des coléoptères, 9(2):776, fn.

Lepturges pictus: Horn, 1880, Trans. Amer. Entomol. Soc., 8:126, pl. 2, fig. 2; Leng and Hamilton, 1896, Trans. Amer. Entomol. Soc., 23:127; Dillon, 1956, Ann. Entomol. Soc. Amer., 49:343, fig. 6; Bayer and Shenefelt, 1969, Univ. Wisc. Res. Bull., 275:29, fig. 37; Rice and Enns, 1981, Trans. Mo. Acad. Sci., 15:101.

Lepturges symmetricus var. *pictus*: Blatchley, 1910, Coleoptera in Indiana, p. 1076, fig. 462; Smith, 1910, N.J. St. Mus. Rept., 1909:334.

Lepturges symmetricus ab. *pictus*: Aurivillius, 1923, Coleop. Cat., 74:416.

Male. Form moderate-sized; integument pale to dark reddish brown, appendages paler, tibiae and femora usually infuscated, antennae unicolorous; pubescence fine, dense, appressed, brownish to grayish. Head with front convex, densely micropunctate, pubescence dense, appressed, margins with several, long, erect setae; genae slightly shorter than lower eye lobes; eyes small, upper lobes separated by about three times their width; vertex deeply impressed; antennae extending about five segments beyond elytra, third segment shorter than fourth, fifth subequal to fourth, suberect setae at apices of segments short. Pronotum broader than long; disk densely micropunctate with large punctures sparsely scattered around middle, at sides, and on basal depression; pubescence fine, grayish, with brownish maculae at sides of middle at apex and before base and usually at bases of lateral tubercles, middle vaguely impressed longitudinally; lateral tubercles placed slightly before base; prosternum finely pubescent; meso- and metasternum finely, densely punctate and pubescent. Elytra about 2.4 times longer than broad; basal gibbosities and impressions shallow; pubescence dense, appressed, grayish except for dark maculae as follows: two pairs of small rounded spots behind basal margin near sides of scutellum and behind, on humeri, an elongate pair at sides of basal one-third but not attaining margins, three elongate maculae behind middle, lateral pair largest and attaining margins, other two pairs discal but not attaining suture, three pairs of small maculae near apex, some maculae often coalescing; punctures moderately coarse, dense, obsolete near apex; apices slightly dehiscent, obliquely rounded. Legs robust, finely pubescent. Abdomen densely micropunctate, finely, densely pubescent; last sternite rounded at apex. Length, 8-9.5 mm.

Female. Form similar. Femora less robust. Abdomen with last sternite elongate, emarginate at apex. Length, 9-10.5 mm.

Type locality. Ohio.
Range. East-central states to Mississippi.
Flight period. May to July.
Host plants. Celtis, Juglans.

This species may be separated from *L. symmetricus* by the absence of dark lateral vittae beginning below the humeri. In *L. pictus* the maculae of the elytra are very distinct and clearly delimited. Also, the genae are only slightly shorter than the lower eye lobes, which are considerably shorter in *symmetricus*.

Lepturges confluens (Haldeman)

Leiopus symmetricus var. *confluens* Haldeman, 1847, Trans. Amer. Philos. Soc., (2)10:50.

Lepturges confluens: Casey, 1913, Memoirs on the Coleoptera, 4:318; Knull, 1946, Ohio Biol. Surv. Bull., 39:252; Fattig, 1947, Emory Univ. Mus. Bull., 5:36; Dillon, 1956, Ann. Entomol. Soc. Amer., 49:340, fig. 4; Dillon and Dillon, 1961, Man. Beetles East. North America, p. 644, pl. 64, no. 6; Gardiner, 1969, Can. Dept. Fish. For. Int. Rept., 0-14:86; Headstrom, 1977, Beetles of America, p. 378; Gosling and Gosling, 1977, Gr. Lakes Entomol., 10:28, fig. 151; Rice and Enns, 1981, Trans. Mo. Acad. Sci., 15:100; Waters and Hyche, 1984, Coleop. Bull., 38:284; Gosling, 1984, Gr. Lakes Entomol., 17:72.

Male. Form small; integument pale to dark reddish brown, appendages often paler; pubescence fine, dense, appressed, brownish to grayish, often irregularly condensed into patches on elytra. Head with front convex, densely micropunctate, pubescence dense, appressed, margins with several long, erect setae; genae two-thirds as long as lower eye lobes; eyes small, upper lobes separated by about three times their width; vertex deeply impressed; antennae extending about six segments beyond body, third segment subequal to fourth, fifth subequal to fourth, suberect setae at apices of segments short. Pronotum broader than long; disk densely micropunctate with coarse punctures sparsely scattered, linear along basal impression; pubescence fine, often with brownish maculae; lateral tubercles placed almost at base; prosternum finely pubescent; meso- and metasternum densely micropunctate, densely pubescent. Elytra about twice as long as broad; basal gibbosities and impressions shallow; punctures coarse, dense, becoming obsolete at apex; surface with paler and dark areas usually coalescing, distinct pattern lacking; pubescence condensed into irregular grayish patches; apices broadly rounded. Legs finely, densely pubescent; femora darker at apices; tibiae broadly pale annulate near bases. Abdomen densely micropunctate, finely, densely pubescent; last sternite subtruncate at apex. Length, 6-9.5 mm.

Female. Form similar. Antennae slightly shorter than in male. Abdomen with last sternite slightly elongate, narrowing and emarginate at apex. Length, 6.5-10 mm.

Type locality. Not indicated (United States).
Range. Eastern United States to Kansas and Texas.
Flight period. May to August.
Host plants. Carya, Cornus, Diospyros, Fagus, Juglans, Liquidambar, Quercus.

The maculation of the elytra in this species is quite variable. Usually there is no distinct pattern, and the paler and dark areas are irregularly coalesced and the entire surface is suffused with grayish pubescence. Occasionally distinct, gray-suffused, brownish maculae are present. The broadly rounded apices of the elytra separate this species from other North American *Lepturges*.

Lepturges angulatus (LeConte)

Liopus angulatus LeConte, 1852, J. Acad. Nat. Sci. Philadelphia, 2:172.
Leiopus angulatus: White, 1855, Cat. Coleop. Ins. British Mus., 8:388; Lacordaire, 1872, Genera des coléoptères, 9(2):776, fn.
Lepturges angulatus: Provancher, 1877, Pet. Fauna Entomol. Can., 1:628; Linsley and Martin, 1933, Entomol. News, 44:182; Dillon, 1956, Ann. Entomol. Soc. Amer., 49:340; Gosling and Gosling, 1977, Gr. Lakes Entomol., 10:28; Rice and Enns, 1981, Trans. Mo. Acad. Sci., 15:100.
Lepturges angulatus angulatus: Dillon, 1956, Ann. Entomol. Soc. Amer., 49:342, fig. 5; Turnbow and Wappes, 1980, J. Ga. Entomol. Soc., 15:345; Rice and Enns, 1981, Trans. Mo. Acad. Sci., 15:100.
Lepturges symmetricus var. *angulatus*: Smith, 1910, N.J. St. Mus. Rept., 1909:334.
Lepturges confluens var. *angulatus*: Leonard, 1928, Cornell Agr. Exp. Sta. Mem., 101:452.
Lepturges confluens angulatus: Alexander, 1958, Proc. Okla. Acad. Sci., 38:47.
Lepturges canus Casey, 1913, Memoirs on the Coleoptera, 4:317. New synonymy.
Lepturges angulatus canus: Dillon, 1956, Ann. Entomol. Soc. Amer., 49:343; Turnbow and Wappes, 1978, Coleop. Bull., 32:370; Hovore and Penrose, 1982, Southwest. Nat., 27:26; Hovore et al., 1987, Proc. Calif. Acad. Sci., 44:319, fig. 19.

Male. Form small; integument pale to dark reddish brown, antennae usually paler; pubescence fine, dense, appressed, grayish and brown, brownish maculae at middle of elytra suffused with gray. Head with front convex, densely micropunctate, pubescence dense, appressed, bottom margin with several long erect hairs; genae about two-thirds as long as lower eye lobes; eyes rather small, upper lobes separated by about three times their width; vertex deeply impressed between eyes; antennae about twice as long as body, third segment subequal to fourth, fifth equal to fourth, third and fourth segments with a few short suberect setae. Pronotum broader than long; disk with four vague calluses at each side of middle; punctures sparse, scattered around middle and at sides; pubescence fine, dense, appressed, each side of middle with a short macula at base and at apex, usually not uniting near middle; lateral tubercles placed at base; prosternum finely pubescent; meso- and metasternum densely clothed with pale appressed pubescence. Elytra a little more than twice as long as broad; basal gibbosities and impressions shallow: basal punctures moderately fine, separated, becoming coarser and denser at middle and obsolete toward apex; surface with dark maculae coalescing, base with a short, usually arcuate pair on each side of scutellum, sides with maculae from humeri to near apex, irregularly expanding onto disk near middle, apices with dark maculae at sides, at least middle maculae suffused with grayish; apices obliquely truncate. Legs finely, densely pubescent; femora narrowly dark annulate at apices; tibiae broadly brownish at apices. Abdomen densely micropunctate, finely densely pubescent; last sternite slightly emarginate at apex. Length, 5.5-8 mm.

Female. Form similar. Antennae slightly shorter. Abdomen with last sternite slightly elongate, narrowing and emarginate at apex. Length, 6-10 mm.

Type locality. of *angulatus*, Georgia; *canus*, Austin, Texas.

Range. Eastern United States to Kansas and Texas, Mexico.

Flight period. March to August.

Host plants. Acacia, Celtis, Ficus, Gleditsia, Juglans, Leucaena, Pithecellobium, Quercus, Ulmus.

This species may be separated from the similar-appearing *L. symmetricus* by the grayish suffused elytral maculations and the more widely separated upper eye lobes. The obliquely truncate elytral apices distinguish *L. angulatus* from *L. confluens*.

We have not seen sufficient series of this species to justify the recognition of subspecies. Most of the host records indicated are for the Texas populations.

Liopinus, new genus

Amniscus Haldeman, 1847, Trans. Amer. Philos. Soc. (2)10:46 (part).

Liopus: LeConte, 1852, J. Acad. Nat. Sci. Philadelphia, 2:170; Horn, 1880, Trans. Amer. Entomol. Soc., 8:123; LeConte and Horn, 1883, Smithson. Misc. Coll., 507:324; Leng and Hamilton, 1896, Trans. Amer. Entomol. Soc., 23:121 (part); Wickham, 1897, Can. Entomol., 29:202; Wickham, 1898, Can. Entomol., 30:37; Blatchley, 1910, Coleoptera in Indiana, p. 1073 (part).

Leiopus: Haldeman, 1847, Trans. Amer. Philos. Soc., (2)10:50; Thomson, 1860, Class. ceram., p. 12; Thomson, 1864, Systema. ceram., p. 26; Lacordaire, 1872, Genera des coléoptères, 9(2):775 (part); Casey, 1913, Memoirs on the Coleoptera, 4:310; Bradley, 1930, Man. Genera Beetles, p. 246; Chagnon, 1933-40, Coleop. Prov. Quebec, p. 272; Knull, 1946, Ohio Biol. Surv. Bull., 39:248; Chagnon and Robert, 1962, Prin. Coleop. Prov. Quebec, p. 272.

Sternidius LeConte, 1873, Smithson. Misc. Coll., 11(264):234 (part); Dillon, 1956 (not LeConte, 1873), Ann. Entomol. Soc. Amer., 49:208; Dillon and Dillon, 1961, Man. Beetles East. North America, p. 640; Arnett, 1962, Beetles U.S., 103:872, 873; Bayer and Shenefelt, 1969, Univ. Wisc. Res. Bull., 275:27; Lewis, 1977, Pan-Pac. Entomol., 62:171; Rice and Enns, 1981, Trans. Mo. Acad. Sci., 15:99, Lewis, 1986, Pan-Pac. Entomol., 62:171.

Form small to moderate-sized, subdepressed; pubescence short, dense, appressed, pronotum and elytra often with small dark spots. Head with front convex, transverse to quadrate; genae subequal to or shorter than lower eye lobes; eyes moderate-sized, deeply emarginate, upper lobes small, widely separated; antennal tubercles prominent, widely divergent; antennae eleven-segmented, 1-1/3 to two times as long as body, third segment longer than first, fourth subequal to or shorter than third. Pronotum broader than long; lateral tubercles acute, placed before base; base broadly impressed transversely, impression extending under lateral tubercles; apex very narrowly impressed; disk convex, with three often-dark, shallow, rounded calluses; punctures fine, dense; prosternum with intercoxal process narrow, one-third to one-sixth as wide as coxal cavities; mesosternal process abruptly declivous anteriorly, much broader than prosternal process; metasternum with episternum narrow, tapering posteriorly. Elytra about twice as long as broad, sides tapering at apical one-third; base shallowly gibbose, basal depressions shallow; disk convex, each elytron with two longitudinal costae; small tufted tubercles often present; punctures small, dense, becoming obsolete toward apex; pubescence short, appressed, usually with dark spots and/or vittae; apices obliquely truncate to rounded. Legs short; femora strongly clavate; tarsi short, first segment about as long as two following together, third segment cleft to base. Abdomen normally segmented.

Type species. Lamia alpha Say.

This genus is characterized by the acute lateral tubercles and uninterrupted basal sulcus of the pronotum, the narrow prosternal process, broader mesosternal process, and by the dark spots and calluses of the pronotum and usual dark spots of the elytra.

Because of what we interpret as a valid type species designation for *Sternidius* by LeConte in 1873, we are forced to propose a new generic name for the species currently recognized as *Sternidius*. In the Smithson. Misc. Coll., 11(264):234, LeConte proposed the name *Sternidius* for a group of species previously included under the name *Liopus*. The key to *Sternidius* listed seven species including *variegatus* Haldeman. In Smithson. Misc. Coll., 11(265):338, also appearing in May-June 1873, LeConte stated, "The new genus *Sternidius* is founded upon *Amniscus variegatus* Hald. and allies, contained in division C of my revision, Journ. Acad. Nat. Sci. Phil., 2nd ser. ii 172"; this statement appears to be a valid subsequent designation of a type species for *Sternidius*. Thus, *Sternidius* is the proper generic name for Dillon's *Astyleiopus,* and the species previously listed as *Sternidius* require a new generic name.

This group is very variable and species are often difficult to define.

We have relied on the recent revision of *Sternidius* by Lewis (1986) for much of the synonymical bibliography. Our approach to the taxonomy of the group has been more conservative, but future studies on the habits and host-plant relationships may confirm Lewis' classification.

Considerations of the Mexican fauna are, unfortunately, beyond the scope of this study. The species occurring in the southwestern United States probably have affinities with Mexico and possibly even with Central America.

We recognize ten species in our fauna.

KEY TO THE NORTH AMERICAN SPECIES OF *LIOPINUS*

1	Pronotum with pubescence concolorous or with three rounded dark spots or longitudinal vittae	2
	Pronotum with pubescence irregularly mottled whitish and brown. Elytra with three dark, transverse maculae. Length, 7-9.5 mm. Southern Texas	*wilti*
2(1)	Epipleura lacking dark maculae, humeri without black spots on top	3
	Epipleura with dark maculae often extending onto disk near middle; if maculae absent, humeri black on top	5
3(2)	Elytra with apices obliquely truncate to emarginate	4
	Elytra with apices rounded, disk with a few dark tufted tubercles along costae, dark vittae absent. Length, 5.6-9 mm. Southern Arizona	*chemsaki*
4(3)	Thoracic sterna and appendages mottled with small rounded, brownish spots; elytra usually with a transverse, slightly oblique whitish fascia behind middle. Length, 3.5-6.5 mm. Eastern United States to Texas	*punctatus*

	Underside and appendages uniformly pubescent; elytra with pale postmedian fascia vague, v-shaped if present. Length, 3.5-7 mm. Eastern United States to Texas .... *misellus*
5(2)	Elytra with apices rounded; if subtruncate, epipleural maculae extending to lateral margins 6
	Elytra with apices truncate, obliquely truncate to obliquely emarginate truncate 9
6(5)	Epipleural maculae extending to lateral margins 7
	Epipleural maculae small, not attaining lateral margins, vaguely rectangular in shape. Elytra with pubescence uniformly grayish except for maculae and tufted spots. Length, 5-8 mm. Southwestern Texas to south central Arizona .... *imitans*
7(6)	Epipleural maculae vaguely rounded, not extending to humeri, usually extending partially onto disk 8
	Epipleural maculae vittiform, extending from humeri to beyond middle of elytra, not or barely extending onto disk. Elytra with postmedian maculae strongly oblique back from suture. Length, 5-6.5 mm. Montane southern Arizona .... *incognitus*
8(7)	Antennae with fourth segment distinctly longer than first. Elytra not or vaguely costate, tufted dark tubercles reduced to flat spots; postmedian dark maculae usually well developed. Length, 4.5-8.5 mm. Arizona .... *decorus*
	Antennae with fourth segment shorter than or subequal to first. Elytra distinctly costate, tufted dark tubercles elevated; postmedian dark maculae usually reduced to sutural spots. Length, 5-8 mm. Southern Arizona to Texas .... *centralis*
9(6)	Hind tarsi with first segment shorter than two following together. Elytra with epipleural maculae not attaining humeri, usually extending onto disk; postmedian maculae slightly oblique. Length, 5-8.5 mm. East-central United States to Florida and Texas, Mexico .... *mimeticus*
	Hind tarsi with first segment equal in length to two following segments together. Elytra with epipleural maculae extending to humeri and onto disk; postmedian maculae, if present, strongly oblique. Length, 4.5-8.5 mm. Eastern North America to New Mexico and Colorado .... *alpha*

Liopinus wilti (Horn), new combination
(Figure 15)

Liopus Wiltii Horn, 1880, Trans. Amer. Entomol. Soc., 8:124.
Liopus wiltii: Leng and Hamilton, 1896, Trans. Amer. Entomol. Soc., 23:122.
Leiopus wiltii: Casey, 1913, Memoirs on the Coleoptera, 4:310; Linsley and Martin, 1933, Entomol. News, 44:182; Vogt, 1949, Pan-Pac. Entomol., 25:181.
Sternidius wiltii: Dillon, 1956, Ann. Entomol. Soc. Amer., 49:209, fig. 8; Turnbow and Wappes, 1978, Coleop. Bull., 32:370; Lewis, 1986, Pan-Pac. Entomol., 62:174, fig. 1; Hovore et al., 1987, Proc. Calif. Acad. Sci., 44:318, fig. 18.

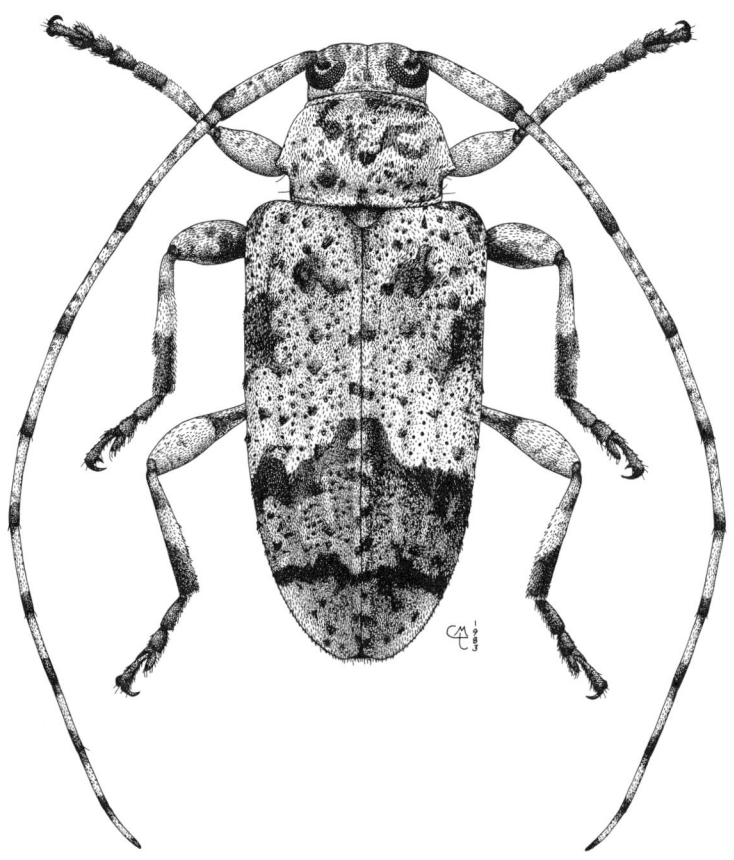

Figure 15. *Liopinus wilti* (Horn), male.

Male. Form moderate-sized, somewhat robust; integument dark reddish brown, appendages paler; pubescence dense, short, appressed, mottled whitish, brown, and black, epipelura not maculate. Head with front convex, pubescence dense, pale brownish with small darker spots interspersed; genae shorter than lower eye lobes; antennae about 1-1/2 times longer than body, segments narrowly dark annulate at apices, segments pale pubescent, basal segments with small, irregular dark spots, third segment longer than first, fourth subequal to first. Pronotum broader than long, lateral tubercles broad; base broadly impressed, apex narrowly impressed; disk vaguely callused; punctures fine, dense; pubescence mottled grayish and brown; prosternum densely pubescent, intercoxal process narrow; mesosternum with intercoxal process broad, about half as broad as coxal cavities; metasternum densely pale pubescent with small brownish spots interspersed. Elytra slightly less than twice as long as broad; basal gibbosities moderate; costae vague, bearing a variable number of dark tufted callosities; punctures coarse at base, becoming obsolete toward apex; pubescence dense, appressed, mottled brownish, gray and black, often with a transverse whitish macula near middle, bordered with black posteriorly; apices obliquely truncate. Legs stout, middle and hind femora dark annulate at middle, tibiae dark annulate at apex and base; tarsi dark. Abdomen densely pale pubescent; last sternite shallowly emarginate at apex. Length, 7.4-9 mm.

Female. Form similar. Abdomen with fifth sternite twice as long as fourth. Length, 8-10 mm.

Type locality. Texas.

Range. Southern Texas.

Flight period. April to June.

Host plants. Acacia farnesiana, Prosopis.

This species may be readily recognized by its rather large, robust form and distinctive color pattern. Often the elytra are vaguely trimaculate with pale pubescence.

Liopinus chemsaki (Lewis), new combination

Sternidius chemsaki Lewis, 1977, Pan-Pac. Entomol., 53: 196; Lewis, 1986, Pan-Pac. Entomol., 62: 180.

Male. Form elongate; integument reddish brown to black; pubescence dense, short, appressed, grayish, brown, and black, epipleura not maculate. Head with front micropunctate, densely pubescent; eyes with lower lobes subequal in length to genae; antennae extending about five segments beyond elytra, segments narrowly dark annulate at apices, basal segments dark mottled, third segment longer than first, fourth shorter than third, slightly longer than first. Pronotum broader than long; base broadly, shallowly impressed, apex very narrowly impressed; disk vaguely tricallused, punctures fine, dense; pubescence dense, appressed, callosities usually dark brownish; prosternum densely, uniformly pubescent; meso- and metasternum densely, uniformly pubescent. Elytra slightly more than twice as long as broad; basal gibbosities broad, moderately elevated; depressions behind gibbosities paler; each elytron tricostate, costae not attaining apices; punctures moderately coarse, dense, becoming absolete toward apex; pubescence dense, short, appressed, grayish, often slightly condensed along costae, dark tufted tubercles sparsely interspersed on costae, usually with a larger pair on basal gibbosities and behind middle; apices rounded. Legs densely grayish pubescent, tibiae vaguely dark biannulate; tarsi dark. Abdomen uniformly densely pubescent; last sternite a little longer than fourth, subtruncate at apex. Length, 6-8 mm.

Female. Form similar. Abdomen with fifth sternite at least twice as long as fourth. Length, 7-9.5 mm.

Type locality. Madera Canyon, Santa Cruz Co., Arizona.
Range. Montane southern Arizona.
Flight period. July to September.
Host plants. Quercus.

The lack of epipleural vittae and presence of rounded elytral apices readily identifies this species. The elytra lack dark or pale transverse vittae, and the tufted dark tubercles are sparse along the costae.

Liopinus punctatus (Haldeman), new combination

Amniscus punctatus Haldeman, 1847, Trans. Amer. Philos. Soc., (2)10:49.
Sternidius punctatus: LeConte, 1873, Smithson. Misc. Coll., 11(264):235; Lewis, 1986, Pan-Pac. Entomol., 62:182.
Liopus punctatus: Horn, 1880, Trans. Amer. Entomol. Soc., 8:124; Chittenden, 1894, Proc. Entomol. Soc. Wash., 3:101; Leng and Hamilton, 1896, Trans. Amer. Entomol. Soc., 23:125; Wickham, 1898, Can. Entomol. 30:37; Blatchley, 1910, Coleoptera in Indiana, p. 1075; Smith, 1910, N.J. St. Mus. Rept., 1909:333; Craighead, 1923, Can. Dept. Agr. Bull., (n.s.) 27:117.
Leiopus punctatus:Leonard, 1910, Cornell Agr. Exp. Sta. Mem., 101:452; Casey, 1913, Memoirs on the Coleoptera, 4:311; Chagnon, 1933-40, Coleop. Prov. Quebec, p. 272; Knull, 1946, Ohio Biol. Surv. Bull., 39:249, pl. 25, fig. 109; Fattig, 1947, Emory Univ. Mus. Bull., 5:35; Chagnon and Robert, 1962, Prin. Coleop . Prov. Quebec, p. 272.
Leiopus maculipennis Blatchley, 1922, Can. Entomol., 54:31.
Sternidius fascicularis maculipennis: Dillon, 1956, Ann. Entomol. Soc. Amer., 49:213.

Male. Form small; integument pale to dark reddish brown; pubescence short, dense appressed, whitish, pale and dark brownish, mottled, epipleura not maculate. Head with front densely micropunctate, pubescence dense, mottled pale and dark brownish; eyes with lower lobes subequal in length to genae; antennae extending about four segments beyond elytra, segments narrowly dark annulate at apices, basal segments mottled with darker brown, third segment longer than first, fourth subequal to third. Pronotum broader than long; base broadly, shallowly impressed, apex very narrowly impressed; disk convex, very finely, densely punctate; pubescence short, dense, mottled, usually with three dark spots; prosternum finely, densely pubescent; meso- and metasternum densely clothed with pale pubescence with brownish spots interspersed. Elytra a little more than twice as long as broad; basal gibbosities small, shallow; each elytron tricostate, costae not attaining apices; punctures moderately coarse, dense, becoming obsolete toward apex; pubescence fine, dense, appressed, pale brownish, dark, tufted tubercles numerous along costae and suture, a whitish, slightly oblique fascia usually present behind middle; apices obliquely truncate to slightly emarginate. Legs densely pubescent, paler pubescence with darker spots interspersed; tibiae dark biannulate; tarsi dark. Abdomen uniformly densely pale pubescent; last sternite slightly longer than fourth, rounded at apex. Length, 5-7 mm.

Female. Form similar. Abdomen with last sternite at least twice as long as fourth. Length, 5-7 mm.

Type locality. Of *punctatus*, none (United States); *maculipennis*, Dunedin, Florida.

Range. Eastern United States from Pennsylvania to Florida and Texas.

Flight period. March to August.

Host plants. Amelanchier, Celtis, Cornus, Diospyros, Parthenocissus, Prunus.

This species is readily recognizable by the mottled-appearing pubescence. The thoracic sterna and appendages are clothed with pale pubescence with darker spots interspersed. The lack of epipleural dark vittae is also distinctive.

Judging by the material available for study, the elytral pattern is quite variable. The tufted dark tubercles along the costae are variable in number and there is often an oblique dark fascia behind the pale one. Often the costae and suture have small, rounded, white spots. Occasionally the pronotal spots are extended into longitudinal vittae. The whitish postmedian fascia of the elytra is often reduced to a series of patches.

In a series from Missouri reared from *Diospyros* by T. C. MacRae, the integument on the basal half of the elytra is dark. This extends over the epipleura, but is not a clearcut epipleural vitta as in *L. alpha*.

Liopinus misellus (LeConte), new combination

Liopus misellus LeConte, 1852, J. Acad. Nat. Sci. Philadelphia, 2: 173.
Leiopus misellus: White, 1855, Cat. Coleop. Ins. British Mus., 8:389; Lacordaire, 1872, Genera des coléoptères, 9(2): 776, fn.; Casey, 1913, Memoirs on the Coleoptera, 4: 313.
Sternidius alpha misellus:Dillon, 1956, Ann. Entomol. Soc. Amer., 49:216; Stein and Tagestad, 1976, USDA For. Serv. Res. Pap., RM-171:41; Rice

and Enns, 1981, Trans. Mo. Acad. Sci., 15:99; Gosling, 1986, Gr. Lakes Entomol., 19:157.
Sternidius misellus: Lewis, 1986, Pan-Pac. Entomol., 62: 183.

Male. Form small; integument light brownish to piceous; pubescence short, fine, appressed, grayish, brownish to black, epipleura not maculate. Head with front densely micropunctate, densely pubescent; eyes with lower lobes subequal in length to genae; antennae extending about five segments beyond elytra, segments narrowly dark annulate at apices, basal segments not mottled, third segment longer than first, fourth subequal to third. Pronotum broader than long; base broadly, shallowly impressed, apex very narrowly impressed; disk convex, vaguely tricallused; punctures fine, dense; pubescence fine, dense, pale brownish, calluses usually dark brownish; prosternum finely, densely pubescent; meso- and metasternum uniformly finely, densely pubescent. Elytra about twice as long as broad; basal gibbosities barely elevated; costae often vague; punctures moderately coarse, dense, becoming obsolete toward apex; pubescence fine, dense, grayish to brownish, small, dark, tufted tubercles usually sparse, postmedian fasciae usually present, pale pubescence variably interspersed before fasciae; apices obliquely truncate to emarginate. Legs finely pubescent; tibiae usually dark biannulate; tarsi dark. Abdomen uniformly densely pubescent; last sternite lightly emarginate at middle. Length, 4-7 mm.

Female. Form similar. Abdomen with fifth sternite twice as long as fourth. Length, 5-7 mm.

Type locality. Illinois.
Range. Eastern United States to Kansas and Texas.
Flight period. April to July.
Host plants. Carpinus, Carya, Castanea, Diospyros, Parthenocissus, Quercus.

This species may be recognized by the relatively uniform pubescence and lack of epipleural vittae.

The elytral pattern can be quite variable, with varying amounts of whitish pubescence and dark, tufted tubercles. The postmedian dark fasciae of the elytra are often reduced.

Liopinus imitans (Knull), new combination

Leiopus imitans Knull, 1936, Entomol. News, 47: 107.
Sternidius imitans: Dillon, 1956, Ann. Entomol. Soc. Amer., 49: 217; Linsley, Knull, and Statham, 1961, Amer. Mus. Nov., 2050: 29; Lewis, 1977, Pan-Pac. Entomol., 53: 198, fig. 3; Lewis, 1979, Pan-Pac. Entomol., 55: 25; Lewis, 1986, Pan-Pac. Entomol., 62: 199, fig. 3.

Male. Form small to moderate-sized, slightly elongate; integument reddish brown to piceous; pubescence short, dense, grayish, appressed, maculae and small tufts black. Head with front densely micropunctate, densely clothed with mottled grayish and brownish pubescence; genae subequal in length to lower eye lobes; antennae extending about 4-1/2 segments beyond elytra, segments narrowly dark annulate at apices, outer segments also narrowly annulate at bases, basal segments with small, dark spots, eleventh segment all dark, third segment longer than first, fourth slightly shorter than third, longer than first. Pronotum longer than broad; base broadly, shallowly impressed, apex very narrowly impressed; disk convex, calluses vague; punctures fine, dense, usually obscured by pubescence; pubescence dense, grayish, usually with two rounded dark spots before middle, median dark spot usually elongate, reaching to basal margin; prosternum finely, densely pubescent; meso- and metasternum densely, gray pubescent, dark spots sparse. Elytra slightly more than twice as long as broad; basal gibbosities very shallow; costae absent; punctures moderately coarse, dense, becoming obsolete toward apex; pubescence dense, grayish, appressed, small, dark, tufted spots numerously interspersed, sides at basal one-third with vaguely rectangular dark fasciae which extend slightly onto disk and one-half the distance to lateral margins down the epipleura, disk with a postmedian, moderate-sized, sutural black fascia, dark fasciae usually outlined with pale grayish pubescence; apices rounded. Legs with femora densely pubescent, pubescence interrupted by small dark spots; tibiae dark biannulate; tarsi black. Abdomen uniformly densely pubescent; last sternite subtruncate to shallowly emarginate at apex. Length, 5-7 mm.

Female. Form slightly more robust. Abdomen with last sternite twice as long as fourth. Length, 6-8 mm.

Type locality. Davis Mts., Texas.

Range. Southwestern Texas to south central Arizona.

Flight period. June to September.

Host plants. Probably *Quercus*.

This species is distinctive by the shape and placement of the elytral vittae. These characters separate it from both *L. decorus* and *L. incognitus*.

Liopinus incognitus (Lewis), new combination

Sternidius incognitus Lewis, 1977, Pan-Pac. Entomol., 53: 20, fig. 5; Lewis, 1986, Pan-Pac. Entomol., 62: 196.

Male. Form small, moderately robust; color brown to piceous, covered with predominantly cinereous pubescence; sides of pronotum and elytra with black markings. Head impunctate, covered with cinereous pubescence; frons transverse. Antennae at least 1-1/2 times as long as body, slender, annulate; scape slightly mottled, third segment much less so, subopaque; pubescence of remaining segments sparse, more or less uniform; fourth segment longer than scape, subequal to third; remainder of segments gradually decreasing in length. Pronotum transverse, widest across acute lateral tubercles, which are placed at basal third; sides gradually divergent to tubercles, then abruptly constricted behind, forming the basal transverse sulcus; basal and apical margins subequal; punctures small, dense, shallow, partially obscured by vestiture; disk with three callosites, more or less coalescent and devoid of cinereous pubescence; one each side of middle just behind apex, one median, elongate, extending to basal margin; sides with a distinct fuscous vitta running the entire length of pronotum. Scutellum triangular to broadly rounded, impunctate, pubescent. Elytra moderately robust, about as long as wide; sides subparallel to middle (scarcely indented behind humeri), then slightly expanded and gradually convergent to apex, apices rounded to subtruncate; punctures of disk dense, subconfluent, much larger than those of pronotum, largest at basal third, then gradually decreasing in size to apex, partially obscured by pubescence; punctation of sides distinct, subconfluent (as large as disk), not obscured by pubescence along vittae; costae evanescent; macular areas as follows: disk with a common black triangular area with its apex at mid-elytra, and sides diverging from suture at an angle of about 45 degrees or greater, not or reaching the lateral border; sides with a lateral vitta, extending from behind middle to basal margin, then along sides of pronotum as above described. The vitta encroaches medially to include less than one-tenth of the disk, and is thus sub-obsolete when viewed from above; pubescence black over vittae, cinereous with tawny components forming a saddle anterior to common nacula, behind (within the demarcated apical area) darker, composed of black, cinereous, and tawny elements in variable combination; rows of small, blackish spots varying in extent and number among individuals present. Underside scarcely mottled; femora with dark spot just proximal to club on some specimens, distal tibiae black, annulate; tarsi black; procoxal process narrow, about one-tenth the width of procoxal cavity; mesocoxal process about half the width of mesocoxal cavity; fifth abdominal sternite subequal to fourth. Length 4.7-6.2 mm.

Female. Fifth abdominal sternite about twice the length of fourth. Length 4.7-6.0 mm. (Original description.)

Type locality. Madera Canyon, Santa Rita Mts., Santa Cruz Co., Arizona.
Range. Montane southern Arizona.
Flight period. July and August.
Host plants. Quercus.

This species differs from *L. alpha* by the rounded elytral apices and lack of costal and postmedian tufts. The shape and placement of the epipleural and discal maculae separate *L. incognitus* from other Arizona species.

Liopinus decorus (Fall), new combination

Liopus decorus Fall, 1907, J. N.Y. Entomol. Soc., 15: 84, Schaeffer, 1908, Brooklyn Mus. Inst. Arts Sci. Bull., 1:345.
Sternidius decorus: Dillon, 1956, Ann. Entomol. Soc. Amer., 49:218 (synonymy); Linsley, Knull, and Statham, 1961, Amer. Mus. Nov., 2050: 29; Lewis, 1977, Pan-Pac. Entomol., 53: 200, fig. 4; Lewis, 1979, Pan-Pac. Entomol. 55: 24; Lewis, 1986, Pan-Pac. Entomol., 62: 197, figs. 4, 16.

Sternidius alpha arizonensis Dillon, 1956, Ann. Entomol. Soc. Amer., 49: 217;
    Linsley, Knull, and Statham, 1961, Amer. Mus. Nov., 2050: 29.
Sternidius centralis: Dillon, 1956, Ann. Entomol. Soc. Amer., 49: 218 (part).

Male. Form small to moderate-sized; integument dark reddish brown, spots and maculae black; pubescence short, dense, appressed, grayish, brownish, and black, epipleura with vaguely rounded maculae well behind humeri. Head with front densely, shallowly micropunctate, densely pubescent; eyes with lower lobes subequal in length to genae; antennae extending about five segments beyond elytra, segments narrowly dark annulate at apices, eleventh segment all dark, basal segments sparsely dark-spotted, third segment longer than first, fourth shorter than third, distinctly longer than first. Pronotum broader than long; disk convex, calluses vague; punctures fine, dense; pubescence dense, appressed, usually mottled grayish and brown, calluses usually dark brownish; prosternum finely, densely pubescent; meso- and metasternum densely pubescent, finely dark spotted at sides. Elytra slightly more than twice as long as broad; basal gibbosities shallow; costae vague or absent; small, dark spots not elevated, usually sparsely interspersed; punctures moderately coarse, dense, becoming obsolete toward apex; pubescence dense, appressed, grayish to mottled grayish and brown, postmedian dark macula inverted V-shaped, extending from suture to about middle of disk, epipleural dark maculae almost rounded, extending from lateral margins partially onto disk, not extending to humeri, base usually with small, dark spots; apices truncate. Legs with femora densely pubescent, dark-spotted; tibiae dark biannulate; tarsi black. Abdomen uniformly densely pubescent; last sternite subtruncate at apex. Length, 4.5-7.5 mm.

Female. Form similar. Abdomen with fifth sternite about twice as long as fourth. Length, 5-8.5 mm.

Type locality. Of *decorus*, Williams, Arizona; *arizonensis*, Cave Creek, Chiricahua Mts., Arizona.

Range. Southern Arizona.

Flight period. June to September.

Host plants. Quercus.

The shape of the elytral and epipleural dark maculae readily separate this species from others occurring in Arizona.

The elytral apices may occasionally seem to be rounded, but examination from all angles will indicate at least a slight truncature.

Liopinus centralis (LeConte), new combination

Liopus centralis LeConte, 1884, Trans. Amer. Entomol. Soc., 12: 24; Leng and
    Hamilton, 1896, Trans. Amer. Entomol. Soc., 23: 123.
Leiopus centralis: Casey, 1913, Memoirs on the Coleoptera, 4: 30.

Sternidius centralis: Dillon, 1956, Ann. Entomol. Soc. Amer., 49: 218, figs. 2, 3; Lewis, 1977, Pan-Pac. Entomol., 53: 196, fig. 2; Lewis, 1986, Pan-Pac. Entomol., 62: 175, fig. 2.

Male. Form small, moderately robust; integument reddish brown to piceous; pubescence dense, short, appressed, grayish, brownish, and black, epipleural maculae rounded, not extending to humeri. Head with front densely micropunctate, densely mottled pubescent; genae subequal to or slightly shorter than lower eye lobes; antennae 4-1/2 to five segments longer than body, segments narrowly dark annulate, basal segments dark spotted, third segment longer than first, fourth equal to first. Pronotum broader than long; base broadly impressed, apex very narrowly impressed; disk tricallused, punctures fine, dense; pubescence mottled grayish and brownish, calluses usually dark brown; prosternum densely pubescent; meso- and metasternum densely gray pubescent, sides brownish spotted. Elytra less than twice as long as broad; basal gibbosities shallow dark-tufted; costae usually distinct; punctures moderately coarse, dense, obsolete toward apex, obscured by pubescence; pubescence dense, appressed, mottled grayish and brown maculae black, small, dark, tufted tubercles rather sparse along costae, postmedian dark macula small, usually diamond-shaped at suture, epiplural maculae extending to lateral margins and partially onto disk, not attaining humeri; apices subtrunctate to rounded. Legs with femora densely dark-spotted; tibiae broadly dark annulate at apex, narrowly at base; tarsi black. Abdomen uniformly densely pubescent; last sternite broadly rounded at apex. Length, 5-7.5 mm.

Female. Form similar. Abdomen with fifth sternite twice as long as fourth, narrowing at apex. Length 6-8 mm.

Type locality. Arizona.
Range. Southern Arizona to Texas.
Flight period. July to September.
Host plants. Prosopis.

The short fourth antennal segment, presence of distinct costae and tufted tubercles on the elytra, and the reduced postmedian maculae separate this species from *L. decorus*. The short epipleural maculae make *L. centralis* distinct from *L. alpha*.

Liopinus mimeticus (Casey), new combination

Leiopus mimeticus Casey, 1891, Ann. N.Y. Acad. Sci. 6: 49; Casey, 1913, Memoirs on the Coleoptera, 4: 315.

Sternidius mimeticus: Dillon, 1956, Ann. Entomol. Soc. Amer., 49: 210; Dillon and Dillon, 1961, Man. Beetles East. North America, p. 640, pl. 63, no. 12; Hovore and Penrose, 1982, Southwest. Nat., 27: 26; Lewis, 1986, Pan-Pac. Entomol., 62: 177, fig. 7; Hovore et al., 1987, Proc. Calif. Acad. Sci., 44: 318.

Liopus crassulus Horn, 1880 (not LeConte, 1873), Trans. Amer. Entomol. Soc., 8:124, 125; Leng and Hamilton, 1896, Trans. Amer. Entomol. Soc. 23:122; Beutenmuller, 1896, J. N.Y. Entomol. Soc.; Townsend, 1902, Trans. Texas Acad. Sci., 5:78; Blatchley, 1910, Coleoptera in Indiana, p. 1074.
Leiopus crassulus Leonard, 1928 (not LeConte, 1873), Cornell Agr. Exp. Sta. Mem., 101:452; Linsley and Martin, 1933, Entomol. News, 44:182; Knull, 1946, Ohio Biol. Surv. Bull., 39:249; Fattig, 1947, Emory Univ. Mus. Bull., 5:35
Sternidius crassulus Kirk, 1970 (not LeConte, 1873), S.C. Agr. Exp. Sta. Tech. Bull., 1038:82.
Liopus schwarzi Hamilton, 1896, Trans. Amer. Entomol. Soc., 23: 124. New synonymy.
Leiopus schwarzi: Casey, 1913, Memoirs on the Coleoptera, 4: 316.
Sternidius schwarzi: Dillon, 1956, Ann. Entomol. Soc. Amer., 49: 212; Turnbow and Hovore, 1979, Entomol. News, 90: 225; Lewis, 1986, Pan-Pac. Entomol., 62: 179, fig. 6.
Leiopus moderator Casey, 1913, Memoirs on the Coleoptera, 4: 314.
Sternidius moderator: Dillon, 1956. Ann. Entomol. Soc. Amer., 49: 212, fig. 9.
Leiopus houstoni Casey, 1913, Memoirs on the Coleoptera, 4: 315; Vogt, 1949, Pan-Pac. Entomol., 25: 182.
Leiopus texanus Casey, 1913, Memoirs on the Coleoptera, 4: 315; Vogt, 1949, Pan-Pac. Entomol., 25:182. New synonymy.
Sternidius texanus: Hovore and Penrose, 1982, Southwest. Nat., 27: 26 (habits); Rice, 1985, J. N.Y. Entomol. Soc., 93:1224; Lewis, 1986, Pan-Pac. Entomol., 62: 186, fig. 8; Hovore et al., 1987, Proc. Calif. Acad. Sci., 44: 318, fig. 18.

Male. Form moderate-sized, robust; integument reddish brown; pubescence dense, appressed, grayish and brownish, maculae black, epipleura occasionally lacking dark maculae, pubescence obscuring surface. Head with front densely micropunctate, densely pubescent; genae subequal in length to lower eye lobes; antennae extending about four segments beyond elytra, segments narrowly dark annulate, basal segments mottled, third segment longer than first, fourth subequal to first. Pronotum broader than long, lateral tubercles robust; base broadly, shallowly impressed, apex very narrowly impressed; disk tricallused, punctures around calluses fine, dense; pubescence dense, mottled grayish and brownish, calluses often with dark, rounded spots; prosternum densely pubescent, dark spots numerous. Elytra slightly less than twice as long as broad; basal gibbosities vague, shallow; costae distinct; small, dark, tufted tubercles sparsely interspersed along costae; punctures moderately coarse, dense, becoming obsolete toward apex; pubescence obscuring surface, usually mottled grayish and pale brownish, often darker appearing on apical one-third, humeri with a dark spot on top, epipleural maculae not attaining humeri, extending slightly onto disk, occasionally absent, postmedian, dark maculae narrow, shallowly oblique back from suture, sides at ends of maculae with short, linear, dark maculae; apices obliquely truncate, occasionally vaguely emarginate. Legs with femora densely pubescent, dark-spotted and vaguely annulate with a larger dark spot; tibiae dark biannulate; tarsi black. Abdomen usually uniformly pubescent, occasionally vaguely spotted; last sternite subtruncate to shallowly emarginate at apex. Length, 5-8 mm.

Female. Form similar. antennae extending about two segments beyond elytra. Abdomen with fifth sternite about twice as long as fourth. Length, 6-8.5 mm.

Type locality. Of *mimeticus*, Texas; *schwarzi*, Key West, Florida; *moderator*, District of Columbia; *houstoni*, Brownsville, Texas; *texanus*, Brownsville, Texas.

Range. Eastcentral United States to Florida and Texas, Mexico.

Flight period. April to October.

Host plants. Acer, Celtis, Diospyros, Ficus, Leucaenia, Lysiloma, Metopium, Morus, Piscidia, Prunus, Rhus, Sesbania, Sophora, Zanthoxylum.

The robust body form, dense, usually grayish pubescence, narrow, oblique postmedian elytral maculae and relatively short antennae will distinguish this species. Occasionally the epipleural maculae are lacking but those individuals always have the dark spot on the tops of the humeri. The area behind the postmedian maculae is often brown pubescent and the dark tufted tubercles along the costae vary in number.

Liopinus alpha (Say), new combination

Lamia alpha Say, 1827, J. Acad. Nat. Sci. Philadelphia, 5:270; LeConte, 1859, Complete Writings T. Say, 2:329.

Amniscus alpha: Haldeman, 1847, Trans. Amer. Philos. Soc., (2)10:48.

Liopus alpha: LeConte, 1852, J. Acad. Nat. Sci. Philadelphia, 2:172; Horn, 1880, Trans. Amer. Entomol. Soc., 8:124; Leng and Hamilton, 1896, Trans. Amer. Entomol. Soc., 23:124; Beutenmuller, 1896, J. N.Y. Entomol. Soc., 4:79; Wickham, 1898, Can. Entomol., 30:37; Smith, 1910, N.J. St. Mus. Rept., 1909:333; Blatchley, 1910, Coleoptera in Indiana, p. 1074; Rosewall, 1920, Can. Entomol., 52:203; Craighead, 1921, J. Agr. Res., 22:215; Craighead, 1923, Can. Dept. Agr. Bull., (n.s.) 27:117.

Leiopus alpha: White, 1855, Cat. Coleop. Ins. British Mus., 8:388; Lacordaire, 1872, Genera des coléoptères, 9(2):776, fn.; Casey, 1913, Memoirs on the Coleoptera, 4:314; Blackman and Stage, 1924, N.Y. St. Coll. For. Tech. Pub., 17:116; Leonard, 1928, Cornell, Agr. Exp. Sta. Mem., 101:452; Chagnon, 1933-40, Coleop. Prov. Quebec, p. 272; Hoffmann, 1940, Bull. Brooklyn Entomol. Soc., 35:59; Hoffmann, 1942, USDA Misc. Pub., 466:11; Loding, 1945, Geol. Surv. Ala. Mon., 11:123; Knull, 1946, Ohio Biol. Surv. Bull., 39:250; Fattig, 1947, Emory Univ. Mus. Bull., 5:36; Chagnon and Robert, 1962, Prin. Coleop. Prov. Quebec, p. 272.

Sternidius alpha: LeConte, 1873, Smithson. Misc. Coll., 11(264):235; Dillon, 1956, Ann. Entomol. Soc. Amer., 49:209; Dillon and Dillon, 1961, Man. Beetles East. North America, p. 640, pl. 63, no. 11; Bayer and Shenefelt, 1969, Univ. Wisc. Res. Bull., 275:29, fig. 37; Stein and Tagestad, 1976, USDA For. Serv. Res. Pap., RM-171:40; Gosling and Gosling, 1976, Gr. Lakes Entomol., 10:26, fig. 147; Laliberte et al., 1977, Fabreries, 3:99; Headstrom, 1977, Beetles of America, p. 37; Lewis, 1986, Pan-Pac. Entomol. 62:188, figs. 11, 14.

Liopus alpha var.: Townsend, 1902, Trans. Texas Acad. Sci., 5:78.

Sternidius alpha alpha: Dillon, 1956, Ann. Entomol. Soc. Amer., 49:214; Gardiner, 1966, Can. J. Zool., 44:204, figs. 30, 51; Gardiner, 1969, Can. Dept. Fish. For. Int. Rept., 0-14:85.

Lamia (Mesosa) fascicularis Harris, 1836, Trans. Hartford Nat. Hist. Soc., 1:68, pl. 1, fig. 9. New synonymy.

Amniscus fascicularis: Haldeman, 1847, Trans. Amer. Philos. Soc., (2)10:48; White, 1855, Cat. Coleop. Ins. British Mus., 8:393.

Leptostylus fascicularis: LeConte, 1852, J. Acad. Nat. Sci. Philadelphia, 2:170; Lacordaire, 1872, Genera des coléoptères, 9(2):272, fn.

Liopus fascicularis: Horn, 1880, Trans. Amer. Entomol. Soc., 8:124; Leng and Hamilton, 1896, Trans. Amer. Entomol. Soc., 23:125; Beutenmuller, 1896, J. N.Y. Entomol. Soc., 4:79; Blatchley, 1910, Coleoptera in Indiana, p. 1074; Rosewall, 1920, Can. Entomol., 52:203.

Leiopus fascicularis: Casey, 1913, Memoirs on the Coleoptera, 4:311; Leonard, 1928, Cornell Agr. Exp. Sta. Mem., 101:452; Knull, 1946, Ohio Biol. Surv. Bull., 39:250; Fattig, 1947, Emory Univ. Mus. Bull., 5:35.

Sternidius fascicularis: Dillon, 1956, Ann. Entomol. Soc. Amer., 49:209; Lewis, 1986, Pan-Pac. Entomol., 62:190, fig. 15.

Sternidius fascicularis fascicularis: Dillon, 1956, Ann. Entomol. Soc. Amer., 49:213, fig. 7; Rice and Enns, 1981, Trans. Mo. Acad. Sci., 15:99; Waters and Hyche, 1984, Coleop. Bull., 38:284.

Amniscus alpha var. *divergens* Haldeman, 1847, Trans. Amer. Philos. Soc., (2):10:48.

Leiopus alpha var. *divergens*: Aurivillius, 1923, Coleop. Cat., 47:408.

Amniscus vicinus Haldeman, 1847, Trans. Amer. Philos. Soc., (2)10:49.

Leiopus vicinus: White, 1855, Cat. Coleop. Ins. British Mus., 8:379; Casey, 1913, Memoirs on the Coleoptera, 4:312; Leonard, 1928, Cornell Agr. Exp. Sta. Mem., 101:452.

Sternidius alpha vicinus: Dillon, 1956, Ann. Entomol. Soc. Amer., 49:215; Turnbow and Franklin, 1980, J. Ga. Entomol. Soc., 15:345; Waters and Hyche, 1984, Coleop. Bull., 38:284.

Liopus cinereus LeConte, 1852, J. Acad. Nat. Sci. Philadelphia, 2:173; Horn, 1880, Trans. Amer. Entomol. Soc., 8:124; Packard, 1881, U.S. Entomol. Comm. Bull., 7:75; Beutenmuller, 1896, J. N.Y. Entomol. Soc., 4:79; Wickham, 1898, Can. Entomol., 30:37.

Leiopus cinereus:White, 1855, Cat. Coleop. Ins. British Mus., 8:388; Lacordaire, 1872, Genera des coléoptères, 9(2):776, fn.; Casey, 1913,

Memoirs on the Coleoptera, 4:314; Leonard, 1928, Cornell Agr. Exp. Sta. Mem., 101:452; Knull, 1946, Ohio Biol. Surv. Bull., 39:250; Fattig, 1947, Emory Univ. Mus. Bull., 5:35.
Sternidius cinereus:LeConte, 1873, Smithson. Misc. Coll., 11(264):235; Kirk, 1970, S.C. Agr. Exp. Sta. Tech. Bull., 1038:82.
Liopus alpha var. *cinereus*: Blatchley, 1910, Coleoptera in Indiana, p. 1075.
Liopus rusticus LeConte, 1852, J. Acad. Nat. Sci. Philadelphia, 2:173. New synonymy.
Leiopus rusticus: White, 1855, Cat. Coleop. Ins. British Mus., 8:388; Lacordaire, 1872, Genera des coléoptères, 9(2):776, fn.
Sternidius rusticus: Lewis, 1986, Pan-Pac. Entomol., 62;195, fig. 12.
Liopus alpha var *floridanus* Hamilton, 1896, Trans. Amer. Entomol. Soc., 23:125. New synonymy.
Leiopus floridanus: Blatchley, 1920, Can. Entomol., 52:69.
Leiopus alpha floridanus: Leng, 1920, Cat. Coleop., p. 282.
Liopus alpha var. *floridanus*:Aurivillius, 1923, Coleop. Cat., 74:408; Fattig, 1947, Emory Univ. Mus. Bull., 5:36.
Sternidius floridanus: Dillon, 1956, Ann. Entomol. Soc. Amer., 49:217; Turnbow and Franklin, 1980, J. Ga. Entomol. Soc., 15:345; Lewis, 1986, Pan-Pac. Entomol., 61:192, fig. 9.
Liopus xanthoxyli Shimer, 1868, Trans. Amer. Entomol. Soc., 2:7; Packard, 1871, 1st Rept. Inj. Ins. Mass., p. 13, pl. 1, fig. 2; Packard, 1881, U.S. Entomol. Comm. Bull., 7:132, fig. 61; Beutenmuller, 1896, J. N.Y. Entomol. Soc., 4:79.
Sternidius xanthoxyli: LeConte, 1873, Smithson. Misc. Coll., 11(264):235.
Leiopus xanthoxyli: Packard, 1881, U.S. Entomol. Comm., 7:251, figs. 97, 99.
Leiopus dentatus Casey, 1913, Memoirs on the Coleoptera, 4:310.
Leiopus testaceus Casey, 1913, Memoirs on the Coleoptera, 4:311.
Leiopus pleuralis Casey, 1913, Memoirs on the Coleoptera, 4:312.
Leiopus timidus Casey, 1913, Memoirs on the Coleoptera, 4:313.
Leiopus obscurellus Casey, 1913, Memoirs on the Coleoptera, 4:313.
Leiopus scapalis Casey, 1913, Memoirs on the Coleoptera, 4:312.
Leiopus nelsonicus Casey, 1913, Memoirs on the Coleoptera, 11:291.
Sternidius alpha nigricans Dillon, 1956, Ann. Entomol. Soc. Amer., 49:216; Gilmour, 1965, Cat. Lam. du Monde, 8:579.
Sternidius alpha coloradensis Dillon, 1956, Ann. Entomol. Soc. Amer., 49:216; Gilmour, 1965, Cat. Lam. du Monde, 8:579
Sternidius suturalis Dillon, 1956, Ann. Entomol. Soc. Amer., 49:218.
Sternidius vittatus Dillon, 1956, Ann. Entomol. Soc. Amer., 49:219; Lewis, 1986, Pan-Pac. Entomol., 62:185, fig. 10. New synonymy.

Male. Form small to moderate-sized; integument reddish brown to piceous; pubescence short, dense, appressed, variably whitish, brownish to black, epipleura with long, vittaform maculae. Head with front densely micropunctate, densely pubescent; eyes with lower lobes subequal to or slightly longer than genae; antennae extending about five segments beyond elytra, segments narrowly dark annulate at apices, basal segments often dark-spotted, third segment longer than first, fourth slightly shorter than third, longer than first. Pronotum broader than long; base broadly, shallowly impressed, apex very narrowly impressed; disk convex, vaguely tricallused; punctures fine, dense; pubescence appressed, dense, variable, often with calluses brownish and sides pale pubescent; prosternum finely pubescent; meso- and metasternum densely pubescent, pubescence usually interrupted by small, rounded spots. Elytra a little more than twice as long as broad, sides subparallel to apical one-third; basal gibbosities shallow, usually tuberculate; costae often vague, small tufted tubercles sparsely interspersed along costae; punctures moderately coarse, dense, becoming finer and sparser toward apex; pubescence short, dense, appressed, variably variegated with brownish and grayish, postmedian dark macula strongly oblique, often reduced, epipleural dark maculae extending from humeri to beyond middle, at most barely extending onto disk, postmedian maculae often anteriorly bordered by a whitish macula; apices obliquely emarginate to truncate. Legs with femora mottled pubescent; tibiae dark biannulate; tarsi dark. Abdomen uniformly densely pubescent; last sternite emarginate at apex. Length, 4.5-7.5 mm.

Female. Form similar. Abdomen with fifth sternite about twice as long as fourth. Length, 5-8.5 mm.

Type locality. Of *alpha*, Pennsylvania; *fascicularis*, North America; *divergens*, Pennsylvania; *lateralis*, Pennsylvania; *vicinus*, none (North America); *cinereus* Georgia; *rusticus*, western New York; *floridanus*, Biscayne Bay, Florida; *xanthoxyli*, Mount Carroll, Illinois; *dentatus*, Illinois; *testaceus*, District of Colombia; *pleuralis*, District of Columbia and Delaware; *timidus*, Pennsylvania; *obscurellus*, Bluff Point, Lake Champlain, New York; *scapalis*, Indiana; *nelsonicus*, Nelson Co., Virginia; *nigricans*, Tajique, New Mexico; *coloradensis*, Colorado Springs, Colorado; *suturalis*, Rockdale, Texas; *vittatus*, Lucedale, Mississippi.

Range. Eastern North America to Arizona and Colorado.

Flight period. March to October.

Host plants. Acer, Alnus, Amelanchier, Ampelopsis, Carya, Castanea, Celastrus, Celtis, Citrus, Diospyros, Ficus, Gleditsia, Juglans, Malus, Morus, Platanus, Quercus, Rhus (including *glabra, hytra, copallina, typhina*), *Robinia, Ulmus, Wisteria, Zanthoxylum.*

Parasites. Cenocoelius ashmeadii Dalle Torre, *C. provancheri* (Rohwer), *Heterospilus liopodis* (Brues) (Braconidae).

This is a very variable species with numerous pubescent and color patterns. The presence of the dark epipleural maculae which extend from the humeri to slightly beyond the middle of the elytra is a good diagnostic character. This, along with the truncate to obliquely emarginate elytral apices and strongly oblique postmedian elytral maculae, make *S. alpha* readily identifiable.

Variation in coloration and pubescent pattern may be, in part, geographical. Specimens from the extreme southern United States tend to be darker, with more dark spots. The tufted tubercles of the elytra vary from many to almost none. The postmedian elytral maculae may often be reduced or broken into a series of spots. The amount of whitish or grayish pubescence suffused over the surface ranges from almost none to forming a distinct oblique macula before the dark postmedian one. The pronotum may often lack dark spots on the dorsal calluses.

The large number of plants utilized as larval hosts may partially explain some of the observed variation in the adults.

Because of the lack of biological information on *S. alpha* in many parts of the distributional range, and lack of definitive series for study, we are recognizing no geographical populations at this time.

Genus *Nyssodrysina* Casey

Nyssodrys Leng and Hamilton, 1896 (not Bates, 1864), Trans. Amer. Entomol. Soc., 23:133; Bradley, 1930, Man. Genera Beetles, p. 246.
Leiopus: Lacordaire, 1872; Genera des coléoptères, 9(2):775 (part).
Nyssodrysina Casey, 1913, Memoirs on the Coleoptera, 4:309; Dillon, 1956, Ann. Entomol. Soc. Amer., 49:160; Arnett, 1962, Beetles U.S., 103:872; Monne and Giesbert, 1992, Insecta Mundi, 6:253.
Nyssodrysola Gilmour, 1962, Beitr. Neotrop. Fauna, 2:256, (Types species: *Nyssodrysola stictica* Gilmour, by original designation).
Sternidurges Gilmour, 1959, Anal. Inst. Biol. Mexico, 31:329. (Type species: *Sternidurges apicalis* Gilmour, by original designation).

Form moderately small, ovoid. Head with front convex, about as long as broad; median line extending onto neck; mandibles feebly arcuate; genae convergent, about half as long as lower eye lobes; eyes moderate, deeply emarginate, lower lobes longer than broad, upper lobes small, separated by less than diameter of antennal scape; antennal tubercles prominent, widely divergent; antennae eleven-segmented, 1-1/2 to two times as long as body, basal segments with very few short, suberect hairs beneath, first segment shorter than third, third longer than fourth, fourth slightly longer than first, fifth shorter than first. Pronotum broader than long, sides with a large, acute tubercle near base; base narrowly impressed, impression not extending beyond lateral tubercles; apex very narrowly impressed; disk convex; prosternum with internal coxal process narrow, about one-eighth as broad as coxal cacities; mesosternum with intercoxal process broad, about two-thirds as broad as coxal cavities; metasternum with episternum narrow, parallel. Elytra less than twice as long as broad, tapering behind middle; basal gibbosities vague, shallow; apices narrowing, obliquely truncate. Legs stout; femora strongly clavate; tarsi moderate, slender, first segment longer than following two together, third segment cleft to base. Abdomen normally segmented, fifth sternite of females about three times longer than fourth.

Type species. Liopus haldemani LeConte (by original designation).

This genus may be recognized by the ovoid body form, acute basal lateral tubercles of the pronotum, shape of the eyes, and by the proportions of the coxal processes.

The above description is drawn from the single North American species of this Neotropical group.

Nyssodrysina haldemani (LeConte)
(Figure 16)

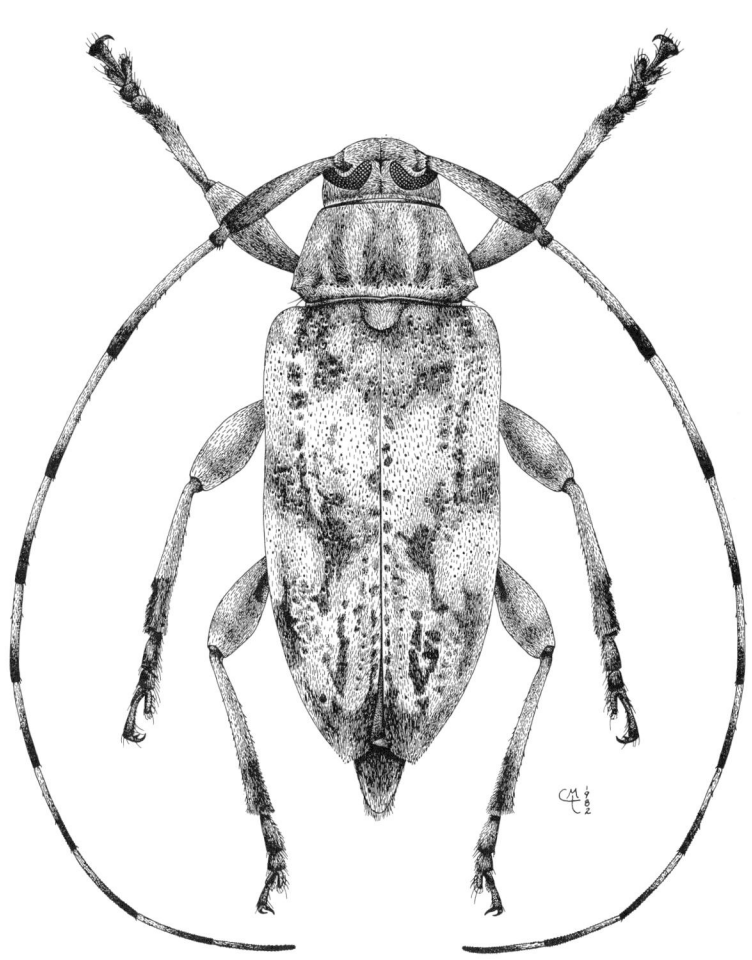

Figure 16. *Nyssodrysina haldemani* (LeConte), female.

Liopus Haldemani LeConte, 1852, J. Acad. Nat. Sci. Philadelphia, 2:173; Horn, 1880, Trans. Amer. Entomol. Soc., 8:124.
Leiopus Haldemani: White, 1855, Cat. Coleop. Ins. British Museum, 8:389; Lacordaire, 1872, Genera des coléoptères, 9(2):776, fn.
Sternidius Haldemani: LeConte, 1873, Smithson. Misc. Coll., 11(264):235.
Nyssodrys haldemani: Horn, 1886, Trans. Amer. Entomol. Soc., 13:xiii; Leng and Hamilton, 1896, Trans. Amer. Entomol. Soc., 23:133; 1923, Aurivillius, 1923, Coleop. Cat., 74:425; Craighead, 1923, Can. Dept. Agr. Bull., (n.s.) 27:121, pl. 12, fig. 7.
Nyssodrysina haldemani: Casey, 1913, Memoirs on the Coleoptera, 4:310, Fattig, 1947, Emory Univ. Mus. Bull., 5:35; Dillon, 1956, Ann. Entomol. Soc. Amer., 49:161; Turnbow and Hovore, 1979, Entomol. News, 90:225.
Nyssodrys contempta Bates, 1864, Ann. Mag. Nat. Hist., (3)13:152 (118); Bates, 1872, Trans. Entomol. Soc. London, 1872:220; Bates, 1881-85, Biol. Centr.-Amer., Coleop., 5:179, 412; Horn, 1886, Trans. Amer. Entomol. Soc., 13:xii.

Male. Form moderately small; integument dark reddish brown; pubescence fine, dense, appressed, grayish and yellowish. Head with front micropunctate, densely pubescent; antennae extending beyond body by five or six segments, segments dark annulate apically, pubescence very fine, appressed. Pronotum with disk impunctate, apical and basal impressions with a row of punctures; disk with a vague, narrow, median line, mostly behind middle; pubescence arranged into longitudinal yellowish vittae, one on each side of middle and a vague interrupted one on each side, areas between yellowish vittae grayish pubescent; prosternum densely pubescent; meso- and metasternum finely, densely clothed with appressed, mottled grayish and yellowish pubescence. Elytra less than twice as long as broad impressions behind vague basal gibbosities shallow; punctures at basal half coarse, dense, punctures becoming obsolete toward apices; pubescence arranged into yellowish, short, elongate, irregular vittae; apices obliquely truncate. Legs finely densely pubescent; tarsi and apices of tibiae dark. Abdomen finely, densely pubescent; last sternite about twice as long as fourth, emarginate at apex. Length, 7-9.5 mm.

Female. Form similar. Antennae slightly shorter. Abdomen with last sternite narrow, moderately elongate. Length, 8-10 mm.

Type locality. Of *haldemani*, Alabama; *contempta*, Mexico.

Range. Eastern United States from Maryland to Florida and Alabama to Panama.

Flight period. April to November.

Host plants. Foresteria, Bursera (in Mexico); *Bursera, Celtis* (United States).

The rather short, ovoid form and yellowish maculations of the pronotum and elytra distinguish this species. It is more common in Mexico and southward than in the United States.

Genus *Urgleptes* Dillon

Urgleptes Dillon, 1956, Ann. Entomol. Soc. Amer., 49:332; Dillon and Dillon, 1961, Man. Beetles East. North America, p. 640; Arnett, 1962, Beetles U.S., 103:873; Bayer and Shenefelt, 1969, Univ. Wisc. Res. Bull., 275:27; Villiers, 1980, Ann. Soc. Entomol. Fr., (n.s.) 16:90, 106; Rice and Enns, 1981, Trans. Mo. Acad. Sci., 15:100.
Leiopus (part): Lacordaire, 1872, Genera des coléoptères, 9(2):775.
Lepturges (part): Leng and Hamilton, 1896, Trans. Amer. Entomol. Soc., 23:127; Blatchley, 1910, Coleoptera in Indiana, p. 1075; Casey, 1913, Memoirs on the Coleoptera, 4:317; Knull, 1946, Ohio Biol. Surv. Bull., 39:251.

Form small, subdepresed; pubescence very fine, appressed. Head with front convex, broader than long; mandibles feebly arcuate; genae slightly convergent, subequal to or half as long as lower eye lobes; eyes small, upper lobes widely separated; antennal tubercles moderate, widely divergent; antennae slender, eleven-segmented, 1-3/4 to 2-1/2 times as long as body, often with several short setae beneath on basal segments, third segment usually longer than first, fourth usually subequal to third. Pronotum broader than long; lateral tubercles acute, placed before extreme base; disk convex, basal impression broad, extending beneath lateral tubercles; apex very narrowly impressed; prosternum with intercoxal process narrow, about one-sixth as broad as coxal cavities; mesosternal process slightly broader than prosternal; metasternum with episternum narrow, parallel-sided. Elytra slightly more than twice as long as broad; base not gibbose; disk shallowly impressed at basal one-third; pubescence short, appressed; apices rounded to obliquely truncate. Legs moderately elongate; posterior femora moderately clavate, slender in females; tarsi slender, first segment of hind pair longer than two following segments combined, third segment cleft to base. Abdomen normally segmented.

Type species. Liopus signatus LeConte (by original designation).

This genus may be separated from *Lepturges* by the basal transverse impression of the pronotum continuing under the lateral tubercles and onto the sides and by the proportions of the first and third antennal segments.

Most species of *Urgleptes* occur in the Neotropical region, and only five are found in North America. Our North American species are small, and only *U. signatus* approaches *Lepturges* in size.

KEY TO THE NORTH AMERICAN SPECIES OF *URGLEPTES*

1 Eyes with genae subequal in length to lower eye lobes 2
   Eyes with genae about half as long as lower eye lobes; postmedian maculae of elytra with very irregular margins, transverse or slightly oblique toward

	posterior. Length, 3.5-5.5 mm. Southern United States .... *foveatocollis*
2(1).	Elytra with postmedian maculae distinct, broad, apices rounded to subtruncate............................. 3
	Elytra with postmedian maculae not well defined, usually with a series of smaller longitudinal maculae, apices shallowly, obliquely emarginate. Length, 3.5-6 mm. Southern Texas .... *celtis*
3(2)	Elytra with apices not broadly maculate 4
	Elytra with apices broadly maculate, postmedian maculae usually tapering toward suture. Length, 3.2-4 mm. Eastern North America................. *facetus*
4(3)	Elytra with postmedian maculae broad, transverse, extending to suture; integument usually dark brownish. Length, 4-6 mm. Eastern North America .... *querci*
	Elytra with postmedian maculae narrow, oblique, not extending to suture; integument usually pale reddish brown. Length 5-9 mm. Eastern North America .... *signatus*

Urgleptes foveatocollis (Hamilton)
(Figure 17)

Liopus foveatocollis Hamilton, 1896, Trans. Amer. Entomol. Soc., 23:125.
Leiopus foveatocollis: Leng, 1920, Cat. Coleop. Amer., p. 283.
Urgleptes foveatocollis: Dillon, 1956, Ann. Entomol. Soc. Amer., 49:338.
Lepturges minutus Champlain and Knull, 1925, Entomol. News, 36:207; Linsley and Martin, 1933, Entomol. News, 44:182. New synonymy.
Urgleptes minutus: Dillon, 1956, Ann. Entomol. Soc. Amer., 49:335; Gilmour, 1965, Cat. Lam. du Monde, 8:588.
Urgleptes kissingeri Dillon, 1956, Ann. Entomol. Soc. Amer., 49:337, fig. 9; Turnbow and Hovore, 1979, Entomol. News, 90:225. New synonymy.
Urgleptes knulli Dillon, 1956, Ann. Entomol. Soc. Amer., 49:337, fig. 8; Hovore et al., 1987, Proc. Calif. Acad. Sci., 44:320. New synonymy.

Male. Form small, subdepressed; integument dark to pale reddish brown, appendages paler, partially infuscated; pubescence fine, dense, appressed, grayish and dark brown. Head with front convex, micropunctate, densely clothed with appressed pubescence; genae about half as long as lower eye lobes; eyes moderate, upper lobes widely separated; antennae about twice as long as body, third segment longer than first, fourth slightly longer than or subequal to third, fifth shorter than fourth. Pronotum broader than long; disk feebly convex, micropunctate; base broadly impressed, impression with a row of punctures; apex very narrowly impressed; pubescence dense, appressed, grayish, brownish maculae usually vague, a longitudinal one on

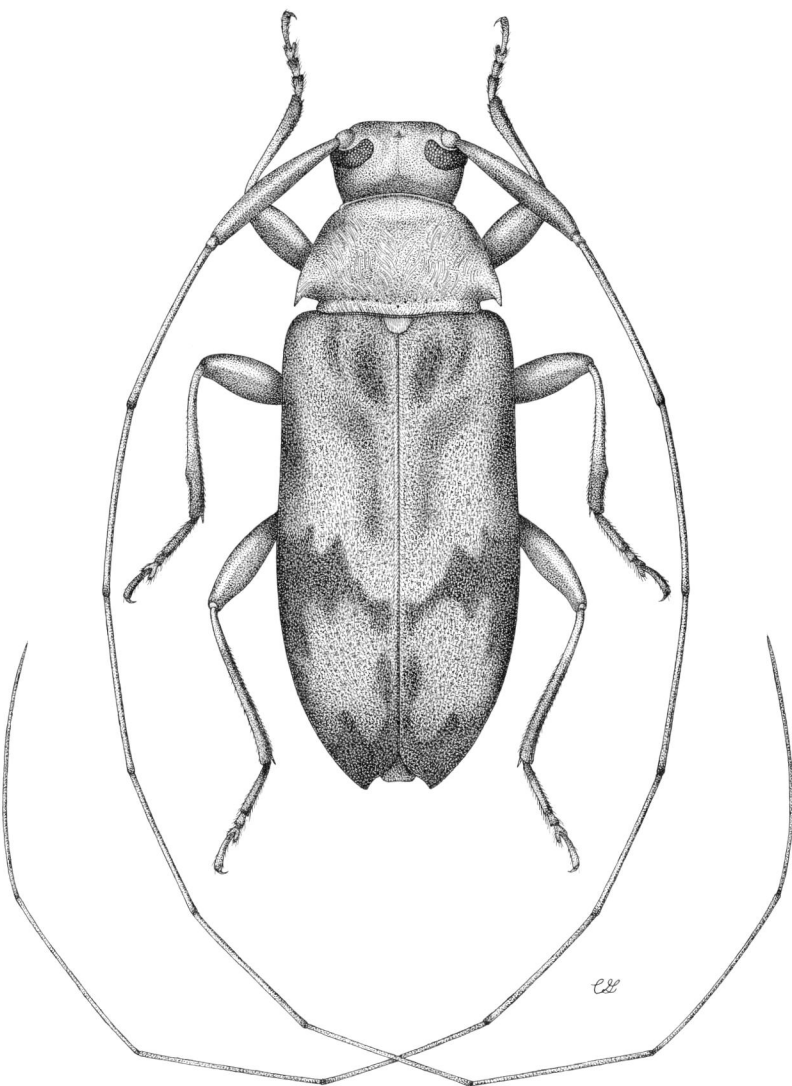

Figure 17. *Urgleptes foveatocollis* (Hamilton), male.

each side of middle and one around base of lateral tubercles; lateral tubercles strongly directed back; prosternum moderately densely pubescent, intercoxal process very narrow; mesosternum plane, intercoxal process about twice width of prosternal process; metasternum finely, densely pubescent. Elytra about twice as long as broad; impressions at basal one-third shallow;

punctures at base moderately coarse, separated, becoming denser near middle and obsolete toward apex; pubescence dense, pale, appressed, maculae pale to dark brownish, base with a subcircular macula around scutellum extending to forward margins of impressions, sides dark maculate to about middle, maculae extending onto disk behind middle forming irregular transverse to slightly oblique bands, apical one-third with small, irregular maculae; apices slightly obliquely truncate. Legs with femora and tibiae partially infuscated, very finely pubescent. Abdomen densely micropunctate, finely, densely pubescent; last sternite rounded at apex. Length, 3.5-5 mm.

Female. Form similar. Antennae slightly shorter. Abdomen with last sternite subtruncate, lightly fringed at apex. Length, 3.5-5.5 mm.

Type locality. Of *foveatocollis*, Biscayne Bay, Florida; *minutus*, Slidell, Louisiana; *kissingeri*, South Miami, Florida; *knulli*, Brownsville, Texas.

Range. Southern United States to Texas.

Flight Period. May to November.

Host Plants. Celtis, Lantana, Lysiloma, Piscidia.

The shorter genae and maculate pattern of the elytra separate this species from other North American *Urgleptes*. It varies geographically, mostly in coloration. The "foveae" of the pronotum described by Hamilton are parts of the dark maculae, and at first glance resemble depressions. The characters used to separate *U. minutus, U. kissingeri,* and *U. knulli* all fit into the normal range of variation.

Urgleptes celtis (Schaeffer)

Lepturges celtis Schaeffer, 1905, Bull. Brooklyn Mus. Ins. Arts Sci., 1:168; Casey, 1913, Memoirs on the Coleoptera, 4:320; Linsley and Martin, 1933, Entomol. News, 44:182.

Urgleptes celtis: Dillon, 1956, Ann. Entomol. Soc. Amer., 49:336; Hovore and Penrose, 1982, Southwest. Nat., 27:26; Hovore et al., 1987, Proc. Calif. Acad. Sci., 44:320.

Male. Form small, subdepressed; integument pale brownish, maculae darker brown; pubescence fine, appressed, grayish and brown. Head with front convex, micropunctate, finely, densely pubescent; genae about as long as lower eye lobes; eyes moderate, upper lobes widely separated; antennae about twice as long as body, third segment longer than first, fourth subequal to third, fifth shorter than fourth. Pronotum broader than long; disk convex, micropunctate, finely, densely pubescent; base broadly impressed, impression punctate; apex very narrowly impressed; darker maculae vague, irregular; prosternum finely pubescent, intercoxal process very narrow; mesosternum with intercoxal process slightly broader than prosternal; metasternum finely, densely pubescent. Elytra slightly more than twice as long as broad; basal impressions shallow; basal punctures separated, becoming denser toward middle and obsolete toward apex; pubescence fine, pale, appressed, brownish on maculae; maculae irregular, usually longitudinal, narrow, base with a pair around scutellum, premedian pair somewhat rectangular, extending from margins to slightly onto disk, postmedian pair oblong, situated toward margins, remainder of surface with often less distinct, partially reticulating, longitudinal vittae; apices shallowly, obliquely emarginate. Legs finely pubescent. Abdomen finely pubescent; last sternite rounded at apex. Length, 3.5-5.5 mm.

Female. Form similar. Abdomen with last sternite subtruncate at apex. Length, 4-6 mm.

Type locality. Esperanza Ranch, Brownsville, Texas.
Range. Southern Texas.
Flight period. March to October.
Host plants. Celtis, Leucaena, Pithecoellobium flexicaule.

The slightly emarginate apices and irregular maculations of the elytra make this species distinct.

Urgleptes facetus (Say)
(Figure 18)

Lamia faceta Say, 1827, J. Acad. Nat. Sci. Philadelphia, 5:271; LeConte, 1857, Compl. Writings T. Say, 2:329.
Amniscus facetus: Haldeman, 1847, Trans. Amer. Philos. Soc., (2)10:49.
Liopus facetus: LeConte, 1852, J. Acad. Nat. Sci. Philadelphia, 2:171; Packard, 1881, U.S. Entomol. Comm. Bull., 7:132, fig. 62.
Leiopus facetus: Haldeman, 1847, Proc. Amer. Philos. Soc., 4:372; White, 1855, Cat. Coleop. Ins. British Mus., 8:387; Fitch, 1859, 4th Rept. Nox. Ben. Ins. N.Y., p. 751; Packard, 1871, Rept. Inj. Ins. Mass., 1:14, pl. 1, fig. 1; Lacordaire, 1872, Genera des coléoptères, 9(2):776, fn.; Packard, 1881, U.S. Entomol. Comm. Bull., 7: 250, fig. 98; Packard, 1883, Guide Study of Insects, p. 712, pl. 15, fig. 1; Packard, 1890, Ins. Inj. For. Shade Trees, p. 913.
Lepturges facetus: Horn, 1880, Trans. Amer. Entomol. Soc., 8:127; Hamilton, 1891, Insect Life, 4:132; Leng and Hamilton, 1896, Trans. Amer. Entomol. Soc., 23:128; Beutenmuller, 1896, J. N.Y. Entomol. Soc., 4:79; Wickham, 1898, Can. Entomol., 30:38; Lugger, 1899, 5th Rept. Minn. Agr. Exp. Sta., p. 125, fig. 131; Lugger, 1899, Minn. Agr. Exp. Sta. Bull., 66:209, fig. 131; Blatchley, 1910, Coleoptera in Indiana, p. 1077, fig. 463c; Smith, 1910, Rept. N.J. St. Mus. Rept., 1909:334; Casey, 1913, Memoirs on the Coleoptera, 4:322; Champlain and Knull, 1925, Entomol. News, 36:141; Knull, 1928, Entomol. News, 39:316; Leonard, 1928, Cornell Agr. Exp. Sta. Mem., 101:452; Knull, 1946, Ohio Biol. Surv. Bull., 39:254; Fattig, 1947, Emory Univ. Mus. Bull., 5:36.
Urgleptes facetus: Dillon, 1956, Ann. Entomol. Soc. Amer., 49:335; Dillon and Dillon, 1961, Man. Beetles East. North America, p. 641, pl. 64, no. 5; Gardiner, 1966, Can. J. Zool., 44:204, fig. 26; Gardiner, 1969, Can. Dept. Fish. For. Int. Rept., 0-14:90 (larva, pupa); Kirk, 1970, S.C. Agr. Exp. Sta.

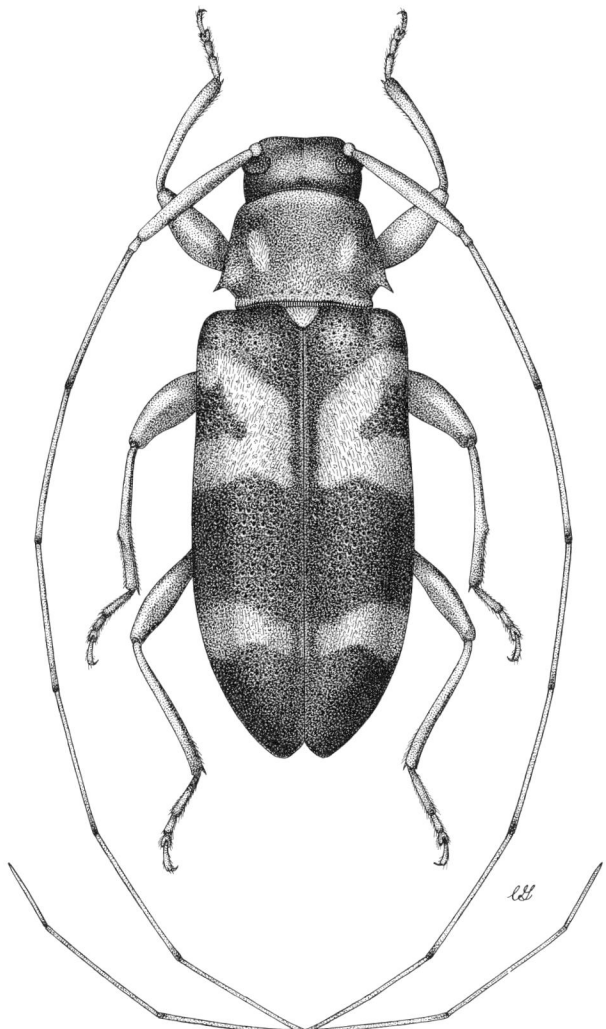

Figure 18. *Urgleptes facetus* (Say), male.

Tech. Bull., 1038:82; Headstrom, 1977, Beetles of America, p. 378; Gosling and Gosling, 1977, Gr. Lakes Entomol., 10:29, fig. 156; Laliberte et al., 1977, Fabreries, 3:100; Rice and Enns, 1981, Trans. Mo. Acad. Sci., 15:100; Gosling, 1984, Gr. Lakes Entomol., 17:73.

Male. Form small, subdepressed; integument pale to dark brownish, antennae often paler; pubescence fine, appressed, grayish and brownish. Head with front convex, micropunctate, finely, moderately densely pubescent; genae about as long as lower eye lobes; eyes moderate, upper lobes widely separated; antennae about twice as long as body, third segment longer than first, fourth subequal to third, fifth shorter than fourth. Pronotum broader than long; disk shallowly convex, micropunctate, finely, densely pubescent; base broadly impressed, impression punctate; apex very narrowly impressed; maculae obsolete; prosternum finely pubescent, intercoxal process very narrow; mesosternal process slightly broader than prosternal; metasternum finely pubescent. Elytra twice as long as broad; basal impressions vague; basal punctures dense, subcontiguous, punctures becoming obsolete toward apex; pubescence fine, pale, dark on maculae; base with macula extending down suture to about middle, sides dark maculate, maculae expanded behind middle onto disk, usually narrowing toward suture, apex with a broad macula, area before postmedian maculae usually with short, oblique maculae toward sides; apices rounded. Legs finely pubescent. Abdomen finely pubescent; last sternite rounded at apex. Length, 3.2-4 mm.

Female. Form similar. Abdomen with last sternite broadly subtruncate at apex. Length, 3.5-4 mm.

Type locality. Not given (North America).
Range. Eastern North America to Minnesota.
Flight period. May to August.
Host plants. Acer, Amelanchier, Carya, Crataegus, Fagus, Maclura, Malus, Morus, Quercus, Rhus.

The two broad, brown maculae of the elytra readily distinguish this species.

Urgleptes querci (Fitch)

Leiopus querci Fitch, 1859, 5th Rept. Nox. Ben. Ins. N.Y., p. 796; Packard, 1881, U.S. Entomol. Comm. Bull., 7:24.

Lepturges querci: Horn, 1880, Trans. Amer. Entomol. Soc., 8:127; Chittenden, 1894, Proc. Entomol. Soc. Wash., 3:101; Leng and Hamilton, 1896, Trans. Entomol. Soc. Amer., 23:128; Beutenmuller, 1896, J. N.Y. Entomol. Soc., 4:79; Wickham, 1898, Can. Entomol., 30:38; Felt, 1906, N.Y . St. Mus. 59th Ann. Rept., 4:439; Blatchley, 1910, Coleoptera in Indiana, p. 1077, fig. 463b; Smith, 1910, N.J. St. Mus. Rept., 1909:334; Casey, 1913, Memoirs on the Coleoptera, 4:321; Craighead, 1923, Can. Dept. Agr. Bull., (n.s.) 27:118; Blackman and Stage, 1924, N.Y. St. Coll. For. Tech. Pub., 17:118; Champlain and Knull, 1925, Entomol. News, 36:141; Leonard, 1928, Cornell Agr. Exp. Sta. Mem., 101:452; Knull, 1930, Entomol. News, 41:102; Barrett, 1932, Univ. Calif. Pub. Entomol., 5:290; Chagnon, 1933-40, Coleop. Prov. Quebec, p. 273; Hoffmann, 1940, Bull. Brooklyn Entomol. Soc., 35:59; Loding, 1945, Geol. Surv. Ala. Mon., 11:123; Knull, 1946, Ohio Biol. Surv. Bull., 38:253; Fattig, 1947, Emory Univ. Mus. Bull., 5:36; Steyskal, 1951, Coleop. Bull., 5:76; Chagnon and Robert, 1962, Prin. Coleop. Prov. Quebec, p. 273; Furth, 1985, Conn. Acad. Arts Sci., 46:192.

Lepturges quercus: Aurivillius, 1923, Coleop. Cat., 74:415; Hoffmann, 1942, USDA Misc. Pub., 466:11.
Urgleptes querci: Dillon, 1956, Ann. Entomol. Soc. Amer., 49:334; Dillon and Dillon, 1961, Man. Beetles East. North America, p. 461, pl. 64, no. 9; Bayer and Shenefelt, 1969, Univ. Wisc. Res. Bull., 275:29, fig. 38; Headstrom, 1977, Beetles of America, p. 378; Gosling and Gosling, 1977; Gr. Lakes Entomol., 10:29, fig. 155; Laliberte et al., 1977, Fabreries, 3:100; Rice and Enns, 1981, Trans. Mo. Acad. Sci., 15:100; Gosling, 1984, Gr. Lakes Entomol., 17:73; Gosling, 1986, Gr. Lakes Entomol., 19:158.
Lepturges tristis Casey, 1913, Memoirs on the Coleoptera, 4:321.

Male. Form small, subdepressed; integument usually dark brownish, antennae paler; pubescence fine, dense, appressed, grayish and brownish. Head with front convex, micropunctate, densely pubescent; genae about as long as lower eye lobes; eyes rather small, upper lobes widely separated; antennae about twice as long as body, third segment longer than first, fourth subequal to third, fifth shorter than fourth. Pronotum broader than long; disk convex, micropunctate, densely pubescent; base broadly impressed, impression with a row of punctures; apex very narrowly impressed; dark maculae vague, usually with one on each side of middle; prosternum finely pubescent, intercoxal process very narrow; mesosternum with intercoxal process about twice as broad as prosternal; metasternum finely, densely pubescent. Elytra about twice as long as broad; basal impressions shallow; basal punctures well separated, becoming slightly more dense near middle and obsolete toward apex; pubescence fine, dense, appressed; maculae dark brownish, base with a vaguely triangular macula behind scutellum narrowing down suture, sides dark maculate to about apical one-third, postmedian maculae broad, transverse, extending to suture, area before postmedian maculae often with small dark spots; apices rounded to truncate. Legs usually infuscated, finely pubescent. Abdomen micropunctate, finely pubescent; last sternite subtruncate at apex. Length, 4-5.5 mm.

Female. Form similar. Abdomen with last sternite rounded at apex. Length, 4-6 mm.

Type locality. Of *querci*, New York; *tristis*, Buena Vista Spring, Franklin Co., Pennsylvania.

Range. Eastern North America to Kansas and Texas.

Flight period. May to September.

Host plants. Acer, Amelanchier, Asimina, Carya, Castanea, Celastrus, Cercis, Juglans, Liriodendron, Malus, Morus, Ostrya, Oxydendron, Quercus, Rhus, Tilia, Ulmus, Wisteria.

Parasites. Calyptus magdalis Cresson (Braconidae).

This species is easily recognizable by the dark integument and single, transverse, dark macula of the elytra.

Urgleptes signatus (LeConte)

Liopus signatus LeConte, 1852, J. Acad. Nat. Sci. Philadelphia, 2:171.
Leiopus signatus: White, 1855, Cat. Coleop. Ins. British Mus., 8:388; Lacordaire, 1872, Genera des coléoptères, 9(2):776, fn.
Lepturges signatus: Horn, 1880, Trans. Amer. Entomol. Soc., 8:127; Chittenden, 1894, Proc. Entomol. Soc. Wash., 3:101; Knobel, 1895, Beetles New England, p. 34, fig. 111; Leng and Hamilton, 1896, Trans. Amer. Entomol. Soc., 23:128; Wickham, 1898, Can. Entomol., 30:38; Blatchley, 1910, Coleoptera in Indiana, p. 1076, fig. 463a; Smith, 1910, N.J. St. Mus. Rept., 1909:334; Casey, 1913, Memoirs on the Coleoptera, 4:320; Blatchley, 1920, Can. Entomol., 52:69; Craighead, 1923, Can. Dept. Agr. Bull., (n.s.) 27:118; Champlain and Knull, 1925, Entomol. News, 36:140; Leonard, 1928, Cornell Agr. Exp. Sta. Mem., 101:452; Knull, 1930, Entomol. News, 41:102; Knull, 1932, Entomol. News, 43:64; Knull, 1946, Ohio Biol. Surv. Bull., 39:253; Fattig, 1947, Emory Univ. Mus. Bull., 5:36; Steyskal, 1951, Coleop. Bull., 5:76; Furth, 1985, Conn. Acad. Arts Sci., 46:192.
Urgleptes signatus: Dillon, 1956, Ann. Entomol. Soc. Amer., 49:333; Dillon and Dillon, 1961, Man. Beetles East. North America, p. 641, pl. 64, no. 8; Headstrom, 1977, Beetles of America, p. 378; Gosling and Gosling, 1977, Gr. Lakes Entomol., 10:29, fig. 154; Gosling, 1984, Gr. Lakes Entomol., 17:73; Gosling, 1986, Gr. Lakes Entomol., 19:158.
Lepturges tenebrosus Casey, 1913, Memoirs on the Coleoptera, 4:320.

Male. Form small to moderate-sized, subdepressed; integument pale reddish brown, maculae brownish; pubescence fine, dense, appressed, grayish and brownish. Head with front convex, micropunctate, densely pubescent; genae subequal in length to lower eye lobes; eyes moderate, upper lobes widely separated; antennae a little less than twice as long as body, third segment longer than first, fourth subequal to third, fifth shorter than fourth. Pronotum broader than long; disk convex, micropunctate, finely, densely pubescent, often infuscated; basal impression broad, with a row of punctures; apical impression very narrow; prosternum finely pubescent, intercoxal process very narrow; mesosternum with intercoxal process slightly broader than prosternal; metasternum finely pubescent. Elytra about twice as long as broad; basal impressions shallow; basal punctures separated, punctures becoming denser toward middle and obsolete toward apex; pubescence fine, pale, brownish on maculae; base with often vague, rounded maculae on each side behind scutellum, sides at basal one-third maculate, postmedian maculae irregular, oblique toward margins and often connecting laterally with narrow, elongate apical maculae; apices narrowly rounded. Legs often infuscated, finely pubescent. Abdomen finely, densely pubescent; last sternite broadly subtruncate at apex. Length, 5-8 mm.

Female. Form similar. Abdomen with last sternite narrowly truncate at apex. Length, 6-9 mm.

Type locality. Of *signatus*, New York; *tenebrosus*, Bluff Point, Lake Champlain, New York.

Range. Eastern North America to Florida and Minnesota.

Flight period. June to August.

Host plants. Acer, Alnus, Carya, Castanea, Cercis, Fagus, Fraxinus, Morus, Quercus, Rhus, Tilia.

The oblique, vaguely semicircular postmedian vittae of the elytra distinguish this species.

Genus *Coenopoeus* Horn

Coenopoeus Horn, 1880, Trans. Amer. Entomol. Soc., 8: 117; LeConte and Horn, 1883, Smithson. Misc. Coll., 507:323; Bates, 1885, Biol. Centr.-Amer., Coleop., 5: 385; Leng and Hamilton, 1896, Trans. Amer. Entomol. Soc., 23:115; Bradley, 1930, Man. Genera Beetles, p. 245; Dillon, 1956, Ann. Entomol. Soc. Amer., 49: 156; Arnett, 1962, Beetles U.S., 103: 872.

Form large, robust, subcylindrical. Head with front short, broader than long, slightly convex; clypeus with apical margin bilobed, emarginate medially; eyes with lower lobes almost rounded, upper lobes separated by a little more than their width; genae subparallel, shorter than lower eye lobes; antennal tubercles prominent, widely separated; antennae slightly longer than body in males, shorter than body in females, scape gradually enlarging toward apex, extending almost to middle of pronotum, third segment shorter than scape, fourth shorter than third, sixth segment apically produced in males, erect setae absent. Pronotum broader than long, sides obtusely tuberculate at middle; disk vaguely callused; basal sulcus broad, extending onto sides; prosternum narrow, deeply excavated, intercoxal process arcuate, medially impressed, about one-fourth as broad as coxal cavities, coxae dorsally carinate; mesosternum with intercoxal process gradually declivous anteriorly, medially excavated anteriorly, more than half as broad as coxal cavities; metasternum with episternum gradually tapering posteriorly. Elytra less than twice as long as broad, tapering at apical one-third; disk vaguely bigibbose near base, erect hairs absent; apices rounded. Legs large, robust; femora strongly clavate, lacking erect setae; tarsi short, broad, third segment cleft to base, claws semidivaricate. Abdomen normally segmented.

Type species: Leptostylus palmeri LeConte (by original designation).

The robust body form, apically produced sixth antennal segment of males and not strictly divaricate tarsal claws which meet one another at a slight angle make this genus distinctive.

Two species of *Coenopoeus* are known, *C. niger* Horn from the Cape Region of Baja California and the following.

Coenopoeus palmeri (LeConte)
(Figure 19)

Leptosylus Palmeri: LeConte, 1873, Smithson. Misc. Coll., 11(264):233.
Coenopoeus palmeri: Horn, 1880, Trans. Amer. Entomol. Soc., 8:118, pl. 2, fig. 1; LeConte and Horn, 1883, Smithson. Misc. Coll., 507:323; Bates, 1885, Biol. Centr.-Amer., Coleop., 5:385; Leng and Hamilton, 1896, Trans. Amer. Entomol. Soc., 23:115; Hunter et al., 1912, USDA Bur. Entomol. Bull. 113:43; Garnett, Can. Entomol., 50:282; Dillon, 1956, Ann. Entomol. Soc. Amer., 49:157, figs. 8, 9; Papp, 1959, Bull. So. Calif. Acad. Sci., 58:91; Linsley, Knull, and Statham, 1961, Amer. Mus. Nov., 2050:29; Mann, 1969, U.S. Nat. Mus. Bull. 256:88 (habits); Raske, 1972, Can. Entomol., 104:121, figs. 1-8 (habits).

Male. Form large, robust; integument black; pubescence very fine, dense, black, usually with a slight violaceous cast, pronotum, elytra, and underside usually maculate with thicker gray-brown pubescence, antennae and tibiae gray annulate. Head with front micropunctate with larger punctures interspersed; vertex punctate as front; pubescence very fine, dark; antennae with segments three to ten broadly gray annulate near bases. Pronotum with disk somewhat flattened medially; punctures sparse, scattered, denser on basal sulcus; pubescence very fine, dark, apical margin fringed with pale pubescence, disk often with scattered patches of pale pubescence, basal margin usually pale pubescent; prosternum finely pubescent; meso- and metasternum micropunctate with a few larger punctures interspersed, pubescence fine, denser at middle of metasternum, often with scattered patches of pale pubescence. Elytra coarsely, rather sparsely punctate near base, punctures extending to about middle; pubescence very fine, dark, often with a large yellow-gray band over basal half and two transverse pale bands over apical half; apices rounded. Legs robust, densely clothed with fine dark pubescence, tibiae usually pale annulate at middle. Abdomen micropunctate; sternites with denser, pale pubescence along apical margins; last sternite narrow, shallowly emarginate at apex. Length, 15-27 mm.

Female. Form similar. Antennae shorter than body, sixth segment not produced at apex. Abdomen with last sternite longer than fourth, truncate at apex. Length, 18-27 mm.

Type locality. Arizona.

Range. Southern Nevada, southern California to Texas and Sinaloa, Mexico (Figure 20).

Flight period. May to September.

Host plants. *Opuntia* spp. (including *versicolor, fulgida, spinosior, bernadina, imbricata, parryi, arbuscula*).

This species may be recognized by the large, robust form, black integument, frequent presence of yellow-gray pubescent patches on the elytra, and by the apically produced sixth antennal segment of the male. *C. niger* from the Cape region of Baja California differs by the smaller average size, lack of pale elytral maculae, and by the denser, coarser punctures of the elytra which extend beyond the middle.

C. palmeri varies considerably in the amount of pale pubescence on the body. In the northern portions of the range, specimens are well marked and the pale pubescence covers most of the elytra. The extent of the pubescence decreases southward, and in southern Arizona and New Mexico all-black individuals are encountered, especially along the coast of Mexico to the state of Sinaloa.

The habits of this species have been observed by Raske (1972). Adults are concealed during the day on the underside of branches of the host. At dusk they climb the plants and feed on the new growth. Mating occurs at night at the top of the cactus. Females deposit eggs on the epidermis of the plant, and larvae enter at the point of attachment. Penetration by the larva stimulates the plant to produce a sticky exudate which is initially mined by the larva. As the larvae grow, they enter and mine the pith. Pupal chambers are formed from frass and plant material at the end of the larval mines. Larvae overwinter in the pupal cell and transform to adults in late spring.

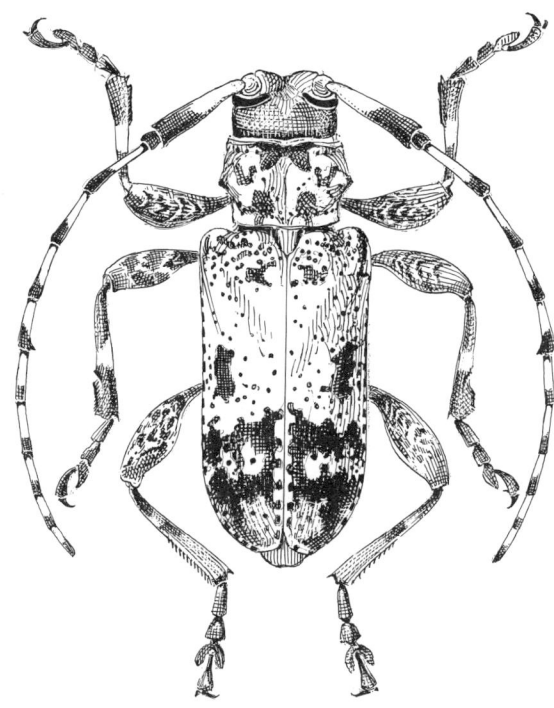

Figure 19. *Coenopoeus palmeri* (LeConte), male.

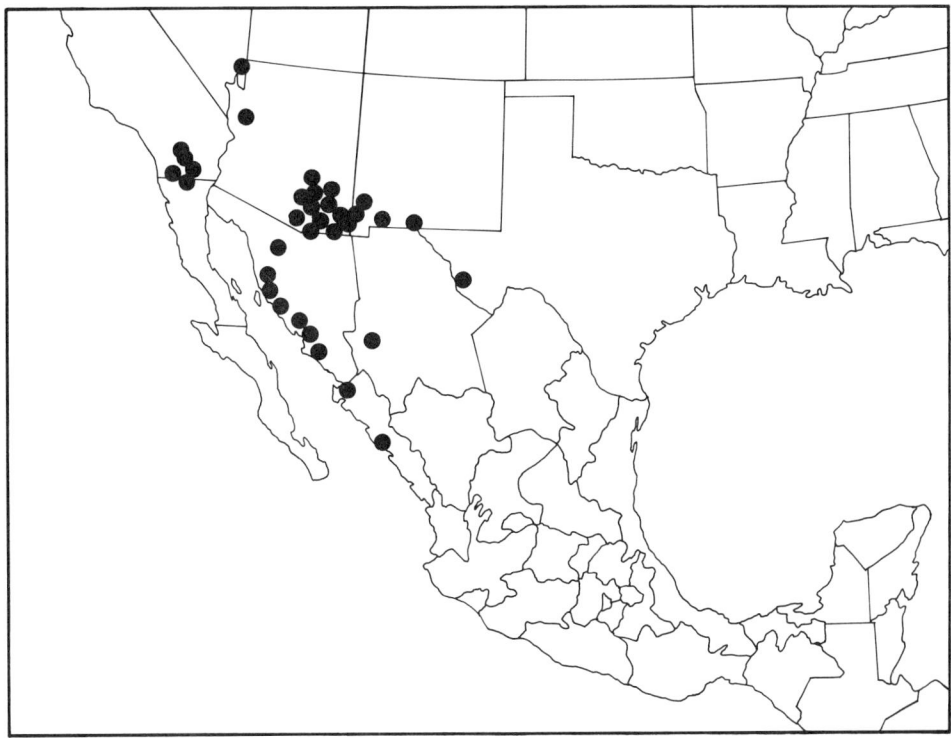

Figure 20. Known geographic range of *Coenopoeus palmeri* (LeConte).

Genus *Styloleptus* Dillon

Styloleptus Dillon, 1956, Ann. Entomol. Soc. Amer., 49: 158; Arnett, 1961, Beetles U.S., 103: 872; Villiers, 1980, Ann. Soc. Entomol. Fr., (n.s.) 16:90, 101; Ivie, 1985, Pan-Pac. Entomol., 61: 315.
Leptostylus: Leng and Hamilton, 1896, Trans. Amer. Entomol. Soc., 23: 116 (part).
Caribbeana Gilmour, 1963, Stud. Fauna Curaçao, Caribbean Is., 18: 97. (Type species: *Caribbeana hebes* Gilmour, by original designation.)

Form small to moderate-sized, subdepressed. Head with front feebly convex, quadrate; mandibles very feebly arcuate; genae convergent, subequal to or slightly shorter than lower eye lobes; eyes moderate, upper lobes widely separated; antennal tubercles prominent, widely divergent; antennae eleven-segmented, extending three or four segments beyond elytra, erect hairs absent beneath, third segment longer than first, fourth subequal to third. Pronotum broader than long, sides with obtuse tumid tubercles just before basal impression; basal impression broad, shallow, narrowly extending to sides; apex abruptly, narrowly constricted; disk convex, tricallused;

prosternum with intercoxal process one-third to one-fourth as broad as coxal cavities; mesosternum with intercoxal process almost as broad as coxal cavities. Elytra twice or less longer than broad; basal gibbosities shallow; costae usually distinct; elevated, tufted tubercles variably interspersed along costae; apices obliquely truncate. Legs robust; femora strongly clavate; tarsi short, first segment of hind pair shorter than following two segments together. Abdomen normally segmented.

Type species: Leptostylus biustus LeConte (by original designation).

The shape of the pronotum with rounded, tumid lateral tubercles near the base and proportions of the intercoxal processes readily identify this genus in our fauna.

Most of the species of *Styloleptus* occur in the West Indies, and the above description is based on the single species found in North America.

Styloleptus biustus (LeConte)
(Figure 21)

Leptostylus biustus LeConte, 1852, J. Acad. Nat. Sci. Philadelphia, 2: 169; Lacordaire, 1872, Genera des coléoptères, 9(2): 772, fn.; LeConte, 1872, Smithson. Misc. Coll., 11(264):233; Horn, 1880, Trans. Amer. Entomol. Soc., 8: 121; Hubbard, 1885, Insects Affect. Orange, p. 174, pl. 14, fig. 2 (habits); Chittenden, 1894, Proc. Entomol. Soc. Washington, 3: 99; Gahan, 1895, Trans. Entomol. Soc. London, 1895: 134; Horn, 1895, Proc. Calif. Acad. Sci., (2)5: 228; Leng and Hamilton, 1896, Trans. Amer. Entomol. Soc., 23: 119; Beutenmuller, 1896, J. N.Y. Entomol. Soc., 4: 79; Wickham, 1897, Can. Entomol., 29: 208; Townsend, 1902, Trans. Texas Acad. Sci., 5: 78; Blatchley, 1910, Coleoptera in Indiana, p. 1072; Craighead, 1923, Can. Dept. Ag. Bull., (n.s.) 27: 116; Leonard, 1928, Cornell Agr. Exp. Sta. Mem., 101: 451; Beaulne, 1932, Nat. Can., 59: 219 (hosts); Knull, 1937, Entomol. News, 48: 42; Linsley, 1942, Proc. Calif. Acad. Sci., (4)24: 70; Loding, 1945, Geol. Surv. Ala. Mon., 11: 122; Fattig, 1947, Emory Univ. Mus. Bull., 5: 34; Duffy, 1960, Mon. Neotrop. Timber Beetles, p. 252.

Amniscus biustus: White, 1855, Cat. Coleop. Ins. British Mus., 8:393.

Exocentrus biustus: Chevrolat, 1862, Ann. Soc. Entomol. Fr., (4)2: 249.

Styloleptus biustus biustus: Dillon, 1956, Ann. Entomol. Soc. Amer., 49: 158; Kirk, 1969, S.C. Ag. Exp. Sta. Bull. 1033: 86; Turnbow and Hovore, 1979, Entomol. News, 90: 224; Rice, 1985, J. N.Y. Entomol. Soc., 93:1224.

Liopus minuens Leng and Hamilton, 1896, Trans. Amer. Entomol. Soc., 23: 123. New synonymy.

Leptostylus minuens: Casey, 1913, Memoirs on the Coleoptera, 4: 316; Aurivillius, 1923, Coleop. Cat., 74: 401.

Styloleptus minuens: Dillon, 1956, Ann. Entomol. Soc. Amer., 49: 159; Turnbow and Hovore, 1979, Entomol. News, 90: 225.

Leptostylus pusillus Blatchley, 1925, Can. Entomol., 42: 167.

Leptostylus scurra var. *dorsalis* Fisher, 1926, Proc. U.S. Nat. Mus., 68: 21.

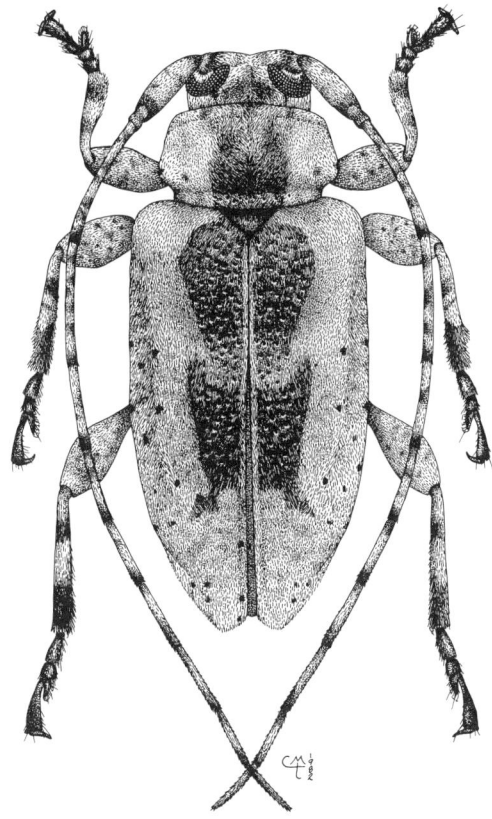

Figure 21. *Styloleptus biustus* (LeConte), male.

Leptostylus bahamicus Champlain and Knull, 1926 (not Fisher, 1925), Entomol. News, 37: 206.

Male. Form small, subdepressed; integument pale to reddish brown; pubescence short, dense, appressed, grayish, brownish and black, elytra and epipleura often black maculate. Head with front densely micropunctate, densely pubescent; genae shorter than or subequal in length to lower eye lobes; antennae extending three or four segments beyond elytra, segments narrowly dark annulate at apices, basal segments dark spotted. Pronotum broader than long; sides with rounded, tumid tubercles near basal margin; disk convex, tricallused; punctures fine, dense; pubescence dense, obscuring surface, grayish to brownish, calluses variably dark maculate, sides occasionally with narrow, interrupted, longitudinal vittae; prosternum finely, densely pubescent; meso- and metasternum irregularly dark-spotted. Elytra twice or less as long as broad; basal gibbosities often vague; punctures dense, slightly coarser than pronotal ones; dark tufted tubercles irregularly interspersed along costae; pubescence dense, appressed, grayish to brownish; epipleura usually dark maculate, maculae often narrowly extending along sides of pronotum to behind eyes; postmedian dark macula often reduced to black spots, occasionally vague dark maculae present behind postmedian pair; apices obliquely truncate. Legs densely mottled pubescent; tibiae with at least middle and hind pairs dark annulate; tarsi dark. Abdomen finely densely pubescent; last sternite subemarginate to rounded at apex. Length, 4.5-9 mm.

Female. form similar. Abdomen with fifth sternite about twice as long as fourth. Length, 5-9 mm.

Type locality. Of *biustus*, southern and middle states; *minuens*, Lake Worth, Florida; *pusillus*, southern Florida; *dorsalis*, Cayamas, Cuba.

Range. Central eastern United States to Florida, west to Texas and Iowa, West Indies.

Flight period. December to July (probably year-round in Florida).

Host plants. Bursera, Cajanus, Carya, Celtis, Citrus, Ficus, Ilex, Liquidambar, Metopium, Mimosa, Morus, Pistacia, Punica, Rhus, Sesbania, Zanthoxylum.

This species is variable in size and coloration. We can find no morphological bases for considering the small forms from Florida as a distinct species. All of the characters previously used to separate *A. biustus* and *A. minuens* are invalid. The color varies from grayish with dark maculae to brownish with only dark spots. The epipleural maculae are occasionally absent in the small Florida specimens. However, when adequate material is available from critical areas it may yet reveal two subspecies.

Some individuals of *S. biustus* suggest, at first glance, a species of *Sternidius*. However, the distinctive shape of the pronotum and broader intercoxal processes will immediately place it in *Styloleptus*.

Genus *Pseudastylopsis* Dillon

Leptostylus: Horn, 1880, Trans. Amer. Entomol. Soc., 8:119 (part); Leng and Hamilton, 1896, Trans. Amer. Entomol. Soc., 23:116 (part).
Pseudastylopsis Dillon, 1956, Ann. Entomol. Soc. Amer., 49:220.
Pseudostylopsis: Arnett, 1962, Beetles U.S., 103:872 (error); Hatch, 1971, Univ. Wash. Pub. Biol., 16:151.

Form moderate-sized, subcylindrical. Head with front quadrate, convex, about as long as broad; vertex deeply impressed between eyes; mandibles shallowly arcuate, pointed at apices; genae convergent, shorter than or longer than lower eye lobes; eyes moderate-sized, separated above by at least diameter of antennal scape; antennal tubercles shallow, broadly divergent; antennae slender, eleven-segmented, slightly longer than body in males, as long as body in females, scape with several long, erect setae beneath, third segment longer than first, fourth shorter than third. Pronotum broader than long, sides obtusely tuberculate at middle; apex shallowly impressed, base broadly impressed; disk with five calluses, median one prominent; punctures dense, often extending onto calluses; prosternum with intercoxal process arcuate, about one-sixth to one-third as broad as coxal cavities; mesosternal process about two-thirds as broad as coxal cavities; metasternum with episternum narrow, parallel-sided, barely tapering. Elytra a little more than twice as long as broad; disk costate, with small tufted tubercles; basal gibbosities shallow; apices rounded to subtruncate. Legs robust; femora pedunculate; middle tibiae with an external sinus; tarsi short, first segment of hind pair about as long as two following segments together. Abdomen normally segmented, last sternite of females longer than fourth.

Type species: Leptostylus nebulosus Horn (by original designation).

This genus is closely related to *Astylopsis*. The two may be separated by the narrower prosternal process, longer third antennal segment, and prominent median callus of the pronotum of *Pseudastylopsis*.

Three species are known.

KEY TO THE NORTH AMERICAN SPECIES OF *PSEUDASTYLOPSIS*

1	Elytra with prominent tufted tubercles at base, apices narrowly subtruncate	2
	Elytra with vague, small, tufted tubercles at base, apices broadly rounded to subtruncate. Pronotum mottled grayish and brown pubescent. Length, 9-14 mm. Oregon to southern Calfornia and western Nevada	*nebulosus*
2(1)	Elytra strongly costate and tuberculate at apical half, impressions behind basal gibbosities deep. Pronotum grayish pubescent around dark vittae. Length, 9-11.5 mm. Southern Arizona to southern Sierra Madre, Mexico	*nelsoni*
	Elytra feebly costate at apical half, tubercles small, dorsal impressions shallow. Pronotum yellow-brown pubescent around dark vittae. Length, 8-12 mm. Southern Arizona to Texas and southern Sierra Madre, Mexico	*pini*

Pseudastylopsis nebulosus (Horn)
(Figure 22)

Leptostylus nebulosus Horn, 1880, Trans. Amer. Entomol. Soc., 8:122; Leng and Hamilton, 1896, Trans. Amer. Entomol. Soc., 23:118; Craighead, 1923,

Can. Dept. Agr. Bull., (n.s.) 27:115; Doane et al., 1936, For. Ins., p. 189; Keen, 1938, USDA Misc. Pub., 273:137; Keen, 1952, USDA Misc. Pub., 273:176; Tyson, 1966, Pan-Pac. Entomol., 42:204.

Pseudastylopsis nebulosus: Dillon, 1956, Ann. Entomol. Soc. Amer., 49:220.

Pseudostylopsis nebulosus: Hatch, 1971, Univ. Wash. Pub. Biol., 16:151 (error).

Male. Form moderate-sized, rather robust, subcylindrical; integument reddish brown, abdomen infuscated; pubescence dense, appressed, mottled grayish, brownish, and black. Head with front micropunctate, densely brown and gray pubescent, long, erect setae present on eye margins, edge of clypeus, and on genae; vertex densely pubescent, minutely punctate, two large punctures present between eyes; genae shorter than lower eye lobes; antennae slightly longer than body, outer segments brownish annulate at apices, third segment longer than first, fourth shorter than third, slightlylonger than first. Pronotum broader than long, sides obtusely tuberculate at middle; disk with calluses prominent, median one usually most

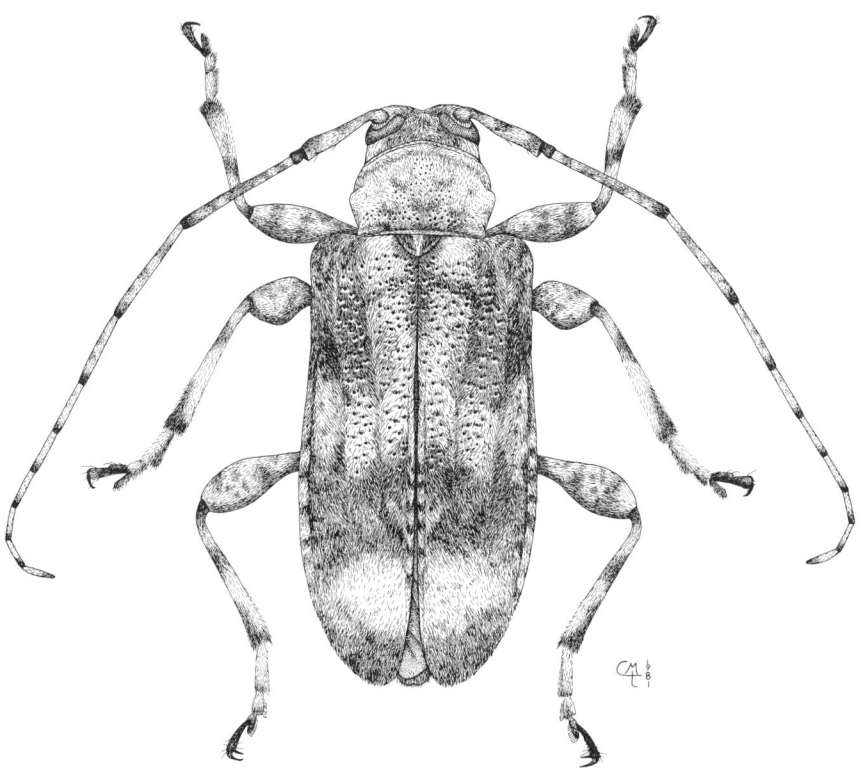

Figure 22. *Pseudastylopsis nebulosus* (Horn), male.

elevated; punctures moderately dense around calluses; pubescence dense, uniform, variegated brownish and gray; prosternum densely pubescent; meso- and metasternum densely grayish pubescent with brownish spots at sides. Elytra about twice as long as broad, slightly expanding at middle; each elytron with four costae, not attaining apex; small tufted tubercles present on costae, especially behind middle; pubescence dense, variegated dark and pale brown, gray, and black, sides with small dark maculae before middle, middle usually with a whitish transverse macula, black spots lightly interspersed over costae; apices rounded. Legs moderate; femora grayish pubescent with brown spots; tibiae biannulate with brownish; tarsi with first two segments pale pubescent. Abdomen densely gray pubescent, spots often vague; last sternite emarginate at apex. Length, 9-14 mm.

Female. Form similar. Antennae about as long as body. Abdomen with last sternite somewhat elongate, narrow and fringed at apex. Length, 11-14 mm.

Type locality. Western Nevada.

Range. Southern Oregon to southern Sierra Nevada, California and western Nevada.

Flight period. June to September.

Host plants. Abies concolor, Pseudotsuga menziesii.

This species may be distinguised by the rather broad form, grayish facies, and very small tufted tubercles of the elytra.

Pseudastylopsis nelsoni, new species

Male. Form moderate-sized, moderately robust; integument reddish brown, head, antennae, and legs partly infuscated; pubescence dense, fine, appressed, mottled brownish, gray, and black. Head with front micropunctate, densely mottled dark brown and grayish pubescent, sides at eye margin with a long, erect seta, genae with three setae on each side, long setae present around mouth-parts; vertex grayish pubescent with small dark spots, two larger dark spots between eyes; genae longer than lower eye lobes; antennae about as long as body, segments dark annulate at apices, outer segments also annulate basally, segments grayish pubescent with dark spots, third segment longer than first, fourth shorter than third, longer than first. Pronotum broader than long, sides subacutely tuberculate behind middle; disk with calluses shallow; median one shallowly elongated to apical margin; punctures dense; pubescence grayish, each side of middle with a medially interrupted dark vitta, often with an inverted v-shaped, dark vitta medially beginning at apical margin, two to three long setae present at sides behind lateral tubercles; prosternum thinly pubescent, intercoxal process about half as broad as coxal cavity; meso- and metasternum variegated pubescent, mesosternal process about two-thirds as broad as coxal cavity. Elytra about twice as long as broad, sides broadening behind middle and tapering toward apex; impressions behind basal gibbosities deep; basal gibbosities moderately crested with tufted tubercles; tufted tubercles numerous on costae behind middle; punctures dense to moderately dense; pubescence dense, appressed, sides with a black vitta extending from humerus to a little behind middle, not reaching lateral margins, middle with a short triangular or inverted v-shaped dark vitta extending back from suture, basal tubercles dark-tipped, a short, dark vitta present across middle of basal gibbosities, surrounding pubescence mottled grayish and brownish, grayish spots usually present at posterior ends of discal impressions and near apex; apices narrowly subtruncate, often slightly dehiscent. Legs mottled grayish pubescent; tibiae dark biannulate, tarsi dark or pale pubescent Abdomen variegated pubescent; last sternite emarginate at apex. Length, 9-11.5 mm.

Female. Form similar. Abdomen with last sternite elongate, fringed at apex. Length, 9.5-11.5 mm.

Holotype male (California Academy of Sciences) and one male paratype from Rustlers Peak, Chiricahua Mts., Arizona, 21 July 1975, on *Pinus ponderosa* (G. H. Nelson).

The prominent tufted tubercles and deep, discal impressions of the elytra make this species distinctive. The coloration and narrow elytral apices separate it from the other known *Pseudastylopsis*.

We are pleased to dedicate this species to G. H. Nelson for his contributions to this project.

A population from the Sierra Madre in the state of Durango, Mexico differs sufficiently to warrant description at this time.

Pseudastylopsis nelsoni australis, new subspecies

Similar to *P. nelsoni nelsoni*. Pronotum with a dark, interrupted vittae on each side of middle. Elytra rather sparsely punctate; median dark vitta narrow, arcuately extending back from suture. Legs with tarsi with pale pubescence. Length, 9-11.5 mm.

Holotype male (California Academy of Sciences) and 19 paratypes (9 males, 10 females) from 5 km W. La Ciudad, Durango, Mexico, 7 August 1983 (E. Giesbert, F. Hovore), 2 August 1983 (F. Hovore). One male and one female paratype, El Salto, Durango, 1 October 1976 (E. Giesbert). One additional female from Durango, Durango, 4 July 1964 (L. A. Kelton), is also assignable to this subspecies.

The absence of a central dark vitta on the pronotum and the more distinct median vittae of the pronotum and of the elytra separate this subspecies from *P. nebulosus*. Additionally, *P. nelsoni nelsoni* has all black tarsi.

Pseudastylopsis pini (Schaeffer)
(Figure 23)

Leptostylus pini Schaeffer, 1905, Brooklyn Mus. Inst. Arts Sci. Bull., 1:165.

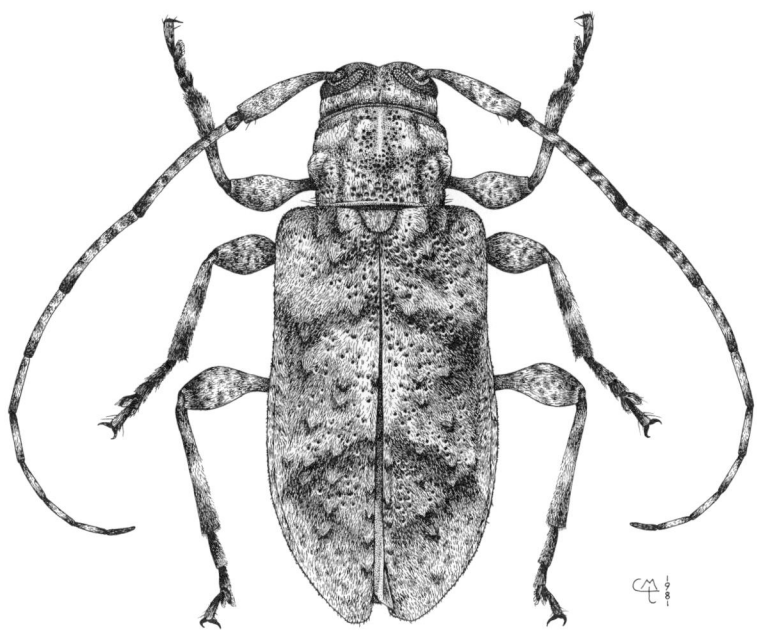

Figure 23. *Pseudastylopsis pini* (Schaeffer), male.

Astylopsis pini: Leng, 1920, Cal. Coleop. North America, p. 282.
Pseudastylopsis pini: Dillon, 1956, Ann. Entomol. Soc. Amer., 49:221; Linsley, Knull, and Statham, 1961, Amer. Mus. Nov., 2050:29.

Male. Form moderate-sized, moderately robust; integument reddish brown, antennae and legs partly infuscated, head dark; pubescence dense, fine, appressed, mottled brownish, gray, and black. Head with front micropunctate, densely mottled gray and brownish pubescent, two long, erect setae present at sides of front on eye margins, two or three on genae, and several above mouth-parts; vertex densely mottled pubescent, minutely punctate, two large, asymmetrical punctures present a little behind eyes; genae subequal in length to lower eye lobes; antennae slightly longer than body, segments dark annulate at apices, eleventh all dark, basal segments dark-spotted, third segment longer than first, fourth slightly shorter than third, longer than first. Pronotum broader than long, sides shallowly, obtusely tuberculate a little behind middle; disk with calluses shallow, median one most prominent; punctures dense, coarse along basal margin; pubescence dense, grayish, pale and dark brown, each side of middle, with a narrow dark vitta usually interrupted at middle, two or three erect setae present on dorsum of lateral tubercles; prosternum rather thinly grayish pubescent; meso- and metasternum moderately densely grayish pubescent, with brownish spots at sides. Elytra less than twice as long as broad, sides slightly expanding behind middle, rather abruptly converging near apex; each elytron with four costae not attaining apex; small, dark, tufted tubercles present on costae and along suture, these more numerous behind middle; pubescence dense, variegated pale and dark brown and gray, middle with a vague, broad, oblique paler fascia, dark pubescence forming an outline of a broad triangle behind middle, scutellum and area around usually with gray pubescence; apices subtruncate. Legs mottled grayish pubescent, femora broadly dark on club, tibiae dark biannulate, tarsi dark. Abdomen finely gray pubescent; last sternite shallowly emarginate at apex. Length, 8-11 mm.

Female. Form similar. Antennae about as long as body. Abdomen with last sternite somewhat elongate, apex narrow and lightly fringed. Length, 9-12 mm.

Type locality. Carr's Peak, Huachuca Mts., Arizona.
Range. Arizona to Texas to Durango, Mexico.
Flight period. June to September.
Host Plants. Pinus.

The coloration and large tufted tubercles of the elytra distinguish this species from *P. nebulosus*. The Mexican species, *P. squamosus* Chemsak and Linsley, differs by the short, scale-like pubescence of the pronotum and elytra and the vague whitish vitta at the apical one-third of the elytra.

Genus *Trichastylopsis* Dillon

Leptostylus: Horn, 1880, Trans. Amer. Entomol. Soc., 8:119 (part); Leng and Hamilton, 1896, Trans. Amer. Entomol. Soc., 23:116 (part).
Trichastylopsis Dillon, 1956, Ann. Entomol. Soc. Amer., 49:148; Arnett, 1962, Beetles U.S., 103:871.

Moderate-sized, cylindrical; pubescence dense, thick, appressed, forming rows of tufts down elytra. Head with front convex, quadrate; median line extending onto neck; mandibles feebly arcuate; genae slightly convergent, about as long as lower eye lobes; eyes small, lower lobes rounded, upper lobes separated by about their width; antennal tubercles prominent, widely divergent; antennae eleven-segmented, slightly longer than body, basal segments with a few short, suberect hairs beneath, third segment longer than first, fourth equal to first, fifth shorter than fourth. Pronotum broader than long, sides with moderate-sized, broad tubercles behind middle; base broadly impressed, impression extending to sides; apex narrowly impressed at middle, impressions broadening toward sides; disk with two calluses behind apical margin and a small median callus behind middle; prosternum with intercoxal process two-thirds as broad as coxal cavity, sides slightly angulate near middle; mesosternum with intercoxal process almost as broad as coxal cavity, abruptly declivous anteriorly; metasternum with episternum narrow, parallel. Elytra about twice as long as broad, tapering near apex; basal gibbosities moderate-sized, tufted; each elytron tricostate, pubescence forming small tufts down costae; pubescence dense, thick, appressed, longer, suberect hairs usually sparsely interspersed at sides and on tufts; apices narrowing, obliquely truncate. Legs stout, long, flying hairs numerous; femora strongly clavate; tarsi short, broad, first segment of hind pair subequal in length to two following segments together, third segment cleft to base. Abdomen normally segmented.

Type species. Leptostylus albidus LeConte (by original designation).

The rather broad, cylindrical form, the long flying hairs of the legs, and the slightly expanded sides of the prosternal process (sometimes vague) make *Trichostylopsis* readily recognizable.

A single species is known from North America.

Trichastylopsis albidus (LeConte)

Leptostylus albidus LeConte, 1852, J. Acad. Nat. Sci. Philadelphia, 2:168; Lacordaire, 1872, Genera des coléoptères, 9(2):772, fn.; LeConte, 1873, Smithson. Misc. Coll., 11(264):233; Horn, 1880, Trans. Amer. Entomol. Soc., 8:121; Leng and Hamilton, 1896, Trans. Amer. Entomol. Soc., 23:119; Craighead, 1923, Can. Dept. Agr. Bull., (n.s.) 27:116.
Amniscus albidus: White, 1855, Cat. Coleop. Ins. British Mus., 8:392.
Trichastylopsis albidus: Dillon, 1956, Ann. Entomol. Soc. Amer., 49:148; Hovore and Giesbert, 1976, Coleop. Bull., 30:358; Hovore, 1983, Coleop. Bull., 37:386; MacKay, Zak, and Hovore, 1987, Coleop. Bull., 41:366.
Leiopus setipes Casey, 1891, Ann. N.Y. Acad. Sci., 6:48; Dillon, 1956, Ann. Entomol. Soc. Amer., 49:219 (Incertae sedis); Lewis, 1986, Pan-Pac. Entomol., 62:171 (synonymy).
Liopus setipes: Leng and Hamilton, 1896, Trans. Amer. Entomol. Soc., 23:126.
Leptostylus falli Linsley, 1934, Entomol. News, 45:182.
Trichastylopsis falli: Linsley, Knull, and Statham, 1961, Amer. Mus. Nov., 2050:29.

Male. Form small to moderate-sized; integument dark reddish brown to fuscous, appendages often paler; pubescence dense, thick, grayish and brown, tufts usually present on elytra and longer, suberect hairs usually interspersed; legs with long, flying hairs. Head with front densely mottled pubescent; antennae extending about three segments beyond elytra, basal segments with brown spots, all segments narrowly brown annulate at apices. Pronotum usually with dorsal calluses darker; punctures moderately dense; pubescence usually obscuring surface, grayish, brown, and often black; prosternum densely pubescent; meso- and metasternum densely pubescent, erect hairs absent. Elytra coarsely punctate at base, punctures becoming finer and sparser toward apex; pubescence dense, thick, disk often with a large whitish macula over basal two-thirds, macula sometimes reduced to narrow oblique fasciae, apical one-third mottled dark pubescent, pubescence along costae and tufts pale brownish, sides with a black integumental macula from behind humeri to a little beyond middle, long erect hairs often interspersed on sides and on tufts; apices slightly obliquely truncate. Legs stout; femora dark at apices, long, suberect hairs sparse; tibiae dark biannulate, long, suberect hairs numerous; tarsi dark, pale pubescent. Abdomen densely clothed with appressed pubescence; last sternite broadly rounded at apex. Length, 7-10 mm.

Female. Form similar. Antennae slightly longer than body. Abdomen with last sternite narrowing at apex, fringed. Length, 7-10.5 mm.

Type locality. Of *albidus*, junction of the Colorado and Gila, Arizona; *setipes*, El Paso, Texas; *falli*, Carr Canyon, Huachuca Mts., Arizona.

Range. Texas to southern California, northern Mexico.

Flight period. April to July.

Host plants. Populus, Acer.

This species is variable in coloration and pubescent pattern. The broad whitish macula of the elytra is reduced in most specimens, and the surface is brownish with a small oblique, whitish macula near the middle. The elytral costae are usually yellow-brown pubescent and the tufts are variable. The erect hairs also vary and are often obsolete.

Genus *Astylidius* Casey

Astylidius Casey, 1913, Memoirs on the Coleoptera, 4:308; Bradley, 1930, Man. Genera Beetles, p. 246; Knull, 1946, Ohio Biol. Surv. Bull., 39:247; Dillon, 1956, Ann. Entomol. Soc. Amer., 49:149; Arnett, 1962, Beetles U.S., 103:872; Rice and Enns, 1981, Trans. Mo. Acad. Sci., 15:98.

Form small, depressed. Head with front convex, about as long as broad; mandibles feebly arcuate; genae convergent, longer than lower eye lobes; eyes small, deeply emarginate, upper lobes small, separated by more than diameter of antennal scape; antennal tubercles prominent, widely divergent; antennae eleven-segmented, at least 1-1/2 times longer than body in males, about four segments longer in females, third segment longer than fourth, fourth subequal to or slightly longer than first. Pronotum broader than long, strongly tuberculate at sides behind middle; disk with three rather large calluses; apex narrowly impressed transversely, base broadly impressed, impression extending to underside; prosternum with intercoxal process barely arcuate, about half as broad as coxal cavities; mesosternal process about as broad as coxal cavities, lightly arcuate anteriorly. Elytra a little more than 1-1/2 times as long as broad, sides slightly emarginate before middle; epipleura vertical; basal gibbosities distinct, lightly crested with tufted tubercles; disk costate; small tufted tubercles interspersed down costae; apices rather narrow, rounded to subacuminate. Legs stout; femora pedunculate; tarsi short, slender, first segment longer than two following together, third segment cleft to base. Abdomen normally segmented, fifth sternite of females twice as long as fourth.

Type species: Leptostylus parvus LeConte (by original designation).

The broad form, rather small size, long antennae, and the presence of three dorsal calluses of the pronotum separate *Astylidius* from related genera. *Leptostylus* differs by the shorter antennae and by having five dorsal calluses on the pronotum.

Dillon (1956) recognized two species of *Astylidius* on the basis of the length-width proportions of the pronotum and shape of the elytral apices. Since these characteristics vary within the same population, we recognize only one.

Astylidius parvus (LeConte)

Leptostylus parvus LeConte, 1873, Smithson. Misc. Coll., 11(264):233; Horn, 1880, Trans. Amer. Entomol. Soc., 8:121; Chittenden, 1894, Proc. Entomol. Soc. Wash., 3:99; Leng and Hamilton, 1896, Trans. Amer. Entomol. Soc., 23:119; Wickham, 1897, Can. Entomol., 29:208; Townsend, 1902, Trans. Texas Acad. Sci., 5:78; Blatchley, 1910, Coleoptera in Indiana, p. 1072; Blatchley, 1919, Can. Entomol., 51:65; Craighead, 1923, Can. Dept. Agr. Bull., (n.s.) 27:116.

Astylidius parvus: Casey, 1913, Memoirs on the Coleoptera, 4:308; Champlain and Knull, 1925, Entomol. News, 36:140; Leonard, 1928, Cornell Agr. Exp. Sta. Mem., 101:452; Linsley and Martin, 1933, Entomol. News, 44:182; Knull, 1946, Ohio Biol. Surv. Bull., 39:247, pl. 22, fig. 90; Steyskal, 1951, Coleop. Bull., 5:76; Dillon, 1956, Ann. Entomol. Soc. Amer., 49:150; Turnbow and Wappes, 1978, Coleop. Bull., 32:370; Turnbow and Wappes, 1981, Southwest. Entomol., 6:77; Furth, 1985, Trans. Conn. Acad. Arts Sci., 46:192; Hovore et al., 1987, Proc. Calif. Acad. Sci., 44:317.

Astylidius versutus Casey, 1913, Memoirs on the Coleoptera, 4:308. New synonymy.

Astylidius versutus versutus: Dillon, 1956, Ann. Entomol. Soc. Amer., 49:149; Rice and Enns, 1981, Trans. Mo. Acad. Sci., 15:98.

Astylidius versutus downiei Dillon, 1956, Ann. Entomol. Soc. Amer., 49:150, fig. 4; Gilmour, 1965, Cat. Lam. du Monde, 8:571; Gosling and Gosling, 1977, Gr. Lakes Entomol., 10:25; Morris, 1987, J. Entomol. Sci., 22:141. New synonymy.

Astylidius leiopinus Casey, 1913, Memoirs on the Coleoptera, 4:308; Vogt, 1949, Pan-Pac. Entomol., 25:181.

Leptostylus monki Knull, 1936, Entomol. News, 47:106.

Male. Form small; integument dark reddish brown; pubescence fine, dense, appressed, grayish and pale and dark brownish. Head with front micropunctate, densely mottled pubescent; antennae extending beyond body by at least five segments, segments dark annulate apically, outer segments also narrowly dark annulate basally, basal segments with dark spots. Pronotum with disk sparsely punctate around calluses, basal impression with a row of punctures; pubescence fine, dense, mottled grayish and brownish, calluses usually brownish; prosternum finely grayish pubescent; meso- and metasternum finely pubescent, spotted at sides. Elytra slightly more than 1-1/2 times as long as broad; impressions behind gibbosities rather deep; tufted tubercles on gibbosities larger than those arranged down costae; pubescence dense, mottled grayish and brownish, tubercles brownish, disk with a narrow whitish vitta behind middle extending obliquely back from suture, epipleura behind humeri with a large, dark spot; apices rounded to lightly acuminate. Legs with pubescence of femora pale, interrupted by dark spots; tibiae dark biannulate; tarsi with first segment pale pubescent at least basally. Abdomen finely pale pubescent; fifth sternite rounded at apex. Length, 5.5-7 mm.

Female. Form similar. Antennae a little shorter. Abdomen with fifth sternite about twice as long as fourth, apex narrow, shallowly emarginate. Length, 6-7 mm.

Type locality. Of *parvus*, western states; *versutus*, District of Columbia; *leiopinus*, Columbus, Texas; *monki*, Donna, Texas; *downiei*, Tippecanoe Co., Indiana.

Range. Middle eastern states to Minnesota, Mississippi, and Texas.

Flight period. May to August.

Host plants. Acer, Albizia, Aesculus glabra, Celastrus, Diospyros, Ficus, Morus, Pithecellobium, Rhus, Ulmus, Zanthoxylum.

This species is distinctive in the small, rather broad body shape, long antennae, broad intercoxal processes, and by the narrow, inverted v-shaped pale vitta of the elytra.

Genus *Lagocheirus* Dejean

Lagocheirus Dejean, 1835, Cat. Coleop. Dejean, 2nd ed., p. 336; Dejean, 1837, Cat. Coleop. Dejean, 3rd ed., p. 322; White, 1855, Cat. Coleop. Ins. British Mus., 8:315; Thomson, 1860, Class. ceramb., p. 9; Lacordaire, 1872, Genera des coléoptères, 9(2):762; Leng and Hamilton, 1896, Trans. Amer. Entomol. Soc., 23:115; Dillon, 1956, Ann. Entomol. Soc. Amer., 49:136; Dillon, 1957, Bull. British Mus., 6(6):7; Duffy, 1960, Mon. Neotrop. Timber Beetles, pp. 25, 29, 41; Arnett, 1962, Beetles U.S., 103:871; Lane, 1973, Stud. Entomol., 16:529; Villiers, 1980, Ann. Soc. Entomol. Fr., (n.s.) 16:90, 92.

Lagochirus Erichson, 1847, Arch. Naturg., 13(1):144; LeConte, 1873, Smithson. Misc. Coll. 11(265):337; Horn, 1880, Trans. Amer. Entomol. Soc., 8:117; Bates, 1880, Biol. Centr.-Amer., Coleop., 5:144; LeConte and Horn, 1883, Smithson. Misc. Coll., 507:323; Casey, 1913, Memoirs on the Coleoptera, 4:303; Aurivillius, 1923, Coleop. Cat., 74:391; Bradley, 1930, Man. Genera Beetles, p. 245; de Zayas, 1975, Rev. Fam. Ceramb., p. 226. (Type species: *Lagochirus plantaris* Erichson, monobasic.)

Lagocheirus Thomson, 1864, Syst. ceramb., pp. 27, 355; Lane, 1973, Stud. Entomol., 16:530. (Type species: *Lagocheirus binumeratus* Thomson, by original designation.)

Form moderate-sized to large, depressed. Head with front quadrate, convex, about as broad as long; mandibles arcuate, acute at apices; genae shorter than lower eye lobes; eyes large, moderately coarsely faceted, deeply emarginate, upper lobes narrow, separated by diameter of antennal scape or less; antennal tubercles prominent, broadly divergent at bases; antennae slender, extending beyond body by four or five segments in males, a little longer than body in females, sixth segment of males with an apical, tufted spur, scape slender, about as long as fifth segment, third longer than first, fourth slightly shorter than third, segments seven to eleven shortened, third segment with a few, longer, suberect hairs beneath. Pronotum much broader than long, sides with robust, median tubercles; disk with five prominent calluses; apex narrowly impressed, base broadly, deeply impressed; prosternum very narrow, intercoxal process arcuate, medially impressed, two-fifths to one-half as broad as coxal cavities, coxal cavities closed behind; mesosternum with intercoxal process about as broad as coxal cavities, almost plane, abruptly declivous anteriorly; metasternum with episternum narrow, tapering posteriorly. Elytra a little more than 1-1/2 times as long as broad; basal gibbosities shallow, somewhat transverse; disk punctate over basal half, costae vague, bearing small, tufted tubercles, especially basally and apically, basal gibbosities often linearly crested; apices truncate to slightly emarginate truncate. Legs robust; femora strongly clavate; tibiae moderate, front pair in males deeply excavated and fringed beneath at apical half; tarsi short, broad, front pair expanded and densely fringed at sides in males, third segment cleft to base. Abdomen normally segmented.

Type species: Cerambyx araneiformis Linnaeus (monobasic).

This genus may be recognized among the North American Acanthocinini by the large, robust size, strongly tuberculate and callused pronotum, the broad intercoxal processes, the apical spur of the sixth antennal segment of males, and the strongly fringed and broadened front tarsi of males.

Dillon (1956) described each of the two species occurring in our fauna as new. We regard both of these as members of the rather widespread Neotropical species, *L. undatus* (Voet). A third species, *procerus* Casey, occurs in Baja California and northwestern mainland Mexico. We are unaware of any records from the United States.

KEY TO THE NORTH AMERICAN SPECIES OF *LAGOCHEIRUS*

Antennae with fourth segment biannulate with ashy pubescence; punctures of pronotum outlined with white. Length, 12-19 mm. Southern Texas to Central America .... *undatus*
Antennae with fourth segment uniformly pubescent; punctures of pronotum not outlined with white. Length, 13-25 mm. Florida and West Indies, Texas to South America .... *araneiformis*

Lagocheirus undatus (Voet)

Cerambyx undatus Voet, 1778, Cat. Coleop., 2:11, pl. 9, fig. 34; Voet, 1794, Cat. Coleop., ed. Panzer, 3:27, pl. 9, fig. 24.
Lagochirus undatus: Aurivillius, 1923, Coleop. Cat., 74:393.
Lagocheirus undatus undatus: Dillon, 1957, Bull. British Mus., 6(6):143; Duffy, 1960, Mon. Neotr. Timber Beetles, p. 239, figs. 141-142.
Lagocheirus undatus: Lane, 1973, Stud. Entomol., 16:530.
Lagocheirus obsoletus Thomson, 1860, Class. ceramb., p. 10; Bates, 1874, Trans. Entomol. Soc. London, 1874:229; Bridwell, Proc. Haw. Entomol. Soc., 4:319; Duffy, 1953, Proc. Haw. Entomol. Soc., 15:140, 154, figs. 24-25.
Lagochirus obsoletus: Bates, 1880-85, Biol. Centr.-Amer., Coleop., 5:145, 383.
Lagochirus longipennis Bates, 1880-85, Biol. Centr.-Amer., Coleop., 5:145, 383; Chemsak and Linsley, 1970, J. Kansas Entomol. Soc., 43:411 (lectotype).
Lagocheirus texensis Dillon, 1956, Ann. Entomol. Soc. Amer., 49:139; Turnbow and Wappes, 1981, Southwest Entomol., 6:77; Hovore et al., 1987, Proc. Calif. Acad. Sci., 44:317, fig. 17. New synonymy.
Lagocheirus undatus mariorum Dillon, 1957, Bull. British Mus., 6(6):144. New synonymy.
Lagocheirus zimmermani Dillon, 1952, Entomol. News, 63:207. New synonymy.
Lagocheirus zimmermani zimmermani: Dillon, 1957, Bull. British Mus., 6(6):145.
Lagocheirus zimmermani aukena Dillon, 1957, Bull. British Mus., 6(6):145. New synonymy.

Male. Form moderate-sized, slightly tapering; integument brownish; pubescence dense, appressed, black, brownish, and white. Head with front micropunctate, larger punctures sparsely interspersed, pubescence dense, appressed, mottled grayish and brown, margins of eyes, genae, and clypeus with long, erect setae; vertex convex, minutely punctate, with a row of coarse punctures behind eyes, pubescence mottled, two dark spots present between eyes; antennae longer than body, sixth segment with a distinct tufted, apical spur, segments four and five distinctly biannulate with ashy pubescence, scape extending to middle of lateral tubercle, third segment longer than first, fourth subequal to third, remaining segments gradually decreasing in length. Pronotum almost twice as long as broad, sides broadly tuberculate at middle; punctures scattered around calluses; pubescence dense, appressed, mottled dark and pale brownish, punctures outlined by white pubescence, two short, black, oblique maculae present behind apical margin, and two linear maculae behind middle, extending to basal margins, a few long, erect setae present near sides behind middle; prosternum densely mottled pubescent; meso- and metasternum densely pubescent, sides brownish with small dark brown spots. Elytra less than twice as long as broad; base moderately coarsely, rather sparsely punctate, punctures becoming finer toward apex; each elytron with four costae, middle three pairs with rows of tufted tubercles which are more distinct near base and behind middle; pubescence dense, appressed, mottled, an indistinct, broadly V-shaped, brown maculae present behind scutellum, sides dark brown from humeri to behind middle, joining a dark lateral spot at middle, whitish pubescence suffused throughout, often forming a macula behind dark median spots, usually with two pairs of short, narrow, dark maculae behind middle near suture; erect hairs often present on basal half; apices truncate to emarginate truncate. Scutellum broadly to narrowly brownish at sides. Legs robust, mottled pubescent. Abdomen densely pale pubescent with numerous small brown spots interspersed; last sternite shallowly emarginate at apex. Length, 12-19 mm.

Female. Form similar. Antennae a little longer than body. Abdomen with last sternite narrowly subtruncate at apex. Length, 14-18 mm.

Type locality. Of *undatus*, Indiis Orientalibus; *obsoletus*, Mexico; *longipennis*, R. Sarston, British Honduras; *texensis*, Dimmit Co., Texas; *mariorum,* Maria Madre Island, Tres Marias Islands, Mexico; *zimmermani*, Honolulu, Hawaii; *aukena*, Aukena Island, Mangareva Islands.

Range. Southern Texas to Costa Rica, Hawaii.

Flight period. March to October (May to September in Texas).

Host plants. Aleurites, Allamanda, Araucaria, Bursera, Euphorbia, Ficus, Forstiera, Hibiscus, Manihot , Plumeria, Pseudopanax.

The biannulate antennal segments and white outlined pronotal punctures distinguish this species from *araneiformis*. The population occurring in southern Texas falls within the range of variation exhibited by *undatus* throughout its range.

The frequent occurrence of erect hairs on the elytra presents an interesting situation. Specimens available from Sinaloa to Oaxaca eastward to Yucatan and north to San Luis Potosí possess this characteristic. We can detect no other differences between individuals with erect hairs and those without. The population occurring in Texas lacks erect hairs and the one introduced into Hawaii possesses them. Based on the known distribution of *undatus* and the occurrence of mainland individuals with erect elytral setae, we are synonymizing *L. zimmermani.*

Lagocheirus araneiformis (Linnaeus)

Cerambyx araneiformis Linnaeus, 1767, Syst. Nat., ed. 12, p. 625.
Lagocheirus araneiformis: Dejean, 1835, Cat. Coleop. Dejean, 2nd ed., p. 336.
Lagochirus araneiformis: Horn, 1880, Trans. Amer. Entomol. Soc., 8:117.

Male. Form moderate-sized to large, tapering posteriorly; integument reddish brown to piceous; pubescence dense, short, appressed, pale and dark brownish and grayish. Head with front micropunctate, larger punctures interspersed, pubescence dense, mottled grayish and brown, margins of front, genae, and clypeus with long, erect setae; vertex minutely punctate with a row of coarse punctures behind eyes, pubescence mottled with two dark spots between eyes; antennae extending five to six segments beyond body, segment six with a tufted apical spur, segments uniformly pale brownish pubescent, scape extending beyond middle of lateral tubercle, third segment longer than scape, fourth shorter than third but longer than scape. Pronotum a little more than 1-1/2 times broader than long, sides broadly tuberculate at middle; punctures sparsely scattered around calluses; pubescence pale brownish and gray, dark maculae present as follows: an oblique pair behind apical margin, a basal pair extending onto dorsal calluses, and a short pair behind lateral tubercles, a few erect setae present at sides behind middle of lateral tubercles; prosternum densely, finely pubescent; meso- and metasternum densely clothed with pale brownish, appressed pubescence interrupted by small darker spots. Elytra less than twice as long as broad; base coarsely, separately punctate, punctures becoming very fine and sparse behind middle; each elytron vaguely tricostate, base with three rows of elevated, tufted tubercles; pubescence pale brownish to grayish, mottled, base narrowly darker, a narrow, short, transverse, dark line usually present behind scutellum, sides dark to middle expanding onto disk to form median dark, semi-rounded maculae, posterior margins of median maculae with narrow, whitish, angulate markings, two small dark spots present behind middle near suture, scattered, small, dark and whitish spots present behind middle; apices truncate. Scutellum darker basally and beneath. Legs robust, mottled pubescent. Abdomen mottled pubescent; last sternite emarginate at apex. Length, 13-25 mm.

Female. Form similar. Antennae a little longer than body. Legs with femora less robust. Abdomen with last sternite narrow at apex, deeply emarginate. Length, 13-25 mm.

Type locality. Jamaica.

Range. Florida and West Indies, Mexico to Brazil.

This common, widespread species was divided into a number of subspecies by Dillon (1957). The nomenclatoral status of these populations is beyond the scope of this work, but we will retain the available name, *stroheckeri* for the Florida components. *L. stroheckeri granulatus* Dillon was based on a single individual allegedly from Texas. We have not seen any other examples from Texas and are placing this name in synonymy with *L. araneiformis ypsilon* (Voet) which occurs from Mexico to Panama.

Lagocheirus araneiformis stroheckeri Dillon

Lagocheirus stroheckeri Dillon, 1956, Ann. Entomol. Soc. Amer., 49:138.
Lagocheirus stroheckeri stroheckeri Dillon, 1956, Ann. Entomol. Soc. Amer., 49:138, fig. 2.
Lagocheirus araneiformis stroheckeri: Dillon, 1957, Bull. British Mus., 6:148; Gilmour, 1968, Stud. Fauna Curaçao, Carib. Is., 25:156.

Integument reddish brown. Pubescence of pronotum and elytra pale brownish interspersed with white. Elytra with whitish postmedian markings usually extending across to suture. Length, 14-25 mm.

Type locality. Miami, Florida.
Range. Southern Florida (and Cuba according to Dillon, 1957).
Flight period. April to November.
Host plants. Bursera simaruba.
Most of the specimens available for study have the integument and pubescence paler and the white streaks of the elytra extend across to the suture.

Genus *Leptostylopsis* Dillon

Leptostylopsis Dillon, 1956, Ann. Entomol. Soc. Amer., 49:144; Arnett, 1962, Beetles U.S., 103:871, 872; Bayer and Shenefelt, 1969, Univ. Wisc. Res. Bull., 275:26; Villiers, 1980, Ann. Soc. Entomol. Fr., (n.s.) 16:90, 96.

Form moderate-sized, usually robust. Head with front convex, about as long as broad; mandibles feebly arcuate; genae slightly convergent, as long as lower eye lobes; eyes moderate, deeply emarginate, upper lobes small, well separated above; antennal tubercles prominent, widely divergent; antennae eleven-segmented, usually extending about four segments beyond elytra, third segment slightly longer than first, fourth subequal to first. Pronotum broader than long, sides tumid to tuberculate slightly behind middle; disk with five or seven obtuse tubercles; base and apex rather deeply impressed transversely; prosternum narrow, intercoxal process lightly arcuate, about two-thirds as broad as coxal cavities; mesosternum with intercoxal process broader than coxal cavities, lightly, narrowly declivous anteriorly. Elytra a little more than 1-1/2 times longer than broad; basal gibbosities shallow; disk costate with varying numbers of tufted tubercles; apices narrowly to obliquely truncate. Legs stout; femora pedunculate; tarsi short, stout, first segment shorter than following two together, third segment cleft to base. Abdomen normally segmented, last sternite of females longer than fourth, narrowing apically.

Type species: Leptostylus terraecolor Horn (by original designation).

This genus is distinctive by the broad intercoxal processes. The longer antennae, less convex body form, and less prominent discal tubercles of the pronotum separate it from *Leptostylus.*

Five species are known in North America.

KEY TO NORTH AMERICAN SPECIES OF *LEPTOSTYLOPSIS*

1	Pronotum with lateral tubercles rounded, obtuse	2
	Pronotum with lateral tubercles prominent, conical; elytra with yellow lines on disk. Length, 6.4 mm. Texas	*luteus*
2(1)	Pronotum sparsely, irregularly punctate around discal tubercles	3

	Pronotum densely punctate around median tubercle, punctures forming an arcuate pattern; pubescence, except for small, dark vittae, usually uniformly brownish. Length, 7-12 mm. Florida ....*terraecolor*
3(2)	Pronotum with seven discal tubercles; tarsi pale pubescent 4
	Pronotum with five shallow discal tubercles; tarsi dark pubescent. Length, 8.5-14 mm. Maryland to Florida west to Louisiana ....*planidorsus*
4(3)	Elytra with a narrow, zigzag, white fascia behind middle, pubescent tufts dark; pronotum with a dark basal fascia on each side. Length, 7-11 mm. Florida, Cuba .... *albofasciatus*
	Elytra usually with a dark fascia behind middle, pubescent tufts pale; pronotum lacking dark basal fasciae. Length, 7-13 mm. Southern Florida, West Indies .... *argentatus*

Leptostylopsis luteus Dillon

Leptostylopsis luteus Dillon, 1956, Ann. Entomol. Soc. Amer., 49:147; Bayer and Shenefelt, 1969, Univ. Wisc. Res. Bull., 275:28, fig. 36; Hovore et al., 1987, Proc. Calif. Acad. Sci., 44:318.

"*Female.* Dark reddish brown, densely covered with ashy pubescence, lightly clouded with brown on head above, on pronotal disk, and on sides of elytra, especially near apex, and faintly tinged with luteous on head and pronotum. Elytra each with four distinct luteous lines interrupted by many long tufts of fuscous hairs, one of which tufts forms an elongate crest on basal gibbosity; at apex a rather large luteous patch, the anterior margin of which is strongly oblique, ascending to the suture, bordered by an indistinct brown fascia which becomes fuscous at the suture, this fascia in turn margined anteriorly by an interrupted bright ashy fascia; on deflexed sides at basal fourth a distinct, rather large, fuscous macula. Body beneath, legs, and antennae dark ferrugineous, sparsely ashy pubescent; femora and first six antennal segments mottled with reddish brown, the former broadly blackish at base; tibiae broadly fuscous-annulate on apical third; tarsi fuscous, the first segment ashy-annulate at base; apices of antennal segments beginning with second broadly fuscous-annulate, the sixth and following narrowly annulate on base as well.

Head slightly impressed behind antennal tubercles, impunctate; front one-fourth again as wide as high, strongly narrowed between eyes, slightly so below, somewhat convex; eyes with lower lobe one-fourth shorter than gena, slightly transverse, upper lobe one-fourth shorter than gena, slightly transverse, upper lobe one-third as broad as interocular space. Pronotum nearly three-fourths again as broad across lateral tubercles as long, sides gradually tapering to apex which is narrower than base; lateral tubercles rather small, but prominent, armed with a robust, obtuse tooth; disk with five tubercles, the two lateralmost feeble, rest prominent, the anterior pair broad, surface finely, indistinctly punctate, with a row of coarse punctures in basal sulcus. Elytra rather coarsely, densely punctate; each disk with four distinct costae which bear small but pronounced, tufted tubercles; basal gibbosity feebly elevated, at apex bearing an elongate, low crest; apices separately, narrowly rounded, obliquely subtruncate at suture, the outer angle briefly dentate. Metafemora distinctly slenderer than mesofemora. Fifth sternite half again as long as fourth. Antennae one and two-fifth times as long as body, beneath with only one or two fringing hairs near apex of scape; scape slightly surpassing middle of pronotum; third segment one-fourth again as long as first; fourth feebly longer than first; rest distinctly diminishing in length.

Length 6.4 mm.; width 2.6 mm." (Original description.)
Type locality. Esper. Ranch, Brownsville, Texas.
Range. Southern Texas.
Flight period. October.
Host plants. Hovore et al. (1987) report one adult beaten from dead *Acacia* and another from dead *Baccharis*.

We have seen only the type of this species. It is distinct by its small size, prominent lateral tubercles of the pronotum, and the yellow coloration of the elytra.

Leptostylopsis terraecolor (Horn)

Leptostylus terraecolor Horn, 1880, Trans. Amer. Entomol. Soc., 8:121; Leng and Hamilton, 1896, Trans. Amer. Entomol. Soc., 23:118; Craighead, 1923, Can. Dept. Agr. Bull., (n.s.) 27:116.
Leptostylopsis terraecolor: Dillon, 1956, Ann. Entomol. Soc. Amer., 49:146; Turnbow and Hovore, 1979, Entomol. News, 90:224.
Leptostylus mutilus Casey, 1913, Memoirs on the Coleoptera, 4:307.

Male. Form moderate-sized; integument dark reddish brown; pubescence short, dense, appressed, pale and dark brownish, usually with gray interspersed. Head with front slightly broader than long, micropunctate, densely clothed with appressed brownish pubescence; genae slightly longer than lower eye lobes; upper eye lobes small, separated by about diameter of antennal scape; antennae extending about four segments beyond elytra, segments from third narrowly dark annulate at bases and apices, eleventh all dark, segments to seventh with small dark spots, third segment longer than first, fourth shorter than third, longer than first. Pronotum broader than long, sides broadly tumid, vaguely, obtusely tuberculate; disk with seven shallow calluses, three median ones more prominent; apical and basal impressions deeply punctate, punctures around median callus dense, arcuately arranged, sides sparsely punctate; pubescence short, dense, appressed, brownish, basal margin usually with three dark spots; prosternum densely pubescent, intercoxal process more than one-half as broad as coxal cavity; meso- and metasternum densely pale brownish pubescent, sides with numerous dark spots, mesosternal process broader than coxal cavity. Elytra slightly more than 1-1/2 times as long as broad, tapering at apical one-fourth; basal gibbosities shallow, impression behind moderate; disk with tufted tubercles extending down costae, tubercles more numerous behind middle; pubescence dense, appressed, brownish, each side with a dark lateral vitta extending from humerus to a little behind middle, disk with a narrow, dark, transverse vitta at apical one-third, tubercles pale brownish, area around scutellum dark; punctures rather sparse; apices narrowly truncate. Legs robust, finely, densely clothed with pale brownish pubescence; tibiae narrowly dark at apices, tarsi pale pubescent. Abdomen rather thinly pubescent, small, dark dots numerous; last sternite rounded at apex. Length, 8-12 mm.

Female. Form similar. Antennae slightly longer than body. Abdomen with last sternite a little elongate, narrow, truncate at apex. Length, 7-12 mm.

Type locality. Of *terraecolor*, Florida; *multilus*, Key Largo, Florida.
Range. southern Florida.
Flight period. February to December (most records are April and May).
Host plants. Coccolobis, Ficus, Forestiera, Rhizophora, Sideroxylon, Vitis.

The arcuate puncture pattern of the pronotum makes this species distinctive. The coloration is mostly brownish with paler tubercles on the elytra. In addition to the narrow, dark, transverse band of the elytra, there are often dark oblique bands near the base and dark spots near the apex.

White in 1855 (Cat. Coleop. Ins. British Mus., 8:391) listed *Amniscus terraecolor* as a Newman manuscript name. This listing did not fulfill the conditions to validate the name which was subsequently used by Horn in 1880.

Leptostylopsis planidorsus (LeConte)
(Figure 24)

Leptostylus planidorsus LeConte, 1873, Smithson. Misc. Coll., 11(264):233; Horn, 1888, Trans. Amer. Entomol. Soc., 8:121; Leng and Hamilton, 1896, Trans. Amer. Entomol. Soc., 23:118; Blatchley, 1910, Coleoptera in Indiana, p. 1073; Fattig, 1947, Emory Univ. Mus. Bull., 5:34.
Leptostylopsis planidorsus: Dillon, 1956, Ann. Entomol. Soc. Amer., 49:147; Turnbow and Hovore, 1979, Entomol. News, 80:224.
Leptostylus lecontei Casey, 1913, Memoirs on the Coleoptera, 4:305.
Leptostylus crescenticus Casey, 1913, Memoirs on the Coleoptera, 4:306.

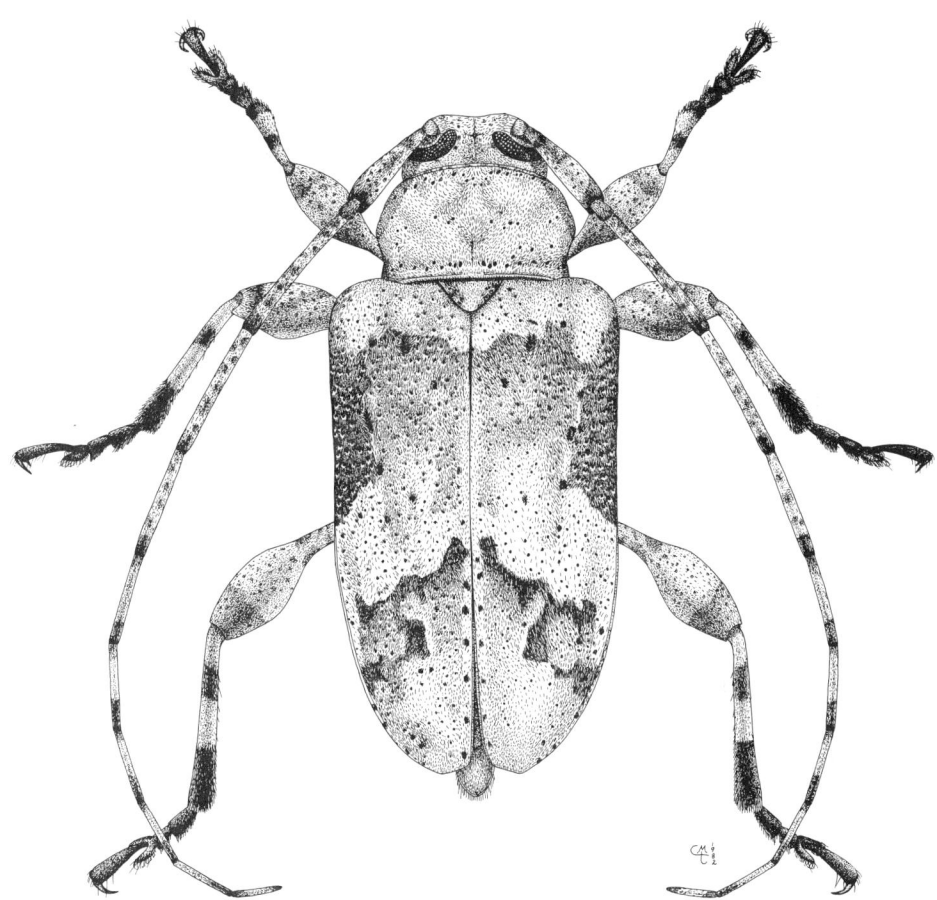

Figure 24. *Leptostylopsis planidorsus* (LeConte), male.

Male. Form moderate-sized; integument reddish-brown, often partially infuscated; pubescence dense, short, appressed, grayish, pale and dark brownish. Head with front about as long as broad, finely, sparsely punctate, moderately densely clothed with fine brownish pubescence, punctures glabrous; genae as long as lower eye lobes; upper eye lobes small, separated by about diameter of antennal scape; antennae extending about four segments beyond elytra, segments dark annulate at apices, segments from sixth dark annulate at bases, basal segments with small dark spots, third segment longer than first, fourth equal to first. Pronotum broader than long, sides shallowly, obtusely tuberculate a little behind middle; disk with five shallow tubercles, three median ones slightly more prominent; apical and basal impressions sparsely punctate; median and lateral punctures sparse, scattered; pubescence fine, dense, usually grayish and brownish mottled, three median tubercles with dark spots; prosternum finely pubescent, intercoxal process more than one-half as broad as coxal cavity; meso- and metasternum finely pubescent, small dark spots numerous at sides, mesosternal process broader than coxal cavity. Elytra a little less than twice as long as broad, tapering at apical one-fourth; basal gibbosities very shallow, impressions behind shallow; dark, tufted tubercles sparse, small, a larger one present on each side near base; pubescence fine, dense, usually mottled grayish and brownish, each side behind middle with an oblique grayish fascia, lateral margins dark to behind middle, lateral fasciae often extending partly onto disk, dark spots usually present behind pale fasciae, apical one-third usually brownish; apices truncate. Legs robust, finely mottled pubescent with small glabrous spots; tibiae dark biannulate; tarsi dark. Abdomen finely pubescent, pubescence interrupted by numerous small spots; last sternite shallowly emarginate at apex. Length, 8.5-13 mm.

Female. Form similar. Antennae a little longer than body. Abdomen with last sternite a little elongate, narrow, truncate at apex. Length, 10-14 mm.

Type locality. Of *planidorsus*, Louisiana; *lecontei*, Lake Worth, Florida; *crescenticus*, Crescent City, Florida.

Range. Maryland to Florida west to Louisiana.

Flight period. May to August.

Host plants. Betula, Cercis.

The five dorsal tubercles of the pronotum and all dark tarsi separate this species from other North American *Leptostylopsis*.

This species is variable in coloration. The postmedian pale fasciae of the elytra are often vague. The lateral dark vittae often extend partially onto the disk, and the pale fasciae are often dark-margined posteriorly. Occasionally the elytral disk is mostly grayish anteriorly.

Leptostylopsis albofasciatus (Fisher)

Leptostylus albofasciatus Fisher, 1926, Proc. U.S. Nat. Mus., 68(22):16; de Zayas, 1975, Rev. Fam. Ceram., p. 232, pl. 29, fig. d.
Leptostylopsis albofasciatus: Gilmour, 1963, Stud. Fauna Curaçao Carib. Is., 17:58.

Male. Form moderate-sized; integument reddish brown; pubescence short, dense, appressed, pale and dark brownish and grayish. Head with front about as long as broad, finely, sparsely punctate, densely clothed with mottled brownish pubescence; genae as long as lower eye lobes; upper eye lobes small, separated by diameter of antennal scape; antennae extending about three segments beyond body, segments brown annulate at bases and apices, basal segments with dark spots, third segment longer than first, fourth equal to first. Pronotum broader than long, sides with small tubercles behind middle; disk with seven calluses, median three more prominent; punctures along apical and basal impressions larger than discal ones; discal punctures fine, sparse, irregularly interspersed; pubescence dense, pale brownish, sides beneath lateral tubercles with a patch of dark brownish, disk with three longitudinal vittae, pair at sides of middle interrupted; prosternum thinly pubescent, intercoxal process more than half as broad as coxal cavity; meso- and metasternum pale pubescent, sides with numerous small spots, mesosternal process broader than coxal cavity. Elytra slightly more than 1-1/2 times as long as broad; basal gibbosities moderately elevated, impression behind moderate; disk with a few dark tufted tubercles along costae; pubescence dense, appressed, mostly light brownish, a dark patch present on each side of basal gibbosities, apical one-third with dark patches, a narrow whitish, zigzag fascia present behind middle, often W-shaped, suture whitish with dark spots, base often with narrow whitish markings; punctures rather fine, sparse; apices obliquely truncate. Legs robust, femora brownish and gray pubescent with small spots; tibiae dark biannulate; tarsi pale pubescent. Abdomen finely pubescent, sides densely spotted; last sternite emarginate at apex. Length, 7-11 mm.

Female. Form similar. Antennae slightly longer than body. Abdomen with last sternite a little elongate, narrow, subtruncate at apex. Length, 11 mm.

Type locality. Cayamas, Cuba.
Range. Southern Florida, Cuba.
Flight period. June.
Host plants. Rhizophora.

The narrow, whitish W-shaped fascia of the elytra make this species distinctive.

We have seen only two specimens from Adam Key, Florida, reared from red mangrove. Zayas (1975) states that *L. albofasciatus* is common in Cuba.

Leptostylopsis argentatus (Jacquelin du Val)

Amniscus argentatus Jacquelin du Val, 1857, in Sagra, Hist. Nat. Cuba, Ins., 7:273.
Leptostylus argentatus: Chevrolat, 1862, Ann. Soc. Entomol. Fr., (4)2:247; Horn, 1880, Trans. Amer. Entomol. Soc., 8:121, 123; Leng and Hamilton, 1896, Trans. Amer. Entomol. Soc., 23:117; Wolcott, 1948, J. Agr. Univ. Puerto Rico, 32:344; Cazier and Lacey, 1952, Amer. Mus. Nov., 1588:51; Gilmour, 1968, Stud. Fauna Curaçao Carib. Is., 25:163; de Zayas, 1975, Rev. Fam. Ceramb., p. 233, pl. 30, fig. a.
Leptostylopsis argentatus: Dillon, 1956, Ann. Entomol. Soc. Amer., 49:145; Morris, 1987, J. Entomol. Sci., 22:141.
Leptostylus taeniatus Casey, 1913, Memoirs on the Coleoptera, 4:306.

Male. Form moderate-sized; integument dark reddish brown; pubescence dense, appressed, grayish, pale and dark brown. Head with front as long as broad, sparsely punctate, densely clothed with grayish pubescence with dark spots interspersed; genae as long as lower eye lobes; upper eye lobes small, separated by about diameter of antennal scape; antennae extending three or four segments beyond elytra, segments narrowly dark annulate at apices and bases, basal segments dark spotted, third segment longer than first, fourth subequal to first. Pronotum broader than long, sides obtusely tuberculate behind middle; disk with seven calluses, median most prominent; punctures sparse, scattered, larger along basal and apical impressions; middle often with a narrow glabrous line extending from median callus to apical margin; pubescence grayish, base with a dark median spot, each side with a short, narrow, dark vitta at apical margin; prosternum gray pubescent, intercoxal process about as broad as coxal cavity; meso- and metasternum gray pubescent, spotted at sides, mesosternal process broader than coxal cavity. Elytra slightly more than 1-1/2 times as long as broad, tapering at apical one-fourth; basal gibbosities moderate, impression behind deep; disk with small tubercles along costae, more numerous behind middle; pubescence dense, appressed, mottled grayish and brownish, a short, dark, transverse vitta present behind middle, usually in the form of an inverted T; punctures fine, sparse; apices shallowly emarginate truncate. Legs robust, femora mottled grayish pubescent; tarsi dark biannulate; tarsi pale pubescent. Abdomen finely pubescent, spotted at sides; last sternite truncate to emarginate at apex. Length, 7-13 mm.

Female. Form similar. Antennae a little longer than body. Abdomen with last sternite a little elongate, apex narrow, emarginate. Length, 8-13 mm.

Type locality. Of *argentatus*, Cuba; *taeniatus*, Lake Worth, Florida.
Range. southern Florida to Georgia, West Indies.
Flight period. March to June (Florida).
Host plants. Zanthoxylum.

This species may be recognized by the short, dark vitta of the elytra and the usual presence of the glabrous median line on the pronotum. Florida specimens are usually grayish brown in overall aspect and occasionally have a pale fascia behind the short dark vitta.

Genus *Leptostylus* LeConte

Leptostylus LeConte, 1852, J. Acad. Nat. Sci. Philadelphia, 2:168; Thomson, 1860, Class. ceram., p. 11; Bates, 1863, Ann. Mag. Nat. Hist., (3)12:101; Thomson, 1864, Syst. ceram., p. 28; Lacordaire, 1872, Genera des coléoptères, 9(2):771; LeConte, 1873, Smithson. Misc. Coll., 11(265):338; Provancher, 1877, Pet. Fauna Entomol. Canada, 1:627; Horn, 1880, Trans. Amer. Entomol. Soc., 8:119 (part); LeConte and Horn, 1883, Smithson. Misc. Coll., 507:323; Leng and Hamilton, 1896, Trans. Amer. Entomol.

Soc., 23:116 (part); Wickham, 1897, Can. Entomol., 29:202, 207 (part); Blatchley, 1910, Coleoptera in Indiana, p. 1071 (part); Casey, 1913, Memoirs on the Coleoptera, 4:305 (part); Craighead, 1923, Can. Dept. Agr. Bull., (n.s.) 27:114 (part); Knull, 1946, Ohio Biol. Surv. Bull., 39:244; Dillon, 1956, Ann. Entomol. Soc. Amer., 49:141; Arnett, 1962, Beetles U.S., 103:872; Bayer and Shenefelt, 1969, Univ. Wisc. Res. Bull., 275:26; Rice and Enns, 1981, Trans. Mo. Acad. Sci., 15:98.

Leptostylis: Bradley , 1930, Man. Gen. Beetles, p. 246 (error).

Amniscus Haldeman, 1847, Trans. Amer. Philos. Soc., (2)10:46 (part).

Form moderate-sized, convex. Head with front almost plane, about as broad as long; mandibles feebly arcuate; genae slightly convergent, as long as lower eye lobes; eyes moderate-sized, deeply emarginate, upper lobes small, well separated above; antennal tubercles prominent, widely divergent; antennae eleven-segmented, a little longer than body in males, segments from fifth short, scape with two setae beneath near apex, third segment longer than scape, fourth shorter than third, shorter than or equal to scape. Pronotum broader than long, sides with obtuse tubercles just behind middle; disk with five prominent calluses, median basal one largest, two vague calluses present at sides above lateral pair; base and apex rather broadly impressed; prosternum narrow, intercoxal process lightly arcuate, more than 1-1/2 as broad as coxal cavities; mesosternum with intercoxal process abruptly delivous anteriorly, as broad as coxal cavities; metasternum with episternum narrow, subparallel. Elytra about twice as long as broad, strongly convex; basal gibbosities prominent, broad; disk costate, costae with rows of pubescent tufts, erect setae absent; apices narrowly, obliquely truncate to rounded. Legs stout; femora pedunculate; tarsi short, stout, first segment shorter than two following segments together. Abdomen normally segmented, last sternite longer than preceding segments.

Type species. Lamia aculifera Say (Thomson designation, 1864).

This genus may be recognized by the convex body form, prominent dorsal tubercles of the pronotum, relatively short antennae, width of the sternal processes, and by the basally gibbose, costate elytra with numerous tufted tubercles.

Three species are known from North America.

KEY TO THE NORTH AMERICAN SPECIES OF *LEPTOSTYLUS*

1   Metatrochanters not apically produced 2
        Metatrochanters apically produced into a sharp spine; elytra with pubescence mostly whitish, with dark vittae laterally behind humeri, medially behind middle, and a transverse band before apex. Length, 9-13 mm. Maryland to Florida and Texas . . . . *asperatus*

2   Antennae with fifth segment one-half as long as first; elytra with basal gibbosities moderate, apices

narrowly truncate; tarsi dark. Length, 6-14 mm.
Eastern North America to Arizona and Kansas and
northeastern Mexico .... *transversus*
Antennae with fifth segment two-thirds as long as
first; elytra with basal gibbosities very prominent,
apices rounded; tarsi pale. Length, 8-11 mm.
Texas to Venezuela .... *gibbulosus*

Leptostylus asperatus (Haldeman), new status
(Figure 25)

Amniscus albescens var. *asperatus* Haldeman, 1847, Trans. Amer. Philos. Soc., (2)10:46.
Leptostylus albescens ab. *asperatus*: Aurivillius, 1923, Coleop. Cat., 74:399.
Leptostylus transversus asperatus: Dillon, 1956, Ann. Entomol. Soc. Amer., 49:143; Gilmour, 1965, Cat. Lam. du Monde, 8:572.

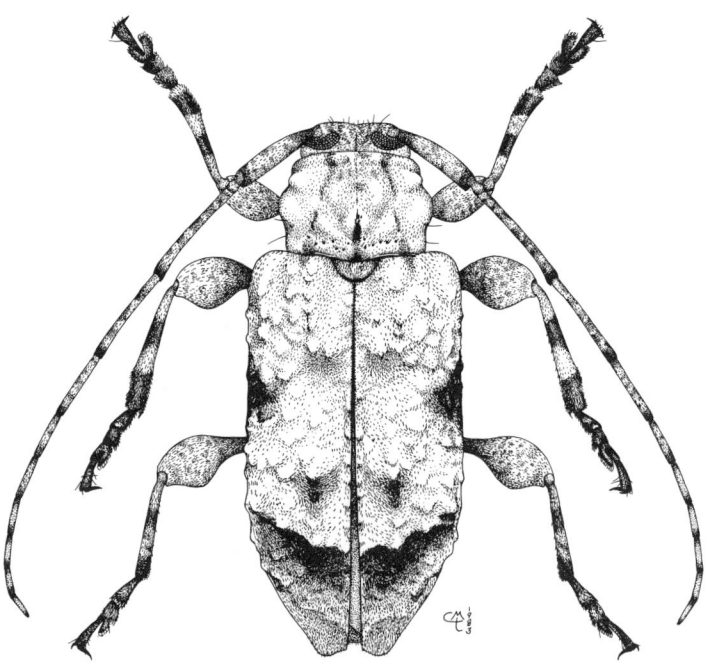

Figure 25. *Leptostylus asperatus* (Haldeman), male.

Leptostylus albescens Dillon, 1956 (not Haldeman, 1847), Ann. Entomol. Soc. Amer., 49:144.

Male. Form moderate-sized; integument piceous to rufopiceous, legs and antennae usually paler; pubescence dense, appressed, grayish, brownish and black, often green. Head with front about as long as broad, minutely, densely punctate, densely clothed with brownish and gray mottled pubescence; genae subequal in length to lower eye lobes; upper eye lobes separated by about diameter of antennal scape; antennae slightly longer than body, segments narrowly dark annulate at apices, scape densely gray pubescent on dorsal surface, segments three to five with pale pubescence interrupted by dark spots, third segment longer than scape, fourth shorter than third, subequal to scape, fifth two-thirds as long as scape, remaining segments short. Pronotum broader than long, sides with obtuse tubercles slightly behind middle; discal calluses strongly elevated; punctures scattered around calluses and on apical and basal impressions; pubescence dense, appressed, grayish with a median black spot on basal margin, sides often with greenish scale-like hairs; prosternum thinly pubescent, intercoxal process more than half as broad as coxal cavity; meso- and metasternum densely, minutely punctate, mottled gray pubescent with greenish tinges at sides, mesosternal process as broad as coxal cavity. Elytra about twice as long as broad, tapering at apical one-fourth; basal gibbosities moderately elevated; impressions behind gibbosities rather deep; disk with numerous tufted tubercles arranged down costae and suture; pubescence dense, appressed, mostly grayish, each side with a broad dark lateral vitta extending from base to behind middle and expanding slightly onto disk before middle, a short transverse vitta usually present behind scutellum, a short arcuate vitta present suturally behind middle, each side with an oblique vitta at apical one-third and another just before apex; sides and disk often tinged with green; punctures coarse, dense, obscured by pubescence; apices narrowly, obliquely truncate. Legs robust; femora pale pubescent near apices; tibiae biannulate with whitish pubescence; tarsi dark pubescent; metatrochanters produced into spines. Abdomen minutely, densely punctate, thinly pubescent; last sternite shallowly rounded at apex. Length, 9-13 mm.

Female. Form similar. Antennae about as long as body. Legs with femora less robust. Abdomen with last sternite about twice as long as fourth, shallowly emarginate at apex. Length, 10-13 mm.

Type locality. New Orleans.
Range. Maryland to Florida and Texas.
Flight period. April to August.
Host plants. Rhus, Quercus.

The distinctive projection of the metatrochanters will immediately identify this species. The dominantly whitish pubescence of the elytra with the distinctive dark markings are characteristic. The sides and disk of the elytra and sides of the sternum are often tinged with bright greenish scale-like hairs.

There has been some confusion concerning the identity of this species. The series of specimens at the Museum of Comparative Zoology, Harvard University, considered to represent *albescens* Haldeman consists of two species. The first specimen has a handwritten label "*Leptostylus aculiferus* Say - *albescens* Haldeman." The next specimen with a type label is *marginellus* Haldeman. The third specimen is *albescens* as defined by Dillon (1956), as are numbers 9 and 12. The twelfth specimen bears a label, "*asperatus* Dej. New Orleans."

Therefore, *albescens* and *marginellus* Haldeman are synonyms of *transversus* Gyllenhal and *asperatus* Haldeman is the correct name for *albescens* Dillon.

Leptostylus transversus (Gyllenhal)

Cerambyx tuberculatus Froelich, 1792 (not Degeer, 1775), Naturforscher, 26:138.
Leptostylus tuberculatus: Aurivillius, 1923, Coleop. Cat., 74:403; Knull, 1946, Ohio Biol. Surv. Bull., 39:245, pl. 20, fig. 78; Fattig, 1947, Emory Univ. Mus. Bull., 5:34.
Lamia transversa Gyllenhal, in Schönherr, 1817, Synon. Ins., 1, 3, append., p. 164.
Amniscus transversus: White, 1855, Cat. Coleop. Ins. British Mus., 8:391.
Leptostylus transversatus: LeConte, 1878, Proc. Amer. Philos., Soc., 17:414 (error).
Leptostylus transversus: Lacordaire, 1872, Genera des coléoptères, 9(2):772, fn.; Dillon, 1956, Ann. Entomol. Soc. Amer., 49:141; Bayer and Shenefelt, 1969, Univ. Wisc. Res. Bull., 275:28, fig. 36; Gosling and Gosling, 1977, Gr. Lakes Entomol., 10:26, fig. 146.
Leptostylus transversus transversus: Dillon, 1956, Ann. Entomol. Soc. Amer., 49:142; Gardiner, 1969, Can. Dept. Fish. For. Int. Rept., 0-14:81; Kirk, 1970, SC Agr. Exp. Sta. Tech. Bull., 1038:82; Turnbow and Franklin, 1980, J. Ga. Entomol. Soc., 15:344; Rice and Enns, 1981, Trans. Mo. Acad. Sci., 15:99; Gosling, 1984, Gr. Lakes Entomol., 17:72.
Lamia aculifera Say, 1823, J. Acad. Nat. Sci. Philadelphia, 3:329; LeConte, 1859, Compl. Writings T. Say, 2:186.
Amniscus aculiferus: Haldeman, 1847, Trans. Amer. Philos. Soc., (2)10:47; White, 1855, Cat. Coleop. Ins. British Mus., 8:392.
Leptostylus aculiferus: LeConte, 1852, J. Acad. Nat. Sci. Philadelphia, 2:168; Fitch, 1857, 3rd Rept. Nox. Ben. Ins. New York, p. 326, pl. 1, fig. 4; Knobel, 1895, Beetles of New England, p. 34, fig. 104; Beutenmuller, 1896, J. NY Entomol. Soc., 4:79; Felt, 1907, N.Y. St. Mus. Mem., 8:461; Smith, 1910, N.J. St. Mus. Rept., 1909:333; Blatchley, 1910, Coleoptera in Indiana, p. 1073, fig. 461; Craighead, 1923, Can. Dept. Agr. Bull., (n.s.) 27:115.
Leptostylus aculifera: Lacordaire, 1872, Genera des coléoptères, 9(2):772, fn.

Leptostylus aculifer: LeConte, 1873, Smithson. Misc. Coll., 11(264):232; Provancher, 1877, Pet. Fauna Entomol. Can., 1:627; Horn, 1883, Trans. Amer. Entomol. Soc., 8:121; Leng and Hamilton, 1896, Trans. Amer. Entomol. Soc., 23:117; Wickham, 1897, Can. Entomol., 29:208, fig. 33; Lugger, 1899, Minn. Agr. Exp. Sta. Bull., 66:208, fig. 130; Champlain and Knull, 1925, Entomol. News, 36:140; Leonard, 1928, Cornell Agr. Exp. Sta. Mem., 101:451; Beaulne, 1932, Nat. Can., 59:219 (hosts); Hoffmann, 1940, Bull. Brooklyn Entomol. Soc., 35:59.

Amniscus marginellus Haldeman, 1847, Trans. Amer. Philos. Soc., (2)10:47; Haldeman, 1847, Proc. Amer. Philos. Soc., 4:372.

Leptostylus aculifera var. *marginellus*: Lacordaire, 1872, Genera des coléoptères, 9(2):772, fn.

Amniscus albescens Haldeman, 1847, Trans. Amer. Philos. Soc., (2)10:46.

Leptostylus aculifera var. *albescens*: Lacordaire, 1872, Genera des coléoptères, 9(2):722, fn.

Leptostylus albescens: Casey, 1913, Memoirs on the Coleoptera, 4:307; Loding, 1945, Geol. Surv. Ala. Mon., 11:122; Fattig, 1947, Emory Univ. Mus. Bull., 5:34; Kirk, 1969, SC Agr. Exp. Sta. Tech. Bull., 1033:86; Turnbow and Hovore, 1979, Entomol. News, 90:224; Waters and Hyche, 1984, Coleop. Bull., 38:284; Furth, 1985, Conn. Acad. Arts Sci., 46:192.

Leptostylus divisus Casey, 1913, Memoirs on the Coleoptera, 4:306; Aurivillius, 1923, Coleop. Cat., 74:401.

Leptostylus transversus dakotensis Dillon, 1956, Ann. Entomol. Soc. Amer., 49:143; Gilmour, 1965, Cat. Lam. du Monde, 8:572. New synonymy.

Leptostylus transversus dietrichi Dillon, 1956, Ann. Entomol. Soc. Amer., 49:144; Gilmour, 1965, Cat. Lam. du Monde, 8:572; Rice, 1985, J. NY Entomol. Soc., 93:1224. New synonymy.

Leptostylus transversus floridellus Dillon, 1956, Ann. Entomol. Soc. Amer., 49:144; Gilmour, 1965, Cat. Lam. du Monde, 8:573; Turnbow and Hovore, 1979, Entomol. News, 90:224. New synonymy.

Male. Form small to moderate-sized; integument piceous and reddish brown; pubescence dense, appressed, grayish, brownish, black, and often tinged with green. Head with front about as long as broad, densely clothed with mottled brownish and grayish pubescence, minutely, densely punctate; genae subequal in length to lower eye lobes; upper eye lobes separated by more than diameter of antennal scape; vertex often with two median dark vittae; antennae about as long as body or slightly shorter, segments brownish annulate at apices, outer segments also brown annulate at bases, segments three and four with dark spots, third segment a little longer than scape, fourth a little shorter than scape, outer segments gradually decreasing, eleventh short. Pronotum broader than long, sides obtusely, rather vaguely tuberculate; discal calluses prominently elevated, median prebasal one largest; punctures scattered around calluses and along apical and basal impressions; pubescence dense, appressed, obscuring surface except occasionally for median callus and a glabrous line extending up to apical margin, color grayish to brownish, usually with a short, linear, dark spot at each side of middle at apex and base, surface occasionally tinged with greenish, sides usually with several long, erect, dark hairs; prosternum thinly pubescent, intercoxal process more than half as broad as coxal cavity; meso- and metasternum densely grayish brown pubescent, mottled with spots of sparse pubescence, mesosternal process about as broad as coxal cavities. Elytra about twice as long as broad, tapering and declivous at apical one-fourth; basal gibbosities moderately elevated, impressions behind gibbosities deep; disk with numerous tufted tubercles linearly arranged down costae and suture; pubescence dense, appressed, mostly brownish, usually with a transverse band of grayish pubescence behind middle, each side with a dark brownish vitta extending from humeri to about middle, where it expands slightly toward disk, usually with a darker band behind pale fascia, often lightly, irregularly tinged with greenish, two pale spots often present anterior to pale fascia; punctures coarse, dense, usually obscured by pubescence; apices narrowly subtruncate. Legs robust; femora brownish and gray mottled pubescent; tibiae biannulate with dark brown. Abdomen densely, finely punctate, moderately densely mottled pubescent; last sternite subtruncate at apex, often lightly notched at middle. Length, 6-12 mm.

Female. Form similar. Antennae usually shorter than body. Legs with femora less robust. Abdomen with last sternite about twice as long as fourth, subtruncate at apex. Length, 7-14 mm.

Type locality. Of *transversus*, America boreali; *tuberculatus*, Virginia; *aculifera*, Missouri; *marginellus*, not indicated (North America); *divisus*, Texas; *albescens*, not indicated (North America); *dakotensis*, Elk Point, South Dakota; *dietrichi*, Lucedale, Mississippi; *floridellus*, Biscayne Bay, Florida.

Range. Eastern North America to Florida and northeastern Mexico, west to Arizona, Kansas and South Dakota.

Flight period. Year round, spring to summer in the northern parts of the range.

Host plants. Aesculus, Amelanchier, Betula, Bursera, Carya, Castanea, Cercis, Cornus, Diospyros, Juglans, Juniperus, Liquidambar, Liriodendron, Maclura, Malus, Pinus, Platanus, Prunus, Quercus, Rhus, Sesbania, Ulmus, Zanthoxylum.

The usual presence of the transverse pale fascia behind the middle of the elytra distinguish this species. The apical one-fourth of the elytra either is darker or a brown band occurs behind the pale fascia. Most specimens have slight greenish tinges scattered over the dorsal surface.

There appears to be little justification for recognizing various subspecies. The more northern forms tend to be more brownish in color with the elytra being similar both basally and apically. Some of the Texas examples tend to be more grayish. In Florida the colors of the pubescence of the elytra are more contrasting with the pale fascia quite distinct.

The subspecies recognized by Dillon (1956) all intergrade and the differences expressed by populations from north to south appear to be clinal. The Florida population could be considered distinct, but the differences diminish westward toward Texas.

Leptostylus gibbulosus Bates
(Figure 26)

Leptostylus gibbulosus Bates, 1874, Trans. Entomol. Soc. London, 1874:230, fn.; Bates, 1885, Biol. Centr.-Amer., Coleop., 5:385; Vogt, 1949, Pan-Pac. Entomol., 25:180 (habits).
Leptostylus gibbulosus gibbulosus: Dillon, 1962, Coleop. Bull., 16:31.
Leptostylus vogti Dillon, 1956, Ann. Entomol. Soc. Amer., 49:141; Dillon, 1962, Coleop. Bull., 16:30. New synonymy.
Leptostylus gibbulosus vogti: Dillon, 1962, Coleop. Bull., 16:31; Hovore et al., 1987, Proc. Calif. Acad. Sci., 44:317.

Male. Form moderate-sized, strongly convex; integument greenish and brownish; pubescence dense, appressed, gray, brownish, and black. Head with front as long as broad, punctures obscured by dense, appressed, variegated brownish and gray pubescence; genae convergent, about as long as lower eye lobes; upper eye lobes separated by about diameter of antennal scape; vertex deeply impressed behind antennal tubercles; antennae slightly longer than body, segments three to eight narrowly brown annulate at apices, segments nine to eleven narrowly brown annulate at apices and bases, basal segments densely grayish pubescent, pubescence mottled with darker spots, segments from fifth uniformly pubescent, third segment longer than scape, fourth subequal to scape, remaining segments gradually decreasing in length, eleventh short. Pronotum broader than long, sides vaguely, obtusely tuberculate at middle; discal calluses strongly elevated, median prebasal onemost prominent; apex rather narrowly impressed, impression interrupted at middle; base more broadly impressed; punctures sparse, scattered around calluses, basal impression with a row of deep punctures; pubescence dense, appressed, variegated grayish and brownish; prosternum densely gray pubescent, intercoxal process about two-thirds as broad as coxal cavity; mesosternum densely pubescent, intercoxal process a little more than 2/3 as broad as coxal cavity; metasternum densely pubescent with small glabrous spots interspersed. Elytra a little more than 1-1/2 times as long as broad, strongly convex, tapering near apex; basal gibbosities prominent, crested by a row of dark tufts; impressions behind gibbosities deep; costae distinct, uniting before apex, two inner pairs with tufted tubercles; punctures very coarse, contiguous, becoming finer at apex; pubescence dense, appressed, brownish on gibbosities and two inner costae, median portion brownish suffused to behind middle, apical one-third with a median whitish, triangular vitta bordered by brownish, remainder of surface grayish pubescent, suture with black spots behind middle; apices narrowly rounded. Legs robust; femora mottled pubescent; tibiae brownish annulate on dorsal surfaces at middle and near apices; tarsi, except extreme apices, pale. Abdomen densely grayish pubescent; last sternite glabrous, dark brownish, pale along apical margin, apex subtruncate, fringed. Length, 8-10 mm.

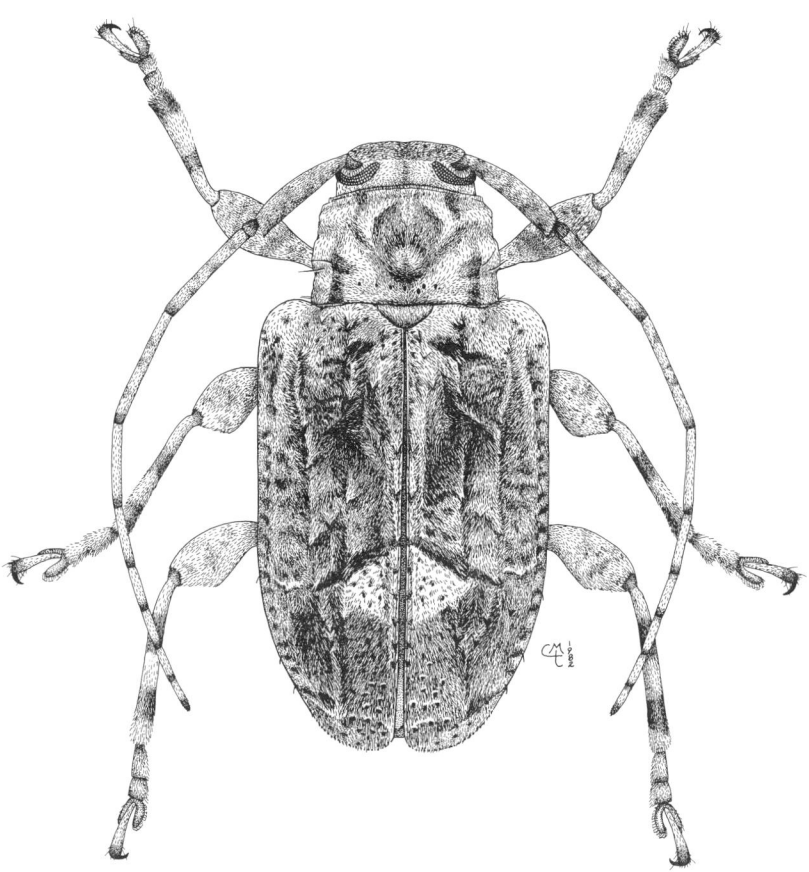

Figure 26. *Leptostylus gibbulosus* Bates, female.

Female. Form similar. Abdomen with last sternite about as long as two preceding segments together, pale area triangular, apex subtruncate, fringed. Length, 8-11 mm.

Type locality. Of *gibbulosus*, Venezuela; *vogti*, S.W. Hidalgo Co., Texas.
Range. Texas to Venezuela.
Flight period. May to November.
Host plants. Sapindus (including *drummondii, saponaria*).
Parasites. Eurytoma sp. (Chalcidae), *Heterospilus* sp. (Braconidae).

This species may be recognized by the partially greenish integument, prominent basal gibbosities, and the color pattern of the elytra. There is little variation in the material we have examined (Texas to Costa Rica). Since specimens from intermediate areas are lacking, we have not recognized subspecies at this time.

Larvae of *L. gibbulosus* feed in the seed of *Sapindus*. Adults are often attracted to light.

TRIBE CYRTININI THOMSON

Thomson, 1864, Syst. ceramb., p. 41 (Cyrtinitae).
Lacordaire, 1872, Genera des coléoptères, 9(2): 818 (Cyrtinides).
LeConte, 1873, Smithson. Misc. Coll., 11(265): 333.
LeConte and Horn, 1883, Smithson. Misc. Coll., 507: 315, 318.
Leng and Hamilton, 1896, Trans. Amer. Entomol. Soc., 23: 107.
Blatchley, 1910, Coleoptera in Indiana, p. 1062.
Aurivillius, 1917, Ark. Zool., 10(23): 48.
Bradley, 1930, Man. Genera Beetles, p. 243.
Knull, 1946, Ohio Biol. Surv. Bull., 39: 232.
Craighead, 1950, USDA Misc. Pub., 657: 242.
Howden, 1959, Can. Entomol., 91: 372.
Arnett, 1960, Beetles of U.S., 103: 869, 888.
Dillon and Dillon, 1961, Man. Beetles East. North America, p. 646.
de Zayas, 1975, Rev. Fam. Ceramb., p. 283.
Villiers, 1980, Ann. Soc. Entomol. Fr., (n.s.) 16: 115.
Rice and Enns, 1981, Trans. Mo. Acad. Sci., 15: 90.

Body small, convex. Head not retractile; front broad; vertex flat or feebly convex between antennae; eyes small, divided; antennae slender, ten or eleven-segmented, scape cylindrical, elongate, slightly enlarging toward apex. Pronotum cylindrical, sides usually unarmed; base constricted; disk convex; prosternum with intercoxal process rather narrow, coxal cavities closed behind, rounded externally; mesosternum with coxal cavities closed externally. Elytra usually depressed and flattened basally, usually with two basal tubercles; suture occasionally fused. Legs short; femora strongly clavate; middle tibiae sulcate; tarsi slender, claws divaricate. Abdomen normally segmented.

The small, ant-like form, convex body, and basal tubercles of the elytra make this tribe easy to recognize among the North American Lamiinae. A single genus occurs in our fauna.

Genus *Cyrtinus* LeConte

Cyrtinus LeConte, 1852, J. Acad. Nat. Sci. Philadelphia, 2: 166; Thomson, 1864, Syst. ceramb., p. 41; Lacordaire, 1872, Genera des coléoptères, 9(2): 819; LeConte, 1873; Smithson. Misc. Coll., 11(265): 333; Horn, 1886, Trans. Amer. Entomol. Soc., 13: xi (synonymy); Wickham, 1897, Can.

Entomol., 29: 202, 204; Blatchley, 1910, Coleoptera in Indiana, p. 1062; Aurivillius, 1917, Ark. Zool., 10(23): 48; Craighead, 1923, Can. Dept. Agr. Bull., (n.s.) 27: 124; Bradley, 1930, Man. Genera Beetles, p. 243; Knull, 1946, Ohio Biol. Surv. Bull., 39: 232; Arnett, 1960, Beetles U.S., 103: 888; Dillon and Dillon, 1961, Man. Beetles East. North America, p. 646; de Zayas, 1975, Rev. fam. Cerambycidae, p. 283; Marinoni, 1977, Dusenia, 10:43; Villiers, 1980, Ann. Soc. Entomol. France, (n.s.) 16: 115.

Myrmolamia Bates, 1885, Biol. Centrali-Amer., Coleop., 5: 363; Horn, 1886, Trans. Amer. Entomol. Soc., 13: xi. (Type species: *Myrmolamia opacicollis* Bates, by present designation.)

Form small, ovoid. Head quadrate, front convex; vertex broad, feebly impressed between antennae; center line extending onto neck; mandibles almost straight, acute at apices; eyes small, finely faceted, divided, lower lobes much shorter than genae; antennae slender, longer than body in males, as long as body in females, segments with at least one long, depressed seta at apices on inside edge, scape longer than third segment, fourth and fifth equal to third. Pronotum cylindrical, apex broader than base; sides unarmed; disk convex; base broadly and deeply impressed transversely; prosternum narrow, narrowly impressed anteriorly, intercoxal process almost plane; meso- and metasternum short. Elytra about twice as long as basal width, broader at middle and tapering toward apex; disk convex, base impressed, often with a large, acute tubercle at each side; wings occasionally absent; apices rounded. Legs short; femora strongly clavate; tibiae slender. Abdomen normally segmented.

Type species. Clytus pygmaeus Haldeman (monobasic).

This genus is characterized by the small ant-like form, ovoid body, and prominent, acute basal tubercles of the elytra.

Two species are known from North America.

KEY TO THE NORTH AMERICAN SPECIES OF *CYRTINUS*

Elytra with basal tubercles arising at basal margins, suture fused, wings absent. Form strongly ovoid, integument dark reddish brown to piceous. Length, 2.3-3.5 mm. Western Texas and northern Mexico ...... *beckeri*
Elytra with basal tubercles arising behind scutellum, wings present. Form more subparallel, integument reddish brown. Length, 2.3-3.5 mm. Eastern North America .... *pygmaeus*

Cyrtinus beckeri Howden

Cyrtinus beckeri Howden, 1960, Can. Entomol., 92: 175, fig. 4; Rice et al., 1985, Coleop. Bull., 39: 23 (habits).

Male. Form small, ovoid; integument dark reddish brown to piceous. Head about as long as broad; front micropunctate with larger punctures interspersed, punctures bearing long, erect setae; vertex micropunctate; antennae slightly longer than body, segments usually darker at apices, scape extending to middle of pronotum, pubescence short, suberect, moderately dense, each segment from second with at least two very long, suberect hairs. Pronotum longer than broad; base broadly impressed; disk strongly convex, densely micropunctate except for vague median callus; punctures sparse, each bearing a short, depressed seta; prosternum sparsely pubescent; meso- and metasternum finely punctate, sparsely pubescent. Elytra fused, about 1.7 times as long as broad, sides expanding from humeri to middle, than narrowing to apices; basal tubercles arising from basal margin, large, acute, directed posteriorly, bearing two-three long setae; impressed area behind tubercles coarsely punctate; disk inflated, sparsely punctate, each puncture bearing a long suberect seta; each elytron with a patch of appressed, white pubescence along suture behind tubercle and extending to basal one-fourth, and an oblique, white pubescent band from margin behind humerus to middle of disk; apices broadly rounded. Legs short, sparsely pubescent; front tibiae with an internal sulcus. Abdomen shining, sparsely punctate and pubescent; last sternite narrowly rounded at apex. Length, 2.3-3.0 mm.

Female. Form similar. Antennae about as long as body. Abdomen with last sternite broadly rounded at apex. Length, 2.5-3.5 mm.

Type locality. Pine Canyon, Chisos Mts., Big Bend National Park, Texas.
Range. Western Texas and northeastern Mexico.
Flight period. May-October.
Host plants. Acer grandidentatum.

This species may be separated from *C. pygmaeus* by the more ovoid body form and darker integument. Additionally, the dorsal tubercles of the elytra arise from the basal margin.

Most known specimens have been beaten from *Acer grandidentatum*, and this plant is assumed to be the host.

Cyrtinus pygmaeus (Haldeman)
(Figure 27)

Clytus pygmaeus Haldeman, 1847, Trans. Amer. Philos. Soc., (2)10: 42.
Cyrtinus pygmaeus: LeConte, 1852, J. Acad. Nat. Sci. Philadelphia, 2: 166; LeConte, 1873, Smithson. Misc. Coll., 11(265): 333; Chittenden, 1894, Proc. Entomol. Soc. Wash., 3: 99 (habits); Leng and Hamilton, 1896, Trans. Amer. Entomol. Soc., 23: 107; Wickham, 1897, Can. Entomol., 29: 204; Blatchley, 1910, Coleoptera in Indiana, p. 1062; Smith, 1910, N.J. Sta. Mus. Rept., 1909: 332; Craighead, 1923, Can. Dept. Agr. Bull., (n.s.) 27: 124, pl. 12, fig. 9, pl. 24, fig. 1, pl. 42 (larva); Leonard, 1928, Cornell Agr. Exp. Sta. Mem., 101: 449; Beaulne, 1932, Nat. Can., 59: 203 (hosts); Knull, 1946, Ohio Biol. Surv. Bull., 39: 232, pl. 19, fig. 77; Fattig, 1947,

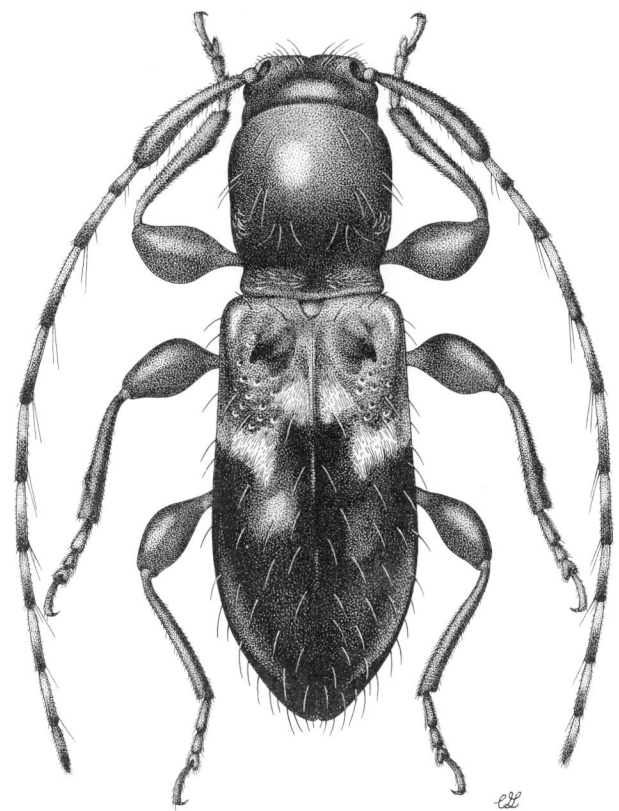

Figure 27. *Cyrtinus pygmaeus* (Haldeman), male.

Emory Univ. Mus. Bull., 5: 30; Vogt, 1949, Pan-Pac. Entomol., 25: 179; Howden, 1959, Can. Entomol., 91: 272; Howden, 1960, Can. Entomol., 92: 176, fig. 3; Arnett, 1960, Beetles U.S., 103: 869, 888; Dillon and Dillon, 1961, Man. Beetles East. North America, p. 646, pl. 64, no. 7; Headstrom, 1977, Beetles of America, p. 380; Rice and Enns, 1981, Trans. Mo. Acad. Sci., 15: 102; Hovore et al., 1987, Proc. Calif. Acad. Sci., 44: 320.

Male. Form small, cylindrical; integument reddish brown, antennae often pale, annulate. Head about as long as broad; front micropunctate with large punctures irregularly interspersed; pubescence long and erect and short and depressed; vertex micropunctate, large punctures sparse; antennae longer than body, segments pale basally; scape extending to about apical one-third of pronotum, pubescence moderate, fine, apical setae long. Pronotum longer than broad; base broadly, deeply impressed; disk strongly convex, vaguely, sparsely micropunctate, larger punctures sparse; pubescence sparse, long and suberect and short and depressed; prosternum sparsely pubescent; meso- and metasternum with a patch of white, appressed pubescence at sides. Elytra about twice as long as basal width, sides moderately expanding to middle and tapering toward apices; wings present; basal tubercles large, tufted, arising from behind basal margin behind scutellum; impressed area coarsely, irregularly punctate; pubescence sparse, long, erect, a short, transverse patch of white, appressed pubescence present suturally behind tubercles and an oblique, pubescent band on each side from lateral margin onto disk but not extending to suture; apices rounded. Legs short, sparsely pubescent; front tibiae lightly sulcate internally. Abdomen sparsely punctate and pubescent; last sternite narrowly rounded at apex. Length, 2.3-3.3 mm.

Female. Form similar. Antennae slightly longer than body. Abdomen with last sternite broadly rounded at apex. Length, 2.5-3.5 mm.

Type locality. Pennsylvania.

Range. Eastern North America to Texas.

Flight period. March to July.

Host plants. Acer, Carya, Cercis, Cornus, Juglans, Liriodendron, Quercus, Robinia, Sapindus.

The narrower body form and paler color separate this species from *C. beckeri.* Also the dorsal elytral tubercles are situated behind the scutellum rather than arising from the basal margin.

TRIBE SAPERDINI MULSANT

Mulsant, 1839, Hist. Nat. Coleop. Fr., Longicornes, p. 181 (Saperdaires).
Thomson, 1864, Syst. ceramb., p. 114 (Saperditae).
Lacordaire, 1872, Genera des coléoptères, 9(2): 831 (Saperdides).
LeConte, 1873, Smithson. Misc. Coll., 11(265): 345.
LeConte and Horn, 1883, Smithson. Misc. Coll., 507: 315, 331.
Harrington, 1899, Ottawa Nat., 13: 62.
Blatchley, 1910, Coleoptera in Indiana, p. 1084.
Bradley, 1930, Man. Genera Beetles, p. 246.
Chagnon, 1933-40, Coleop. Prov. Quebec, p. 275.
Doane et al., 1936, For. Ins., p. 191.
Knull, 1946, Ohio Biol. Surv. Bull., 39: 267.
Breuning, 1952, Entomol. Arb. Mus. Frey, 3: 107.
Duffy, 1953, Mon. British Timber Beetles, p. 281 (larvae).
Dillon and Dillon, 1961, Man. Beetles East. North America, p. 646.
Arnett, 1962, Beetles U.S., 103: 873.
Chagnon and Robert, 1962, Prin. Coleop. Prov. Quebec, p. 275.
Bayer and Shenefelt, 1969, Univ. Wisc. Res. Bull., 275: 30.
Hatch, 1971, Univ. Wash. Pub. Biol., 16: 154.
Rice and Enns, 1981, Trans. Mo. Acad. Sci., 15: 91.

Form small to large. Head with front usually quadrate, plane; eyes finely faceted, deeply emarginate; palpi short, apical segments not acute; antennal tubercles prominent to depressed; antennae slender, about as long as body in males, shorter in females, scape usually slender, subcylindrical, basal segments with a few, short, suberect hairs beneath. Pronotum cylindrical, sides unarmed; prosternum narrow, coxae prominent, intercoxal process narrow, expanded at apex, coxal cavities usually closed behind; mesosternum with intercoxal process gradually arcuate, coxal cavities open to epimeron; metasternum with episternum broad anteriorly, narrowing posteriorly. Elytra lacking carinae; apices usually rounded, occasionally produced. Legs moderately long; femora linear; middle tibiae usually with a dorsal sinus; tarsi short, claws divaricate. Abdomen normally segmented.

In our fauna this tribe may be recognized by the divaricate tarsal claws, barely fringed antennae, and by the broad, posteriorly narrowing metepisternum.

The group is best represented in the Holarctic regions, but also in the Oriental. The single genus occurring in our fauna contains several species of economic importance.

Genus *Saperda* Fabricius

Saperda Fabricius, 1775, Syst. Entomol., p. 184; Fabricius, 1781, Species Ins., 1: 230; Fabricius, 1787, Mantissa Ins., 1: 147; Fabricius, 1792, Entomol. Syst., 1: 307; Fabricius, 1801, Syst. Eleuth., 2: 317; Audinet-Serville, 1835, Ann. Soc. Entomol. Fr., 4: 45; Mulsant, 1839, Hist. Nat. Coleop. Fr., Longicornes, p. 185; Westwood, 1840, Intr. Ins., 2, synopsis, p. 41; Haldeman, 1847, Trans. Amer. Philos. Soc., 2(10): 55; LeConte, 1852, J. Acad. Nat. Sci. Philadelphia, 2: 161; Emmons, 1854, Agr. N.Y., 5: 119; Thomson, 1864, Syst. ceramb., p. 115; Lacordaire, 1872, Genera des coléoptères, 9(2): 832; LeConte, 1873, Smithson. Misc. Coll., 11(265): 346; Provancher, 1877, Pet. Fauna Entomol. Canada, 1: 632; LeConte and Horn, 1883, Smithson. Misc. Coll., 507: 331; Leng and Hamilton, 1896, Trans. Amer. Entomol. Soc., 23: 146; Wickham, 1897, Can. Entomol., 29: 203; Wickham, 1898, Can. Entomol., 29: 203; Felt and Joutel, 1904, N.Y. St. Mus. Bull., 74: 4; Blatchley, 1910, Coleoptera in Indiana, p. 1084; Casey, 1913, Memoirs on the Coleoptera, 4: 358; Craighead, 1923, Can. Dept. Agr. Bull., (n.s.) 27: 127; Mutcher and Weiss 1923, N.J. Dept. Agr. Circ., 58: 19; Bradley, 1930, Man. Genera Beetles, p. 247; Chagnon, 1933-40, Coleop. Prov. Quebec, p. 276; Doane et al., 1936, For. Ins., p. 191; Knull, 1946, Ohio Biol. Surv. Bull., 39: 267; Craighead, 1950, USDA Misc. Pub., 657: 263; Breuning, 1952, Entomol. Arb. Mus. Frey, 3: 141; Breuning, 1960, Frust. Entomol., 3: 14; Dillon and Dillon, 1961, Man. Beetles East. North America, p. 646; Arnett, 1962, Beetles U.S., 103: 873; Chagnon and Robert, 1962, Prin. Coleop. Prov. Quebec, p. 276; Abdullah and Abdullah, 1966, Proc. Roy. Entomol. Soc., (B) 35: 87; Bayer and Shenefelt, 1969, Univ. Wisc. Res. Bull., 275: 30; Hatch, 1971, Univ. Wash. Publ. Biol., 16: 154; Baker, 1972, USDA Misc. Pub., 1175: 184; Headstrom, 1977, Beetles of America, p. 380; Marinoni, 1977, Dusenia, 10: 49; Rice and Enns, 1981, Trans. Mo. Acad. Sci., 15: 102; Drooz, 1985, USDA For. Serv. Misc. Publ., 1426: 295.

Saperda (*Saperda*): Breuning, 1952, Entmol. Arb. Mus. Frey, 3: 142; Gilmour, 1965, Cat. Lam. du Monde, 8: 668.

Anaerea Mulsant, 1839, Hist. Nat. Coleop. Fr., Longicornes, p. 184; Haldeman, 1847, Trans. Amer. Philos. Soc., (2)10: 55; Marinoni, 1977, Dusenia, 10: 49. (Type species: *Cerambyx carcharis* Linnaeus, monobasic).

Saperda (*Anaerea*): Breuning, 1952, Entomol. Arb. Mus. Frey, 3: 142; Gilmour, 1965, Cat. Lam. du Monde, 8: 667.

Compsidia Mulsant, 1839, Hist. Nat. Coleop. Fr. Longicornes, p. 182; Haldeman, 1847, Trans. Amer. Philos. Soc., (2)10: 55. (Type species: *Cerambyx populneus* Linnaeus, monobasic).

Saperda (*Argalia*) Mulsant, 1863, Hist. Nat. Coleop. Fr., Longicornes, p. 381.

Form small to large, subparallel to slightly tapering. Head with front convex to plane, usually broader than long; eyes deeply emarginate, upper lobes small; genae shorter than lower eye lobes; antennae slender, about as long as body in males, shorter in females, basal segments sparsely fringed with fairly short, suberect hairs, scape slender, cylindrical, third segment longer than first, fourth subequal to or slightly longer than first. Pronotum usually broader than long, sides rounded; prosternum narrow, intercoxal process narrow, expanded apically, coxal cavities usually closed behind; mesosternum with coxal cavities open to epimeron; metasternum with episternum broad, tapering posteriorly. Legs moderately long; tarsi short, claws divaricate, usually with a process on anterior and/or middle pair in males. Abdomen normally segmented.

Type species. Cerambyx carcharias Linnaeus (Westwood designation, 1840).

Historically, *Saperda* has been subject to various definitions. Since the group is basically Holarctic, most of the proposed classifications have primarily involved the Palearctic species.

Subsequent to the proposal of the genus *Saperda* by Fabricius in 1775, Mulsant (1839) proposed the genera *Compsidea* and *Anaerea* to be included with *Saperda* in his group "Saperdaires". *Amilia* and *Argalia* were added to this group by Mulsant in 1863. *Argalia* was listed in a footnote on page 381 for species with non-annulated antennae, presumably including *Saperda perforata* (Dallas). Type species for most of the proposed generic names have been fixed by monotypy or subsequent designation, but the type species of *Argalia* is uncertain. Pic (1910) listed *Argalia* as a subgenus of *Saperda* and included two Siberian species. This usage was followed by various Palearctic workers but no type species was designated. This name does not apply to our fauna and we make no attempt to resolve the problem of a valid type species for *Argalia*. North American workers have mostly ignored these names and have preferred to include all of our species in one genus: *Saperda*. We concur in this usage.

Since a number of the North American species of *Saperda* have some economic importance as pests of various fruit and shade trees, there exists a large volume of literature (much of it early) dealing with life histories, injury to hosts, and control. We have made no attempt to include all of such references, since they bear little relevance to the classification of the species. Felt and Joutel (1904) list most of the titles published prior to that date.

This genus is easily recognizable by the divaricate tarsal claws and broad metepisternum. Most species occur in the eastern half of North America and all breed in living trees. Fifteen species are known.

KEY TO THE NORTH AMERICAN SPECIES OF *SAPERDA*

1   Elytra with apices spined or narrowing; head deeply impressed between antennal tubercles; antennae annulate 2
        Elytra with apices rounded to subtruncate; head not or barely impressed between antennal tubercles; antennae usually concolorous 6

2(1)	Elytra with apices spined or acuminate	3
	Elytra with apices narrowed	4
3(2)	Elytra with apices strongly acuminate; color brownish with oblique fasciae on elytra. Length, 14-19 mm. Eastern North America	*obliqua*
	Elytra with apices moderately spined at suture; color grayish to yellowish brown, variegated. Length, 18-31 mm. California to eastern North America	*calcarata*
4(2)	Head with front as long as or longer than broad; elytra sparsely to moderately sparsely punctate, punctures well separated	5
	Head with front broader than long; elytra coarsely, confluently punctate; smaller species. Length, 7-13 mm. Northeastern North America to Pacific Coast	*populnea*
5(4)	Elytra with punctures large, crater-like, irregularly dispersed over surface; antennae with scape slender, grayish pubescent. Length, 10.5-20 mm. Pacific Coast to Utah	*horni*
	Elytra with punctures moderately large, not crater-like, moderately dense; antennae with scape broad, dark pubescent. Length, 11-17 mm. Northeastern North America to North Dakota	*mutica*
6(1)	Elytra with longitudinal and/or oblique vittae	7
	Elytra concolorous or with pubescent white or dark	11
7(6)	Elytra with discal, sutural, or lateral vittae only	8
	Elytra with oblique or transverse vittae	10
8(7)	Elytra with sutural and/or lateral vittae	9
	Elytra with white vittae down middle; front of head and underside densely white pubescent. Length, 13-21 mm. Eastern North America to Saskatchewan	*candida*
9(8)	Elytra with pale sutural and lateral vittae; pronotum with four dark spots, front of head with a median dark spot. Length, 8-12 mm. Eastern North America to Kansas	*puncticollis*
	Elytra with orange lateral vittae; pronotum with lateral vittae only. Length, 9-14 mm. Eastern North America to Texas and North Dakota	*lateralis*
10(7)	Elytra lacking submarginal carinae, basal discal vittae oblique. Length, 10-16 mm. Northeastern North America	*imitans*
	Elytra with submarginal carinae; basal vittae transverse, then oblique near suture, dark spots usually present on each side of basal bands and behind apical bands. Length, 10-17 mm. Eastern North America to Kansas	*tridentata*

11(6)	Elytra with pubescent pale or dark spots	12
	Elytra concolorous	15
12(11)	Elytra with white spots	14
	Elytra usually with six black spots, two median before middle, two near middle, and two median behind middle; pubescence dense, uniformly olive brown. Length, 12-23 mm. Eastern North America	*vestita*
13(12)	Pronotum with longitudinal white vittae at sides	14
	Pronotum uniformly brownish pubescent; elytra with a transverse, irregular pale vitta across middle, occasionally with spots before and behind vitta. Length, 9-17 mm. Eastern North America	*discoidea*
14(13)	Elytra with large, white, irregularly rectangular spots near middle and two smaller spots before apex, pronotal vittae not extending onto elytra at base. Length, 15-21 mm. Northeastern North America	*cretata*
	Elytra with two narrow, elongate spots near suture at middle and two smaller spots near suture near apex, pronotal vittae extending onto base of elytra. Length, 9-15 mm. Northwestern North America	*fayi*
15(11)	Antennae uniformly pubescent; pronotum with two broad, brownish vittae. Legs with middle and front tarsal claws with a large process. Length, 9-13 mm. Eastern North America	*discoidea*
	Antennae with segments from third dark annulate at apices; pronotum grayish pubescent, vittae if present, grayish. Legs with tarsal claws lacking a process. Length, 8-13 mm. Eastern North America to Idaho and Arizona	*inornata*

Saperda obliqua Say

Saperda obliqua Say, 1827, J. Acad. Nat. Sci. Philadelphia, 5: 274; LeConte, 1852, J. Acad. Nat. Sci. Philadelphia, 2: 162; LeConte, 1859, Compl. Writings T. Say, 2: 332; Lacordaire, 1872, Genera des coléoptères, 9(2): 834, fn; LeConte, 1873, Smithson. Misc. Coll., 11(264): 238; LeConte, 1873, Smithson. Misc. Coll., 11(265): 346; Provancher, 1877, Pet. Fauna Entomol. Can., 1: 633; LeConte and Horn, 1883, Smithson. Misc. Coll., 507: 331; Packard, 1890, U.S. Entomol. Comm. 5th Rept., p. 623, fig. 204; Knobel, 1895, Beetles of New England, p. 34, fig. 120; Knab, 1896, Entomol. News, 7: 113; Beutenmuller, 1896, J. N.Y. Entomol. Soc., 4: 80; Leng and Hamilton, 1896, Trans. Amer. Entomol. Soc., 23: 148; Wickham 1898, Can. Entomol., 30: 40; Lugger, 1899, Minn. Agr. Exp. Sta. Bull., 66: 215; Felt and Joutel, 1904, N.Y. St. Mus. Bull., 74: 18, pl. 5, figs. 3, 6 (habits); Chagnon, 1905, Nat. Can., 32: 43; Felt, 1907, Ins. Affect. Trees, p.

480; Blatchley, 1910, Coleoptera in Indiana, p. 1085, fig. 468; Smith, 1910, N.J. St. Mus. Rept., 1909: 335; Johannsen, 1911, Maine Agr. Exp. Sta. Bull., 187: 7; Mutchler and Weiss, 1923, N.J. Dept. Agr. Circ., 58: 4; Craighead, 1923, Can. Dept. Agr. Bull., (n.s.) 27: 129; Leonard, 1928, Cornell Agr. Exp. Sta. Mem., 101: 455; Chagnon, 1933-40, Coleop. Prov. Quebec, p. 278; Procter, 1938, Biol. Surv. Mt. Desert, 6: 153; Knull, 1946, Ohio Biol. Surv. Bull., 38: 268; Fattig, 1947, Emory Univ. Mus. Bull., 5: 41; Craighead, 1950, USDA Misc. Pub., 657: 267 (habits); Beal, 1952, Duke Univ. Sch. For. Bull., 14: 56 (habits); Dillon and Dillon, 1961, Man. Beetles East. North America, p. 647, pl. 65, no. 6; Chagnon and Robert, 1962, Prin. Coleop. Prov. Quebec, p. 278; Abdullah and Abdullah, 1966, Proc. Roy. Entomol. Soc. London, (B) 35: 92; Bayer and Shenefelt, 1969, Univ. Wisc. Res. Bull., 275: 31, fig. 39; Baker, 1972, USDA Misc. Pub., 1175: 188 (habits); Chamberland, 1976, Fabreries, 2: 89; Gosling and Gosling, 1976, Gr. Lakes Entomol., 10: 32, fig. 168; Laliberte et al., 1977, Fabreries, 3: 98; Headstrom, 1977, Beetles of America, p. 381, fig. 531; Drooz, 1985, USDA Misc. Pub., 1426: 298 (habits); Gosling, 1986, Gr. Lakes Entomol., 19: 157.

Saperda (Saperda) obliqua: Breuning, 1952, Entomol. Arb. Mus. Frey, 3: 172; Gilmour, 1965, Cat. Lam. du Monde, 8: 670.

Anaerea obliqua: Haldeman, 1847, Trans. Amer. Philos. Soc., (2)10: 55.

Male. Form moderate-sized, slightly tapering posteriorly; integument reddish brown, antennae with segments from third broadly paler annulate; pubescence dense, short, appressed, mostly orange-brown. Head with front slightly convex, longer than broad; interantennal area deeply impressed; eyes with lower lobes oblong, much longer than genae; antennae about as long as body, segments from third paler than scape, narrowly darker at apices, scape conical, enlarged, third segment longer than first, fourth subequal to first. Pronotum broader than long, apex narrower than base; disk convex, middle with a vague longitudinal carina; punctures coarse, subconfluent; pubescence forming broad longitudinal vittae at middle and at sides; prosternum moderately densely pubescent; meso- and metasternum densely pubescent, pubescence interrupted by glabrous seta-bearing punctures. Elytra about 2-1/2 times as long as broad; punctures moderately coarse, becoming sparser toward apex; pubescence forming oblique vittae, usually with four sparsely pubescent interspaces; apices acuminate, spines long. Legs with pubescence interrupted by small, glabrous punctures; front and middle tarsal claws with small processes. Abdomen densely pubescent, pubescence interrupted by glabrous punctures; apex of last sternite rounded. Length, 14-16 mm.

Female. Form more robust. Antennae shorter than body. Abdomen with last sternite linearly impressed, apex fringed. Length, 15-19 mm.

Type locality. Missouri.
Range. Eastern North America (Figure 28).
Flight period. June to August.
Host plants. Living *Alnus, Betula.*

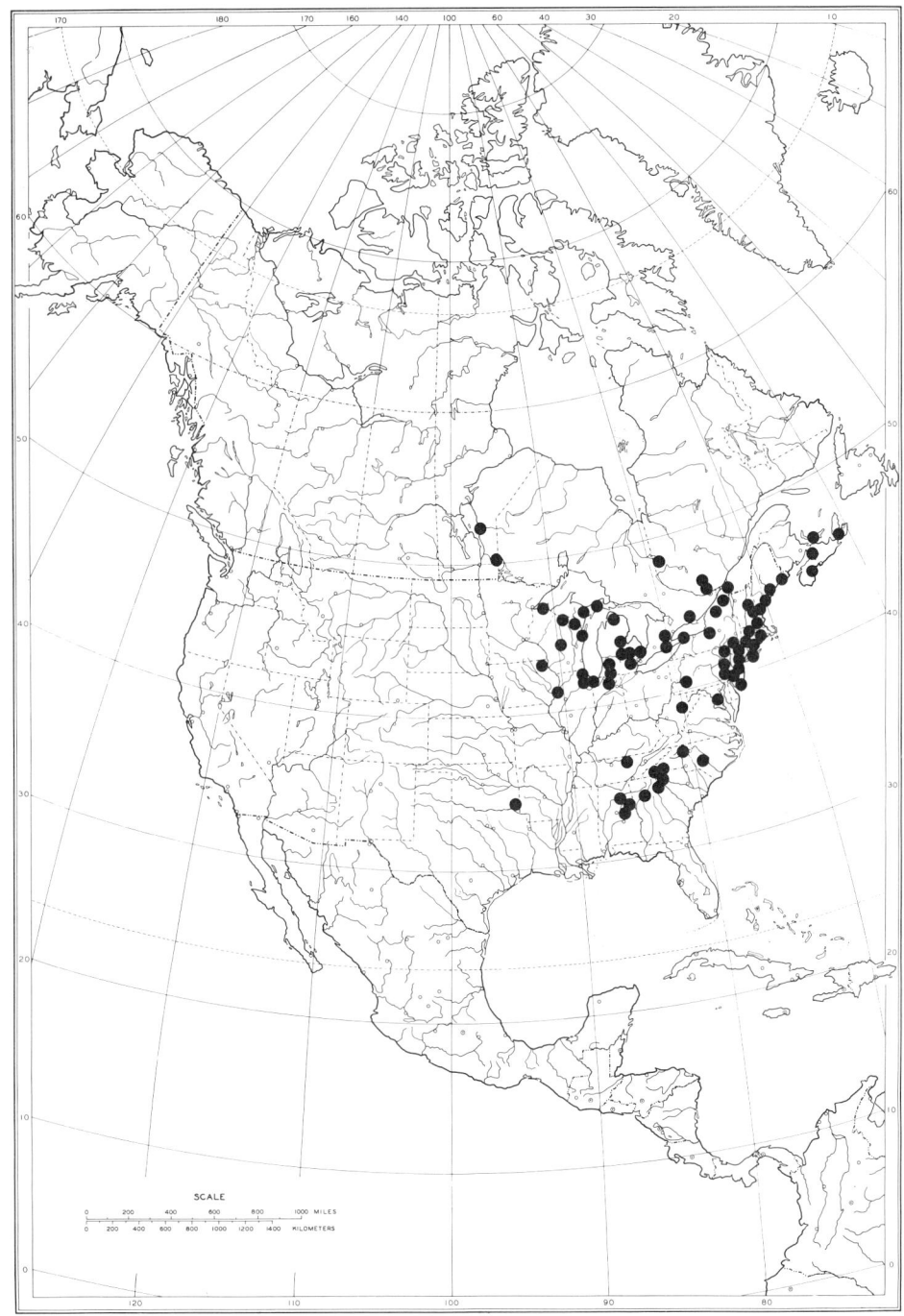

Figure 28. Known geographic range of *Saperda obliqua* Say.

The oblique, pubescent vittae and long apical spine of the elytra immediately identify this species.

Saperda calcarata Say
(Figure 29)

Saperda calcarata Say, 1823, J. Acad. Nat. Sci. Philadelphia, 3: 408; Harris, 1838, Repts. Comm. Zoo. Surv. Sta. (Mass.), p. 90 (habits); Harris, 1841, Rept. Ins. Mass. Inj. Veg., p. 88 (habits); Fitch, 1845, Amer. Quar. J. Agr. Sci., p. 252, pl. 3, fig. 8; LeConte, 1852, J. Acad. Nat. Sci. Philadelphia, 2: 162; Emmons, 1854, Nat. Hist. N.Y. Agr., 5: 121, pl. 16, fig. 1; LeConte, 1859, Compl. Writings T. Say, 2: 190; Fitch, 1859, 5th Rept. Nox. Ben. Ins. N.Y., p. 844; Lacordaire, 1872, Genera des coléoptères, 9(2): 834, fn.; LeConte, 1873, Smithson. Misc. Coll., 11(264): 238; LeConte, 1873, Smithson. Misc. Coll., 11(265): 346; Provancher, 1877, Pet. Fauna Entomol. Can., 1: 633; Packard, 1881, U.S. Entomol. Comm. Bull., 7: 115, 117, fig. 56; LeConte and Horn, 1883, Smithson. Misc. Coll., 507: 331; Knobel, 1895, Beetles of New England, p. 34, fig. 121; Beutenmuller, 1896, J. N.Y. Entomol. Soc., 4: 80; Leng and Hamilton, 1896, Trans. Amer. Entomol. Soc., 23: 148; Wickham, 1898, Can. Entomol., 30: 40; Harrington, 1899, Ottawa Nat., 13: 62; Lugger, 1899, Minn. Agr. Exp. Sta. Bull., 66: 215; Felt and Joutel, 1904, N.Y. St. Mus. Bull., 74: 39, pl. 2, figs. 1-3, pl. 11, fig. 1, pl. 12; Chagnon, 1905, Nat. Can., 32: 43; Felt, 1907, Ins. Affect. Trees, p. 98, pl. 6, fig. 2; Smith, 1910, N.J St. Mus. Rept., 1909: 335; Blatchley, 1910, Coleoptera in Indiana, p. 1086, fig. 469; Gee, 1912, J. Econ. Entomol., 5: 336; Comstock, 1913, Ins. Life, pl. 4, fig. 7; Houser, 1918, Ohio Agr. Exp. Sta. Bull., 332: 318 (habits); Hofer, 1920, USDA Farm. Bull., 1154: 3 (habits); Kotinsky, 1921, USDA Farm. Bull., 1169: 58, figs. 35, 36 (habits); Craighead, 1923, Can. Dept. Agr. Bull., (n.s.) 27: 128, pl. 30, fig. 3, pl. 29, fig. 4; pl. 36, figs. 3, 4; Mutchler and Weiss, 1923, N.J. Dept. Agr. Circ., 58: 4; Hardy and Preece, 1926, Pan-Pac. Entomol., 3: 66 (habits); Hardy, 1926, Rept. Prov. Mus., 1925: 10; Hardy, 1927, Rept. Prov. Mus., 1926: C37; Leonard, 1928, Cornell Agr. Exp. Sta. Mem., 101: 455; Felt and Rankin, 1932, Ins. Dis. Ornam. Trees Shrubs, pp. 431, 481, fig. 214 (habits); Chagnon, 1933-40, Coleop. Prov. Quebec, p. 278, pl. 19, fig. 8; Doane et al., 1936, For. Ins., p. 191, fig. 102 (habits); Procter, 1938, Biol. Surv. Mt. Desert, 6: 153 (habits); Keen, 1938, USDA Misc. Pub. 273:137 (habits); Fenton, 1939, Okla. Agr. Exp. Sta. Circ., 84: 19; Langford and Cory, 1939, Univ. Maryland Ext. Serv. Bull., 84: 52 (habits); Loding, 1945, Geol. Surv. Ala. Mon., 11: 125; Knull, 1946, Ohio Biol. Surv. Bull., 39: 269; Fattig, 1947, Emory Univ. Mus. Bull., 5: 41; Peterson, 1948, Entomol. Soc. Ontario Rept., 1947: 56; Craighead, 1950, USDA Misc. Publ., 657: 264, fig. 56A (habits); English, 1950, Ill. Nat. Hist. Surv. Circ., 47: 59 (habits); Keen, 1952, USDA Misc. Pub., 273: 174

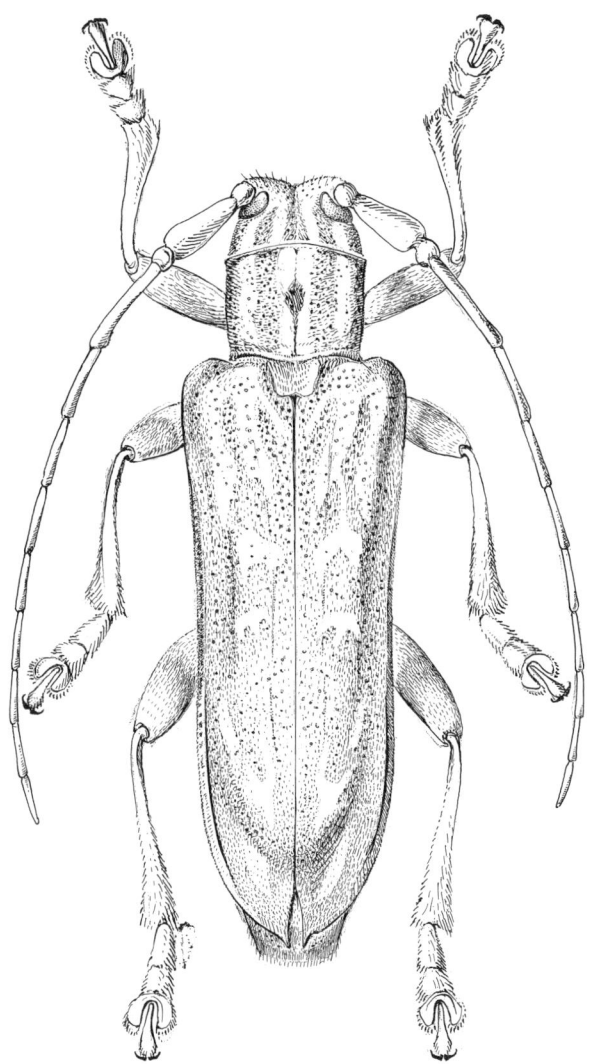

Figure 29. *Saperda calcarata* Say, female.

(habits); Duffy, 1953, Mon. British Timber Beetles, p. 291; Essig, 1958, Ins. West. North America, p. 461; Dillon and Dillon, 1961, Man. Beetles East. North America, p. 647, pl. 65, no. 1; Chagnon and Robert, 1962, Prin. Coleop. Prov. Quebec, p. 278, pl. 19, fig. 8; Abdullah and Abdullah, 1966, Proc. Roy. Entomol. Soc. London, (B) 35: 91; Bayer and Shenefelt,

1969, Univ. Wisc. Res. Bull., 275: 31, fig. 39; Gardiner, 1969, Can. Dept. Fish. For. Int. Rept., 0-14: 104; Solomon, 1969, Ann. Entomol. Soc. Amer. 62: 1214; Hatch, 1971, Univ. Wash. Pub. Biol., 16: 155, pl. 18, fig. 2; Baker, 1972, USDA Misc. Publ., 1175: 185, fig. 60 (habits); Chamberland, 1976, Fabreries, 2: 89; Gosling and Gosling, 1976, Gr. Lakes Entomol., 10: 30, fig. 159; Stern and Tagestad, 1976, USDA For. Serv. Res. Pap., RM-171: 34; Headstrom, 1977, Beetles of America, p. 381; Solomon, 1977, Can. Entomol., 109: 298; Laliberte et al., 1977, Fabreries, 3: 97; Rice and Enns, 1981, Trans. Mo. Acad. Sci., 15: 103; Drooz, 1985, USDA For. Serv. Misc. Pub., 1426: 295, fig. 132; Kukor and Martin, 1986, Oecolog., 7: 138.

Anaerea calcarata: Haldeman, 1847, Trans. Amer. Philos. Soc., 2(10): 55.

Saperda (Anaerea) calcarata: Emmons, 1854, Nat. Hist. N.Y., Agr., 5: 121, pl. 16, fig. 1; Breuning, 1952, Entomol. Arb. Mus. Frey, 3: 153, pl. 4, fig. 12; Gilmour, 1965, Cat. Lam. du Monde, 8: 668.

Saperda adspersa LeConte, 1850, Coleop. Lake Superior, Agassiz and Cabot, p. 234; LeConte, 1852, J. Acad. Nat. Sci. Philadelphia, 2: 162; Lacordaire, 1872, Genera des coléoptères, 9(2): 834, fn.; Gardiner, 1969, Can. Dept. Fish. For. Int. Rept., 0-14: 104 (larva).

Saperda calcarata var. *adspersa:* Felt and Joutel, 1904, N.Y. St. Mus. Bull., 74: 41, pl. 7, fig. 1; Aurivillus, 1923, Coleop. Cat., 74: 481; Leonard, 1928, Cornell Agr. Exp. Sta. Mem., 101: 455; Chagnon, 1933-40, Coleop. Prov. Quebec, p. 278; Abdullah and Abdullah, 1966, Proc. Roy. Entomol. Soc. London, (B) 35: 91.

Saperda (Anaerea) calcarata m. *adspersa*: Breuning, 1952, Entomol. Arb. Mus. Frey, 3: 154; Gilmour, 1965, Cat. Lam. du Monde, 8: 668.

Saperda calcarata ab. *adspersa*: Hatch, 1971, Univ. Wash. Pub. Biol., 16: 155.

Male. Form large, tapering posteriorly; integument black to reddish brown; pubescence dense, short, appressed, grayish mottled with pale brownish. Head with front slightly convex, longer than broad; interantennal area deeply impressed; eyes with lower lobes almost rounded, slightly longer than broad, much larger than genae; antennae slightly longer than body, uniformly grayish pubescent; scape conical, slightly enlarged, third segment longer than first, fourth slightly longer than first. Pronotum broader than long, apex slightly broader than base; disk convex, center with a linear carina behind middle; punctures coarse, confluent to separate; pubescence dense, short, appressed, usually with three longitudinal, brownish vittae, one median and two lateral; prosternum densely pubescent; meso- and metasternum densely pubescent. Elytra about 2-1/2 times as long as broad; punctures rather sparse, shallowly asperate at basal one-third; pubescence dense, usually grayish with two rows of irregular brownish spots; apices narrow, spined at suture. Legs densely pubescent; front and middle tarsal claws with small processes. Abdomen densely pubescent, small denuded spots numerous; last sternite rounded at apex, shallowly emarginate at middle. Length, 18-25 mm.

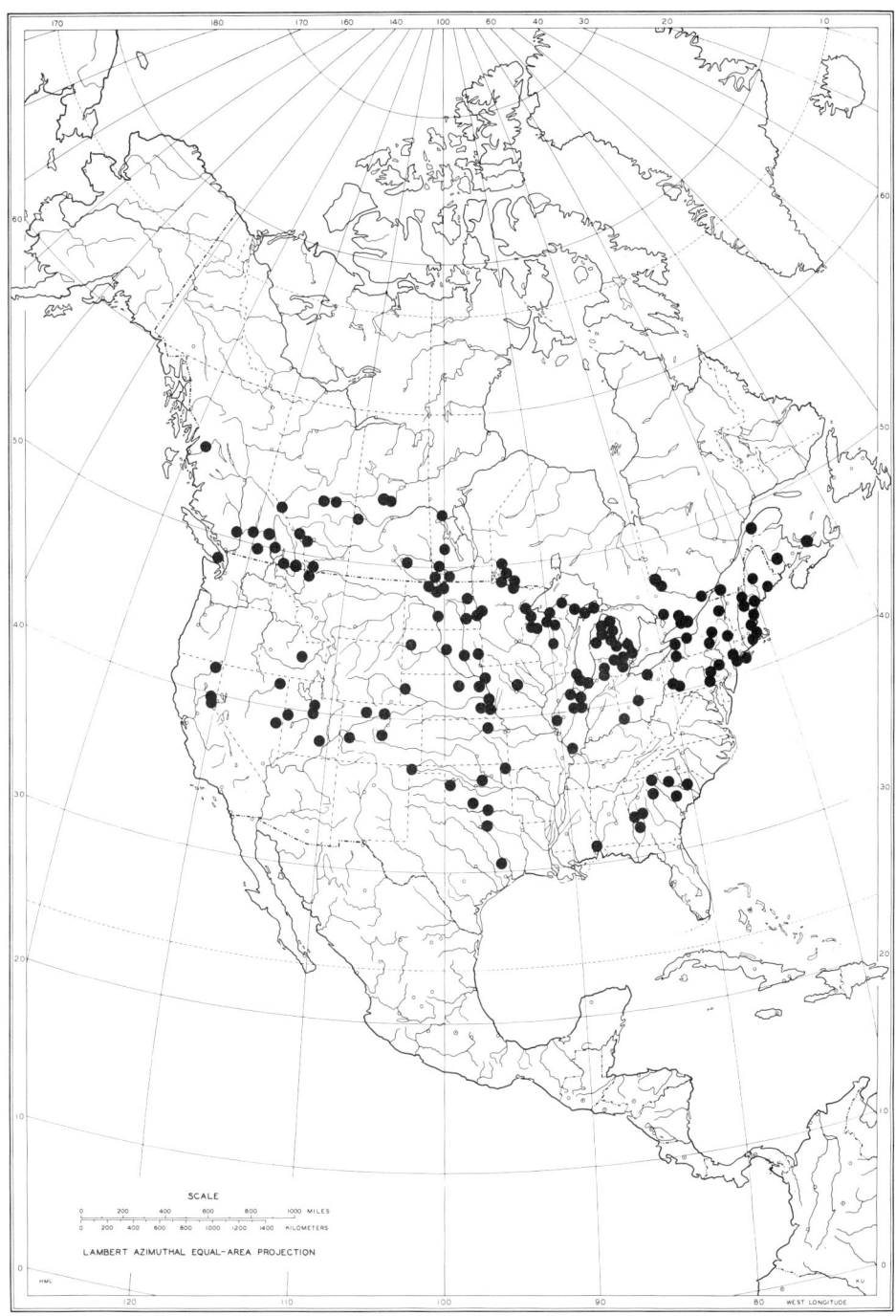

Figure 30. Known geographic range of *Saperda calcarata* Say.

Female. Form more robust. Antennae slightly shorter than body. Abdomen with last sternite linearly impressed at middle, apex densely fringed. Length, 22-31 mm.

Type locality. Of calcarata, Missouri Territory; *adspersa*, Lake Superior.

Range. Eastern North America to California (Figure 30).

Flight period. June to September.

Host plants. Populus, Salix.

Parasites. Eutheresia canescens (Walker), *Ichneumon* sp., *Campaplex sulcatellus* Viereck, *Cremastus* sp.

This is the largest species of *Saperda* in our fauna. The size, pubescent pattern, and spined elytral apices make it easily recognizable.

The coloration of the integument varies from black to pale reddish brown, and the pubescence, particularly the brownish spots of the elytra, varies in size and hue. Western specimens tend to have the integument darker and the pubescence more grayish.

Saperda populnea (Linnaeus)

Cerambyx populneus Linnaeus, 1758, Syst. Nat., ed. 10, p. 394.

Saperda populnea: Fabricius, 1775, Syst. Entomol., p. 186; Felt and Joutel, 1904, N.Y. St. Mus. Bull., 74: 68, pl. 7, fig. 4; Garnett, 1918, Can Entomol., 50: 283; Hardy, 1927, Rept. Prov. Mus., 1926: C37; Doane et al., 1936, For. Ins., p. 192; Keen, 1952, USDA Misc. Pub., 273: 48; Duffy, 1953, Mon. British Timber Beetles, p. 285, fig. 277 (larva); Clark, 1956, Proc. Entomol. Soc. Br. Col., 52: 42; Essig, 1958, Ins. West. North America, p. 462; Hatch, 1971, Univ. Wash., Pub. Biol., 16: 154, pl. 18, fig. 7.

Male. Form small to moderate-sized; subparallel; integument black; pubescence moderately dense to very dense, grayish and orange-brown. Head with front convex, interantennal area shallowly impressed; pubescence dense to moderately dense, long, erect hairs numerous; eyes with lower lobes longer than broad, much longer than genae; antennae about as long as body, segments from third annulate, scape slender, subconical, third segment longer than first or fourth, fourth longer than third. Pronotum slightly broader than long; disk convex, median line vague; base shallowly impressed; punctures moderately coarse, irregularly subreticulate; pubescence sparse to dense, usually with orange-brown or gray lateral vittae and often a narrow median vitta, long, erect hairs numerous; prosternum densely pubescent; meso- and metasternum densely pubescent, pubescence often variegated with yellow brown. Elytra about 2-1/2 times as long as broad; punctures coarse, deep, subconfluent to contiguous; pubescence moderately dense to very dense, grayish to yellow-brown, often with yellow-brown spots, long, erect hairs more numerous at base; apices narrowing, rounded. Legs moderately densely pubescent; tarsal claws lacking a process. Abdomen densely pubescent; last sternite rounded at apex. Length, 9-13 mm.

Female. Form more robust. Antennae shorter than body. Elytra with pubescence often much denser. Abdomen with last sternite linearly impressed at middle. Length, 9-13 mm.

Type locality. Europe.

Range. Europe, Asia, North America.

In North America this species exhibits a complex of phenotypes. All differ from the Eurasian forms, primarily in size and elytral pubescent pattern. Based on the North American material available to us, we have recognized two major populations, one northern and eastern, the other western.

The coarse, deep punctures, yellow-brown pubescence, and narrower elytral apices separate the subspecies of *S. populnea* from *S. inornata* Say.

Only references pertinent to this study are included above. For a more complete listing of synonyms and references for the Palearctic species see Aurivillius (1923).

Saperda populnea moesta LeConte

Saperda moesta LeConte, 1850, Coleop. Lake Superior, Agassiz and Cabot, p. 234; LeConte, 1852, J. Acad. Nat. Sci. Philadelphia, 2: 163; Lacordaire, 1872, Genera des coléoptères, 9(2): 834, fn.; LeConte, 1873, Smithson. Misc. Coll., 11(264): 239; LeConte, 1873, Smithson. Misc. Coll., 11(265): 346; Provancher, 1877, Pet. Fauna Entomol. Can., 1: 635; Packard, 1881, U.S. Entomol. Comm. Bull., 7: 118; LeConte and Horn, 1883, Smithson. Misc. Coll., 507: 331; Beutenmuller, 1896, J. N.Y. Entomol. Soc., 4: 80; Leng and Hamilton, 1896, Trans. Amer. Entomol. Soc., 23: 151; Wickham, 1898, Can. Entomol., 30: 42; Harrington, 1899, Ottawa Nat., 13: 68; Felt and Joutel, 1904, N.Y. St. Mus. Bull., 74: 71, pl. 7, fig. 5; Chagnon, 1905, Nat. Can., 32: 43; Blatchley, 1910, Coleoptera in Indiana, p. 1085; Craighead, 1923, Can. Dept. Agr. Bull., (n.s.) 27: 129; Chagnon, 1933-40, Coleop. Prov. Quebec, p. 276; Craighead, 1950, USDA Misc. Pub., 657: 266; Chagnon and Robert, 1962, Prin. Coleop. Prov. Quebec, p. 276; Gardiner, 1969, Can. Dept. Fish. For. Int. Rept., 0-14: 102 (larva); Baker, 1972, USDA Misc. Pub., 1175: 188 (habits); Drooz, 1985, USDA Misc. Pub., 1426: 298 (habits).

Saperda populnea moesta: Felt and Joutel, 1904, N.Y. St. Mus. Bull., 74: 68; Doane et al., 1936, For. Ins., p. 192; Wong and McLeod, 1965, Can. Dept. For. Bimon. Prog. Rept., 21: 3 (habits); Gosling and Gosling, 1976, Gr. Lakes Entomol., 10: 32, fig. 169; Laliberte et al., 1977, Fabreries, 3: 98.

Saperda populnea var. *moesta*: Aurivillius, 1923, Coleop. Cat., 74: 486; Mutchler and Weiss, 1923, N.J. Dept. Agr. Circ., 58: 11; Leonard, 1928, Cornell Agr. Exp. Sta. Mem., 101: 456; Knowlton and Thatcher, 1936, Utah Acad. Sci. Arts Letters, 13: 281; Knull, 1946, Ohio Biol. Surv. Bull., 39: 273; Clark, 1956, Proc. Entomol. Soc. Brit. Col., 52: 42; Bayer and Shenefelt, 1969, Univ. Wisc. Res. Bull., 275: 32, fig. 40.

Saperda (Saperda) populnea moesta: Breuning, 1952, Entomol. Arb. Mus. Frey, 3: 157; Gilmour, 1965, Cat. Lam. du Monde, 8: 668.
Saperda populnea ab. *moesta*: Hatch, 1971, Univ. Wash. Pub. Biol., 16: 154.

Form small to moderate-sized; pubescence grayish, rather sparse. Head uniformly, sparsely pubescent, long, erect hairs numerous; antennae usually broadly dark annulate. Pronotum usually with grayish lateral vittae, median vittae usually absent; punctures moderately coarse, irregular, shallow. Elytra with punctures coarse, confluent; pubescence sparse, often with small grayish pubescent patches; apices, especially in females, not very narrow. Length, 8-11 mm.

Type locality. Lake Superior.
Range. Northern North America to Idaho.
Flight period. June and July.
Host plants. Populus, Salix.

The reduced pubescence and shinier integument which makes the punctures of the elytra appear larger separate this subspecies. Examples from the more western portion of the range are more densely, uniformly pubescent.

Saperda populnea tulari Felt and Joutel

Saperda populnea tulari Felt and Joutel, 1904, N.Y. St. Mus. Bull., 74: 70, pl. 7, fig. 6; Doane et al., 1936, For. Ins., p. 192.
Saperda populnea var *tulari*: Garnett, 1918, Can. Entomol., 50: 283; Aurivillius, 1923, Coleop. Cat., 74: 487.
Saperda (Saperda) populnea tulari: Breuning, 1952, Entomol. Arb. Mus. Frey, 3: 157; Gilmour, 1965, Cat. Lam. du Monde, 8: 668.
Saperda populnea form *tulari*: Essig, 1958, Ins. West. North America, p. 462.
Saperda populnea ab. *tulari*: Hatch, 1971, Univ. Wash. Pub. Biol., 16: 154.

Form moderate-sized; pubescence grayish and orange-brown, usually dense, especially in females. Head usually with vittae behind eyes in males, densely pubescent in females; antennae about as long as body in males, shorter in females, segments from third narrowly dark annulate at apices. Pronotum with lateral vittae and usually a narrow median vitta in males, females usually much more densely pubescent, vittae often coalesced with other pubescence; punctures moderately coarse, irregular, dense. Elytra with punctures coarse, deep, subconfluent; pubescence in males moderately dense, usually variegated with orange-brown, females usually very densely pubescent; apices narrowly rounded, slightly more broadly in females. Length, 7-13 mm.

Type locality. Tulare County, California.
Range. Washington to California and Nevada.
Flight period. March to November.
Host plants. Populus spp.

California populations from coastal areas, the Central Valley, and Sierra foothills appear to be relatively uniform in pubescence. The males have the elytra variegated with pubescent spots and the females are usually densely orange-brown pubescent. On the east side of the Sierra Nevada, the pubescence appears denser in both sexes, and in western Nevada is greatly reduced. Examples from Oregon and Washington also display reduced pubescence.

The range of this subspecies is difficult to define because of the lack of adequate material from areas adjacent to California. For convenience, all of the western forms are here treated as *S. populnea tulari*.

Saperda horni Joutel
(Figure 31)

Saperda hornii Joutel, 1902, Entomol. News, 13: 33, pl. 2, figs. 1-5; Felt and Joutel, 1904, N.Y. St. Mus. Bull., 74: 22, pl. 7, fig. 3; Garnett, 1918, Can. Entomol., 50: 283; Craighead, 1923, Can. Dept. Agr. Bull., (n.s.) 27: 128, pl. 23, fig. 9; Hopping, 1931, Can. Entomol., 63: 73; Abdullah and Abdullah, 1966, Proc. Roy. Entomol. Soc. London, (B) 35: 91; Hatch, 1971, Univ. Wash. Pub. Biol., 16: 154.
Saperda horni: Linsley, 1936, Pan-Pac. Entomol., 12: 119; Doane et al., 1936, For. Ins., p. 192 (habits); Essig, 1958, Ins. West. North America, p. 462; Chemsak, 1958, Pan-Pac. Entomol., 34: 41 (habits); Tyson, 1966, Pan-Pac. Entomol., 42: 206.
Saperda (Saperda) horni: Breuning, 1952, Entomol. Arb. Mus. Frey, 3: 161.
Saperda uteana Casey, 1924, Memoirs on the Coleoptera, 11: 294.

Male. Form moderate-sized, tapering posteriorly; integument black; pubescence dense, short, appressed, grayish and mottled orange-brown, interrupted by small glabrous spots. Head with front convex, about as broad as long; vertex usually with two yellowish vittae extending onto neck; interantennal area deeply impressed; eyes with lower lobes longer than broad, longer than genae; antennae shorter than body, segments dark annulate at apices, scape cylindrical, not enlarged, third segment longer than first and fourth, fourth longer than first. Pronotum slightly broader than long; disk with a vague median, longitudinal line, usually shallowly impressed at middle before center; punctures coarse, moderately dense to sparse, separated; pubescence dense, grayish with a median and two lateral yellowish vittae, long, erect hairs numerous; prosternum densely clothed with short appressed pubescence and long, flying hairs; meso- and metasternum densely pubescent with small glabrous spots interspersed. Elytra about 2-1/2 times as long as broad; punctures rather sparse, crater-like; pubescence dense, appressed, interrupted by punctures, usually grayish with irregular yellow-orange spots interspersed; apices narrow, obtuse. Legs densely pubescent; front and middle tarsal claws with a small process. Abdomen densely pubescent; last sternite lightly emarginate at middle at apex. Length, 10.5-16 mm.

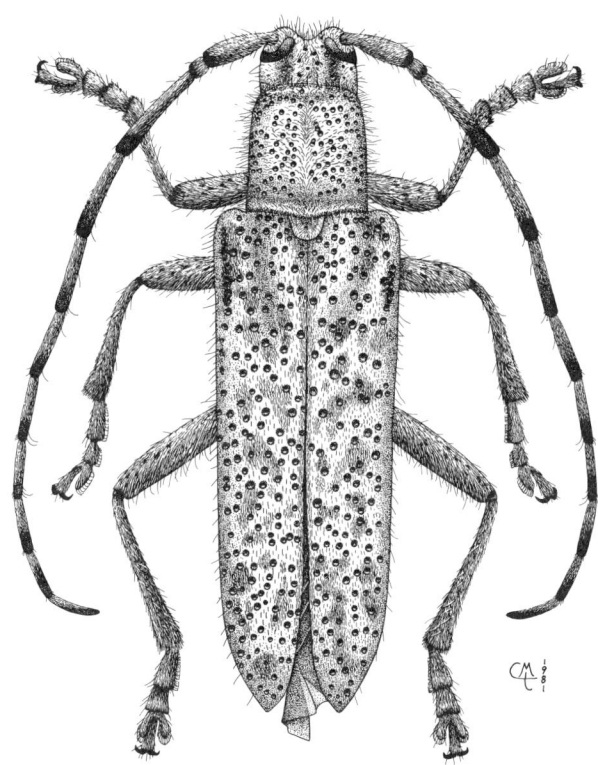

Figure 31. *Saperda horni* Joutel, male.

Female. Form more robust, subparallel. Antennae extending to about fourth abdominal segment. Abdomen with last sternite narrowly longitudinally impressed medially. Length, 13-20 mm.

Type locality. Of horni, Yosemite, California; *uteana*, Vineyard, Utah.

Range. Pacific Coast from British Columbia to southern California and Utah.

Flight period. May to October.

Host plants. Salix.

This species may be recognized by the usually grayish pubescence interrupted by crater-like punctures and yellow-orange spots. The larger size and much sparser punctation of the elytra separate it from *S. populnea* and the larger, sparser punctures of the elytra from *S. mutica*.

Saperda mutica Say

Saperda mutica Say, 1824, J. Acad. Nat. Sci. Philadelphia, 3: 409; LeConte, 1852, J. Acad. Nat. Sci. Philadelphia, 2: 162; LeConte, 1859, Compl. Writings T. Say, 2: 191; Lacordaire, 1872, Genera des coléoptères, 9(2): 834, fn; LeConte, 1873, Smithson. Misc. Coll., 11(264): 238; Leng and Hamilton, 1896, Trans. Amer. Entomol. Soc., 23: 148; Wickham, 1898, Can. Entomol., 30: 40; Harrington, 1899; Ottawa Nat., 13: 67; Felt and Joutel, 1904, N.Y. St. Mus. Bull., 74: 21, pl. 7, fig. 2; Chagnon, 1905, Nat. Can., 32: 43; Blatchley, 1910, Coleoptera in Indiana, p. 1085; Smith, 1910, N.J. St. Mus. Rept., 1909: 335; Mutchler and Weiss, 1923, N.J. Dept. Agr. Circ., 58: 5; Leonard, 1928, Cornell Agr. Exp. Sta. Mem., 101: 455; Chagnon, 1933-40, Coleop. Prov. Quebec, p. 276, pl. 19, fig. 9; Knull, 1946, Ohio Biol. Surv. Bull., 39: 268; Chagnon and Robert, 1962, Prin. Coleop. Prov. Quebec, p. 276, pl. 19, fig. 9; Abdullah and Abdullah, 1966, Proc. Roy. Entomol. Soc. London, (B) 35: 92; Bayer and Shenefelt, 1969, Univ. Wisc. Res. Bull., 275: 31, fig. 39; Hatch, 1971, Univ. Wash. Pub. Biol., 16: 154; Baker, 1972, USDA Misc. Pub., 1175: 188; Stein and Tagestad, 1976, For. Serv. Res. Pap. RM-171: 37; Gosling and Gosling, 1976, Gr. Lakes Entomol., 10: 32, fig. 167; Laliberte et al., 1977, Fabreries, 3: 98; Drooz, 1985, USDA Misc. Pub., 1426: 298.

Anaerea mutica: Haldeman, 1847, Trans. Amer. Philos. Soc., (2)10: 55.

Saperda (Saperda) mutica: Breuning, 1952, Entomol. Arb. Mus. Frey, 3: 160, pl. 4, fig. 14; Gilmour, 1965, Cat. Lam. du Monde, 8: 669.

Male. Form moderate-sized, tapering posteriorly; integument black; pubescence dense, short, appressed, grayish and yellowish to orange. Head with front convex, quadrate; interantennal area deeply impressed; pubescence dense, mottled grayish and orange, neck with a broad orangish, median vitta; eyes with lower lobes about as long as genae; antennae about as long as body, segments except first dark annulate at apices, scape robust, subconical, third segment longer than first and fourth, fourth slightly longer than first. Pronotum slightly broader than long, apex narrower than base; disk convex, median line more prominent over basal half; punctures coarse, subconfluent; pubescence moderately dense, with three broad, longitudinal, orange vittae, one median and two lateral, long, erect hairs sparse; prosternum densely orange pubescent; meso- and metasternum densely, irregularly variegated orange and gray pubescent. Elytra less than 2-1/2 times as long as broad; punctures moderately coarse, rather dense, separated; pubescence short, dense, appressed, grayish, with irregular denser patches of orange interspersed; apices narrow, obtuse. Legs grayish pubescent, pubescence interrupted by glabrous punctures; front and middle tarsal claws with a small process. Abdomen with pubescence interrupted by glabrous punctures; last sternite lightly emarginate medially at apex. Length, 11-13 mm.

Female. Form more robust. Antennae shorter than body. Abdomen with last sternite linearly impressed longitudinally, apex truncate. Length, 14-17 mm.

Type locality. Missouri Territory.
Range. Northeastern North America to North Dakota.
Flight period. June and July.
Host plants. Dead *Salix.*

The more robust scape of the antennae and denser, smaller punctures of the elytra separate this species from *S. horni.*

Saperda candida Fabricius

Saperda candida Fabricius, 1787, Mant. Ins., 1: 147; Fabricius, 1792, Entomol. Syst., 2: 307; Fabricius, 1801, Syst. Eleuth., 2: 319; Haldeman, 1847, Trans. Amer. Philos. Soc., (2)10: 55; LeConte, 1852, J. Acad. Nat. Sci. Philadelphia, 2: 163; Emmons, 1854, Nat. Hist. N.Y., Agr., 5: 121, pl. 16, fig. 3; Thomson, 1857, Arch. Entomol., 1: 392; Lacordaire, 1872, Genera des coléoptères, 9(2): 834, fn.; LeConte, 1873, Smithson. Misc. Coll., 11(264): 238; Cook, 1874, Mich. St. Bd. Agr. Rept., p. 21, fig. 20 (habits); Provancher, 1877, Pet. Fauna Entomol. Can., 1: 633, fig. 48; Koons, 1886, Conn. Bd. Agr. Sec. Rept., p. 15 (habits); Lintner, 1891, Inj. Ins. N.Y. 7th Rept., p. 313, fig. 31; Knobel, 1895, Beetles of New England, p. 34, fig. 122; Beutenmuller, 1896, J. N.Y. Entomol. Soc., 4: 80; Leng and Hamilton, 1896, Trans. Amer. Entomol. Soc., 23: 148; Stedman, 1898, Mo. Agr. Exp. Sta. Bull., 44: 14, fig. 6 (habits); Chittenden, 1898, USDA Div. Entomol. Circ., 32: 1, fig. 1 (habits); Wickham, 1898, Can. Entomol., 30: 41, fig. 5; Harrington, 1899, Ottawa Nat., 13: 62; Lugger, 1899, Minn. Sta. Exp. Sta. 5th Ann. Rept., p. 126, figs. 133, 134 (habits); Lugger, 1899, Minn. Agr. Exp. Sta. Bull., 66: 210, figs., 133, 134 (habits); Felt, 1900, N.Y. St. Mus. Bull., 6: 577; Webster, 1900, J. Columbus Hort. Soc., 15: 5, fig. 5; Banks, 1902, USDA Div. Entomol. Bull., 34 (n.s.): 39, fig. 36; Pettit, 1904, Mich. St. Coll. Agr. Exp. Sta. Spec. Bull., 24: 6, fig. 2 (habits); Felt and Joutel, 1904, N.Y. St. Mus. Bull., 74: 23, pl. 1, figs. 1-8, pl. 8, figs. 1-5, pls. 9, 10, figs. 1-3; Chagnon, 1905, Nat. Can., 32: 43; Chittenden, 1907, USDA Bur. Entomol. Circ., 32: 1, fig. 1 (habits); Hitchings, 1907, Maine Sta. Entomol. 3rd Ann. Rept., p. 96, fig. 16 (habits); Garcia, 1908, New Mex. Agr. Exp. Sta. Bull., 68: 51, fig. 53; Blatchley, 1910, Coleoptera in Indiana, p. 1087, fig. 470; Smith, 1910, N.J. St. Mus. Rept., 1909: 336, figs. 134, 135; Lamson, 1912, Conn. Agr. Exp. Sta. Bull., 71: 61, fig. 6; Baldwin, 1912, Indiana Sta. Entomol. 5th Ann. Rept., p. 58 (habits); Brooks, 1915, USDA Farm. Bull., 675: 1 (habits); Caesar, 1917, Ontario Dept. Agr. Bull., 250: 33, fig. (habits); Becker, 1917, J. Econ. Entomol., 10: 66; Brooks, 1920, USDA Bull., 847: 1 (habits); Haseman, 1920, Mo. Agr. Exp. Sta. Bull., 176: 5, fig. 3 (habits); Patch, 1921, Maine Agr. Exp. Sta. Bull., p. 4, fig. 1 (habits); Wellhouse, 1922, Cornell Agr. Exp. Sta. Mem., 56: 1102;

Mutchler and Weiss, 1923, N.J. Dept. Agr. Circ. 58: 5; Guyton and Knull, 1924, Pa. Dept. Agr. Gen. Bull., 386: 3, 1 pl. (habits); Britton and Zappe, 1927, Conn. Agr. Exp. Sta. Bull., 292: 131, fig. 19 (habits); Leonard, 1928, Cornell Agr. Exp. Sta. Mem., 101: 455; Petch, 1930, Can. Dept. Agr. Entomol. Br. Circ., 73: 1, 3 figs.; Pettit and Hutson, 1931, Mich. St. Coll. Agr. Exp. Sta. Circ. Bull., 137: 52; Felt and Rankin, 1932, Ins. Dis. Ornam. Trees Shrubs, p. 144, fig. 32 (habits); Chagnon, 1933-40, Coleop. Prov. Quebec, p. 276, pl. 19, fig. 6; Doane et al., 1936, For. Ins., p. 191; Kaston, 1937, Conn. Agr. Exp. Sta. Bull., 396: 355; Procter, 1938, Biol. Surv. Mt. Desert, 6: 153; Langford and Cory, 1939, Univ. Md. Ext. Serv. Bull., 84: 50, fig. 32; Hess, 1940, N.Y. St. Agr. Exp. Sta. Bull., 688: 5 (habits); Chandler and Flint, 1942, Ill. Nat. Hist. Surv. Circ., 40: 1, figs. 1, 2; Loding, 1945, Geol. Surv. Ala. Mon., 11: 124; Knull, 1946, Ohio Biol. Surv. Bull., 39: 269; Fattig, 1947, Emory Univ. Mus. Bull., 5: 41; Craighead, 1950, USDA Misc. Pub., 657: 264; Anon., 1950, USDA Leaflet, 274: 1; Essig, 1958, Ins. West. North America, p. 461; Dillon and Dillon, 1961, Man. Beetles East. North America, p. 647, pl. 65, no. 4; Chagnon and Robert, 1962, Prin. Coleop. Prov. Quebec, p. 276, pl. 19, fig. 6; Abdullah and Abdullah, 1966, Proc. Roy. Entomol. Soc. London, (B) 35: 91; Bayer and Shenefelt, 1969, Univ. Wisc. Res. Bull., 275: 31, fig. 39; Hatch, 1971, Univ. Wash. Pub. Biol., 16: 155, pl. 18, fig. 3; Baker, 1972, USDA Misc. Pub., 1175: 187 (habits); Chamberland, 1976, Fabreries, 2: 89; Gosling and Gosling, 1976, Gr. Lakes Entomol., 10: 31, fig. 160; Laliberte et al., 1977, Fabreries, 3: 97; Headstrom, 1977, Beetles of America, p. 380, fig. 530; Rice and Enns, 1981, Trans. Mo. Acad. Sci., 15: 103; Drooz, 1985, USDA Misc. Pub., 1426: 297 (habits); Gosling, 1986, Gr. Lakes Entomol., 19: 157.

Saperda (Saperda) candida: Breuning, 1952, Entomol. Arb. Mus. Frey, 3: 171; Gilmour, 1965, Cat. Lam. du Monde, 8: 670.

Saperda bivittata Say, 1824, J. Acad. Nat. Sci. Philadelphia, 3: 409; Harris, 1838, Repts. Comm. Zool. Surv. Sta. (Mass.), p. 90 (habits); Harris, 1841, Rept. Ins. Mass. Inj. Veg., p. 89; Fitch, 1855, Rept. Nox. Ben. Ins. N.Y., p. 715 (habits); Fitch, 1857, 3rd Rept. Nox. Benef. Ins. N.Y., p. 321, pl. 1, fig. 2 (habits); LeConte, 1859, Compl. Writings T. Say, 2: 190; Riley, 1869, 1st Ann. Rept. Nox. Ben. Ins. Mo., p. 42, fig. 14 (habits); Packard, 1881, U.S. Entomol. Comm. Bull., 7: 136, 137.

Saperda bipunctata Hopping, 1925, Can. Entomol., 57: 208; Abdullah and Abdullah, 1966, Proc. Roy. Entomol. Soc. London, (B) 35: 90. New synonymy.

Saperda (Saperda) candida m. *bipunctata*: Breuning, 1952, Entomol. Arb. Mus. Frey, 3: 172, pl. 5, fig. 21; Gilmour, 1965, Cat. Lam. du Monde, 8: 670.

Saperda candida bipunctata: Stein and Tagestad, 1976, USDA For. Serv. Res. Pap., RM-171: 35; Chemsak and Linsley, 1982, Checklist of Ceram., p. 107.

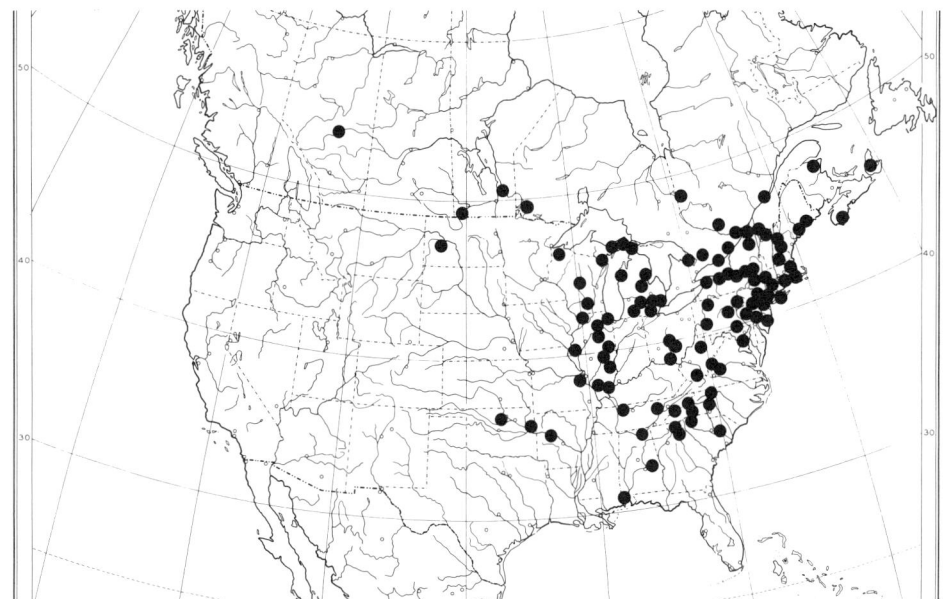

Figure 32. Known geographic range of *Saperda candida* Fabricius.

Male. Form moderate-sized, tapering posteriorly; integument brownish to black, appendages often reddish brown; pubescence very dense, short, appressed, obscuring surface, brownish and white, head, pronotum, and elytra with two longitudinal white vittae, underside densely white pubescent. Head with front convex, quadrate, very densely white pubescent, longer, dark, erect hairs numerous; area between antennal tubercles shallowly impressed; eyes with lower lobes rounded, much longer than genae; antennae slightly longer than body, uniformly clothed with very short, grayish pubescence; scape slender, third segment longer than first, fourth shorter than first. Pronotum slightly broader than long; disk convex with a narrow, longitudinal carina down middle; punctures moderately coarse, dense, usually obscured by pubescence; pubescence dense, appressed, white vittae extending back from head onto elytra; prosternum densely pubescent; meso- and metasternum densely white pubescent. Elytra usually less than 2-1/2 times as long as broad; white vittae usually entire; apices rounded. Legs densely white pubescent; front and middle dorsal claws with a process. Abdomen densely white pubescent; last sternite subtruncate at apex. Length, 13-16 mm.

Female. Form more robust, subparallel. Antennae shorter than body. Abdomen with last sternite medially linearly impressed. Length, 15-21 mm.

Type locality. Of *candida*, none; *bivittata*, United States; *bipunctata*, Aweme, Manitoba.

Range. Eastern United States to Saskatchewan, Canada (Figure 32).

Flight period. May to August.

Host plants. Amelanchier, Amydalus, Araria, Aronia, Cotoneaster, Crataegus, Cydonia, Malus, Prunus, Pyracantha, Pyrus, Sorbus.

Parasites. Cenocoelius saperdae (Ashmead) (Braconidae).

This is one of the most distinctive species of Cerambycidae in our fauna. The brownish coloration with two white vittae extending from the head down the elytra makes *S. candida* easily recognizable. Occasionally the elytral vittae are interrupted medially, and at times a small brownish spot occurs on the vittae near the base of the elytra.

The habits of this species have been well documented because of its economic importance to orchard and ornamental trees. Felt and Joutel (1904) presents a detailed account.

Saperda puncticollis Say

Saperda puncticollis Say, 1824, J. Acad. Nat. Sci. Philadelphia, 3: 406; LeConte, 1852, J. Acad. Nat. Sci. Philadelphia, 2: 164; LeConte, 1859, Compl. Writings T. Say, 2: 189; Lacordaire, 1872, Genera des coléoptères, 9(2): 834, fn; LeConte, 1873, Smithson. Misc. Coll., 11(264): 239; Zimmerman, 1878, Can. Entomol., 10: 220; Harrington, 1890, Entomol. Soc. Ontario, 20th Rept., 1889: 52; Hamilton, 1895, Trans. Amer. Entomol. Soc., 22: 370; Knobel, 1895, Beetles New England, p. 34, fig. 128; Beutenmuller, 1896, J. N.Y. Entomol. Soc., 4: 81; Leng and Hamilton, 1896, Trans. Amer. Entomol. Soc., 23: 151; Wickham, 1898, Can. Entomol., 30: 41; Lugger, 1899, Minn. Agr. Exp. Sta. Bull., 66: 215; Harrington, 1899, Ottawa Nat., 13: 62; Dury, 1902, J. Cincinnati Soc. Nat. Hist., 22: 163; Felt and Joutel, 1904, N.Y. St. Mus. Bull., 74: 66, pl. 6, figs. 5-9; Chagnon, 1905, Nat. Can., 32: 43; Felt, 1907, Ins. Affect. Trees, p. 478, pl. 6, figs. 17-23; Smith, 1910, N.J. St. Mus. Rept., 1909: 336; Blatchley, 1910, Coleoptera in Indiana, p. 1087; Craighead, 1923, Can. Dept. Agr. Bull., (n.s.) 27: 128; Mutchler and Weiss, 1923, N.J. Dept. Agr. Circ., 58: 11; Leonard, 1928, Cornell Agr. Exp. Sta. Mem., 101: 456; Fletcher, 1929, Can. Entomol., 61: 259; Chagnon, 1933-40, Coleop. Prov. Quebec, p. 276, pl. 19, fig. 10; Knull, 1946, Ohio Biol. Surv. Bull., 39: 273; Fattig, 1947, Emory Univ. Mus. Bull., 5: 42; Dillon and Dillon, 1961, Man. Beetles East. North America, p. 651, pl. 64, no. 14; Chagnon and Robert, 1962, Prin. Coleop. Prov. Quebec, p. 276, pl. 19, fig. 10: Abdullah and Abdullah, 1966, Proc. Roy. Entomol. Soc. London, (B) 35: 92; Bayer and Shenefelt, 1969, Univ. Wisc. Res. Bull., 275: 32, fig. 40; Laliberte et al., 1977, Fabreries, 3: 98; Rice and Enns, 1981, Trans. Mo. Acad. Sci., 15: 103; Gosling, 1984, Gr. Lakes Entomol., 17: 73; Furth, 1985, Conn. Acad. Arts Sci., 46: 193.

Saperda (Saperda) puncticollis: Breuning, 1952, Entomol. Arb. Mus. Frey, 3: 170, pl. 4, fig. 20; Gilmour, 1965, Cat. Lam. du Monde, 8: 670.

Compsidea? puncticollis: Haldeman, 1847, Trans. Amer. Philos. Soc., (2)10: 55.
Saperda trigeminata Randall, 1838, Boston J. Nat. Hist., 2: 43; LeConte, 1852, J. Acad. Nat. Sci. Philadelphia, 2: 164 (synonymy).

Male. Form small, subparallel; integument black to dark reddish brown; pubescence fine, dense, appressed, grayish and greenish yellow. Head with front convex, broader than long; interantennal area barely impressed; pubescence dense, greenish yellow with a glabrous spot at middle below vertex and a triangular spot at middle on neck; eyes with lower lobes slightly transverse, much longer than genae; antennae about as long as body, uniformly, finely, gray pubescent, scape slender, conical, third segment longer than first, fourth longer than first, shorter than third. Pronotum broader than long; disk convex, densely punctate; pubescence dense, appressed, greenish yellow, interrupted by four large glabrous discal spots and a spot on each side; prosternum finely, densely gray pubescent; meso- and metasternum finely, densely gray pubescent, middle of metasternum with an ovoid brownish spot. Elytra a little more than twice as long as broad; punctures coarser and denser just before middle, becoming fine and sparse toward apex; lateral margins and suture with broad, densely pubescent, longitudinal vittae joining at apex, discal pubescence much finer, darker; apices rounded. Legs densely pubescent; middle tarsal claws with a short process. Abdomen densely grayish pubescent; last sternite subtruncate to shallowly emarginate at apex. Length, 8-11 mm.

Female. Form more robust. Antennae slightly shorter than body. Abdomen with last sternite longitudinally impressed at middle. Length, 9-12 mm.

Type locality. Of *puncticollis*, Arkansas; *trigeminata*, Cambridge, Massachusetts.

Range. Eastern North America to Kansas.

Flight period. May to July.

Host plants. Parthenocissus, Rhus, Vitis.

This species may be easily recognized by the small size, vittate elytra, and the four black spots of the pronotum.

Saperda lateralis Fabricius
(Figure 33)

Saperda lateralis Fabricius, 1775, Syst. Entomol., p. 185; Fabricius, 1781, Spec. Ins., 1: 233; Fabricius, 1787, Mant. Ins., 1: 149; Fabricius, 1792, Entomol. Syst., Emend. et Auct., 1(2): 312; Olivier, 1795, Entomol., 4: 17, pl. 4, fig. 41; Fabricius, 1801, Syst. Eleuth., 2: 323; LeConte, 1852, J. Acad. Nat. Sci. Philadelphia, 2: 164; Fitch, 1859, 5th Rept. Nox. Ben. Ins. N.Y., p. 60; Lacordaire, 1872, Genera des coléoptères, 9(2): 834, fn.; LeConte, 1873, Smithson. Misc. Coll., 11(264): 239; Provancher, 1877, Pet. Fauna Entomol. Can., 1: 635; Packard, 1881, U.S. Entomol. Comm. Bull., 7: 59,

141; Knobel, 1895, Beetles of New England, p. 34, fig. 127; Beutenmuller, 1896, J. N.Y. Entomol. Soc., 4: 80; Leng and Hamilton, 1896, Trans. Amer. Entomol. Soc., 23: 150; Wickham, 1898, Can. Entomol., 30: 41; Lugger, 1899, Minn. Agr. Exp. Sta. Bull., 66: 215; Felt and Joutel, 1904, N.Y. St. Mus. Bull., 74: 59, fig. 6, pl. 7, fig. 8; Chagnon, 1905, Nat. Can., 32: 43; Blatchley, 1910, Coleoptera in Indiana, p. 1088; Smith, 1910, N.J. St. Mus. Rept., 1909: 336; Craighead, 1923, Can. Dept. Agr. Bull., (n.s.) 27: 130, pl. 16, fig. 9; Champlain and Knull, 1925, Entomol. News, 36: 141; Leonard, 1928, Cornell Agr. Exp. Sta. Mem., 101: 456; Chagnon, 1933-40, Coleop. Prov. Quebec, p. 278, pl. 19, fig. 7; Doane et al., 1936, For. Ins., p. 192; Hoffman, 1942, USDA Misc. Pub., 466: 11; Loding, 1945, Geol. Surv. Ala. Mon., 11: 125; Knull, 1946, Ohio Biol. Surv. Bull., 39: 272; Fattig, 1947, Emory Univ. Mus. Bull., 5: 42; Breuning, 1960, Frust. Entomol., 3: 12; Dillon and Dillon, 1961, Man. Beetles East. North America, p. 651, pl. 64, no. 17; Chagnon and Robert, 1962, Prin. Coleop. Prov. Quebec, p. 278, pl. 19, fig. 7; Abdullah and Abdullah, 1966, Proc. Roy. Entomol. Soc. London, (B) 35: 92; Bayer and Shenefelt, 1969, Univ. Wisc. Res. Bull., 275: 31, fig. 39; Baker, 1972, USDA Misc. Pub., 1175: 188 (habits); Perry, 1975, Coleop. Bull., 29: 59; Chamberland, 1976, Fabreries, 2: 89; Laliberte et al., 1977, Fabreries, 3: 98; Gosling and Gosling, 1977, Gr. Lakes Entomol., 10: 32, fig. 166; Rice and Enns, 1981, Trans. Mo. Acad. Sci., 15: 103; Waters and Hyche, 1984, Coleop. Bull., 38: 285; Gosling, 1984, Gr. Lakes Entomol., 17: 73; Furth, 1985, Conn. Acad. Arts. Sci., 46: 193; Drooz, 1985, USDA For. Serv. Misc. Pub., 1426: 298; Gosling, 1986, Gr. Lakes Entomol., 19: 157.

Saperda (Saperda) lateralis: Breuning, 1952, Entomol. Arb. Mus. Frey, 3: 164; Gilmour, 1965, Cat. Lam. du Monde, 8: 669.

Compsidea lateralis: Haldeman, 1847, Trans. Amer. Philos. Soc., (2)10: 55; Haldeman, 1847, Proc. Amer. Philos. Soc., 4: 373.

Saperda lateralis lateralis: Stein and Tagestad, 1976, USDA For. Serv. Res. Pap., RM-171: 36.

Saperda lateralis var. *abbreviata* Fitch, 1859, 5th Rept. Nox. Ben. Ins. N.Y., p. 841.

Saperda lateralis ab. *abbreviata*: Aurivillius, 1923, Coleop. Cat., 74: 486.

Saperda (Saperda) lateralis m. *abbreviata*: Breuning, 1952, Entomol. Arb. Mus. Frey, 3: 165; Gilmour, 1965, Cat. Lam. du Monde, 8: 669.

Saperda lateralis var. *suturalis* Fitch, 1859, 5th Rept. Nox. Ben. Ins. N.Y., p. 841.

Saperda lateralis ab. *suturalis*: Aurivillius, 1923, Coleop. Cat., 74: 486.

Saperda (Saperda) lateralis m. *suturalis*: Breuning, 1952, Entomol. Arb. Mus. Frey, 3: 165; Gilmour, 1965, Cat. Lam. du Monde, 8: 669.

Saperda lateralis var. *connecta* Felt and Joutel, 1904, N.Y. St. Mus. Bull., 74: 60, fig. 6h, i, pl. 7, fig. 9; Chagnon, 1905, Nat. Can., 32: 43; Smith, 1910, N.J. St. Mus. Rept., 1909: 336; Leonard, 1928, Cornell. Agr. Exp. Sta. Mem., 101: 456; Abdullah and Abdullah, 1966, Proc. Roy. Entomol. Soc. London, (B) 35: 92.

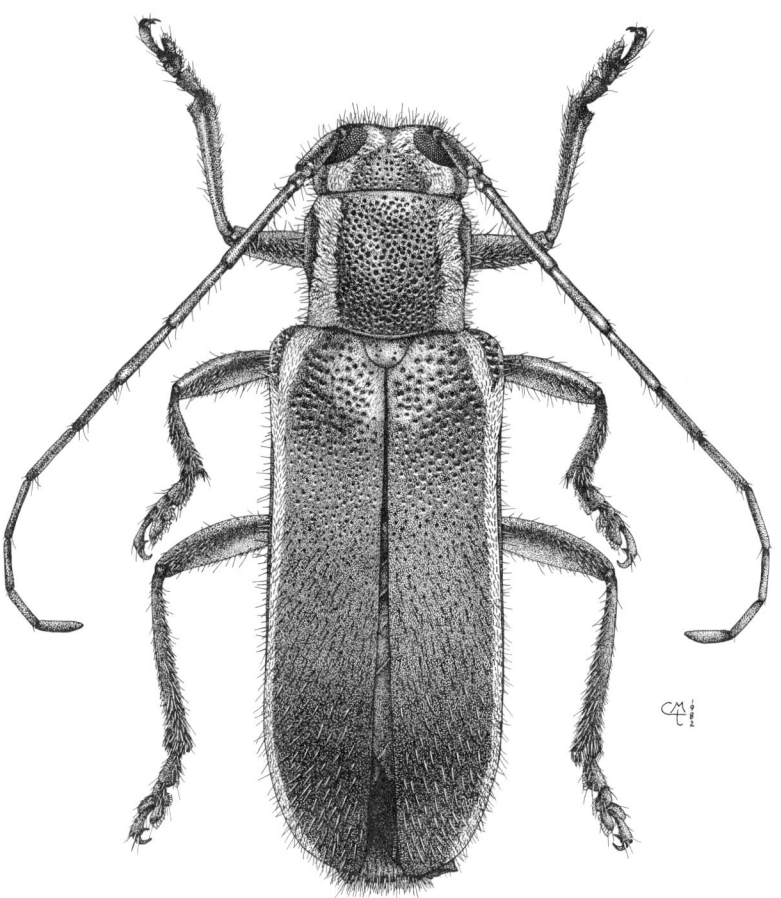

Figure 33. *Saperda lateralis* Fabricius, female.

Saperda lateralis ab. *connecta*: Aurivillius, 1923, Coleop. Cat., 74: 486.
Saperda (Saperda) lateralis m. *connecta*: Breuning, 1952, Entomol. Arb. Mus. Frey, 3: 165; Gilmour, 1965, Cat. Lam. du Monde, 8: 669.
Saperda imitans var. *connecta*: Mutchler and Weiss, 1923, N.J. Dept. Agr. Circ., 58: 11.
Saperda (Saperda) lateralis m. *transeuns* Breuning, 1952, Entomol., Arb. Mus. Frey, 3: 165, pl. 4, fig. 17; Gilmour, 1965, Cat. Lam. du. Monde, 8: 669.

Male. Form small to moderate-sized, subparallel; integument black; pubescence very fine, short, appressed, head, pronotum, and elytra with broad, longitudinal, orange pubescent vittae, occasionally transverse vittae present on elytra. Head with front convex, about as broad as long; vertex with two orange pubescent vittae beginning at bases of antennal tubercles, extending along eye margins and joining pronotal vittae; eyes with lower lobes large, slightly elongate, much longer than genae; antennae longer than body, segments uniformly pubescent, grayish beneath, brownish above, scape slender, cylindrical, third segment longer than first, fourth subequal to first. Pronotum slightly broader than long; disk convex, moderately coarsely, densely punctate; pubescence fine, dense, each side with a broad orange pubescent vitta, sides below vittae with two dark spots, long, erect hairs numerous; prosternum finely, densely, gray pubescent; meso- and metasternum finely, densely, gray pubescent. Elytra less than 2-1/2 times longer than broad; basal punctures moderately coarse, dense, becoming finer and sparser toward apex; each side with a broad, orange, sublateral pubescent vitta, suture often with a narrow vitta, transverse oblique vittae occasionally present at middle and near base; apices broadly rounded. Legs finely pubescent; front and middle tarsal claws with small processes. Abdomen finely, densely pubescent; last sternite shallowly emarginate at apex. Length, 9-12 mm.

Female. Form more robust. Antennae shorter than body. Abdomen with last sternite linearly impressed at middle. Length, 10-14 mm.

Type locality. Of *lateralis*, *abbreviata*; *suturalis*, New York; *connecta*, none; *transeuns*, Cornwells, Pennsylvania

Range. Eastern North America to Texas and North Dakota.

Flight period. May to August.

Host plants. Acer, Alnus, Carpinus, Carya, Fraxinus, Prunus, Quercus, Rhus, Sambucus, Tilia, Ulmus.

The orange pronotal and elytral vittae make this species distinctive. Occasionally the elytra will have a median transverse fascia, and additionally an occasional oblique one at the basal one-third.

Saperda imitans Felt and Joutel

Saperda imitans Felt and Joutel, 1904, N.Y. St. Mus. Bull., 74: 50, fig. 5b, pl. 3, fig. 4; Chagnon, 1905, Nat. Can., 32: 43; Blatchley, 1910, Coleoptera in Indiana, p. 1088; Mutchler and Weiss, 1923, N.J. Dept. Agr. Cir., 58: 10; Craighead, 1923, Can. Dept. Agr. Bull., (n.s.) 27: 131; Blackman and Stage, 1924, N.Y. St. Coll. For. Tech. Pub., 17: 120; Leonard, 1928, Cornell Agr. Exp. Sta. Mem., 101: 456; Chagnon, 1933-40, Coleop. Prov. Quebec, p. 278, pl. 19, fig. 4; Knull, 1946, Ohio Biol. Surv. Bull., 39: 271; Chagnon and Robert, 1962, Prin. Coleop. Prov. Quebec, p. 278, pl. 19, fig. 4; Abdullah and Abdullah, 1966, Proc. Roy. Entomol. Soc. London, (B) 35: 91; Baker, 1972, USDA Misc. Pub., 1175: 188; Gosling and Gosling, 1976, Gr.

Lakes Entomol., 10: 31, fig. 164; Laliberte et al., 1977, Fabreries, 3: 98; Rice and Enns, 1981, Trans. Mo. Acad. Sci., 15: 103; Drooz, 1985, USDA Misc. Pub., 1426: 298.

Saperda (Saperda) imitans: Breuning, 1952, Entomol. Arb. Mus. Frey, 3: 164, Gilmour, 1965, Cat. Lam. du Monde, 8: 669.

Male. Form moderate-sized, subparallel; integument black, antennae often paler; pubescence fine, appressed, grayish, head, pronotum, and elytra with orange pubescent vittae. Head with front convex, quadrate; interantennal area shallowly impressed; pubescence fine, grayish, orange vittae extending from vertex, around eyes, and joining pronotal vittae; eyes with lower lobes slightly elongate, genae very short, antennae slightly longer than body, uniformly pubescent, scape slender, conical, third segment larger than first, fourth subequal to first. Pronotum about as long as wide; disk convex, densely punctate; pubescence fine, dense, grayish, each side with a broad orange pubescent vitta, middle often with a thin grayish vitta, each side with two black spots; prosternum densely gray pubescent; meso- and metasternum densely gray pubescent. Elytra more than 2-1/2 times longer than broad; basal punctures moderately coarse, contiguous, punctures becoming finer and sparser toward apex; pubescence fine, grayish, each elytron with a broad, orange pubescent vitta extending from humeri to apex but not attaining lateral margins, basal one third with oblique orange vittae, middle with oblique orange vittae, apical one-third with short transverse vittae arising from lateral vittae; apices rounded. Legs finely, densely pubescent; front and middle tarsal claws with small processes. Abdomen densely pubescent; last sternite shallowly emarginate at apex. Length, 10-15 mm.

Female. Form more robust. Head with front plane; antennae about as long as body. Abdomen with last sternite longitudinally impressed at middle. Length, 14-16 mm.

Type locality. Near New York City.

Range. Northeastern North America.

Flight period. May to July.

Host plants. Carya, Prunus, Salix, Tilia.

This species can be separated from *S. tridentata* by the oblique vittae at the basal one-third of the elytra and lack of elytral carinae.

Saperda tridentata Olivier

Saperda tridentata Olivier, 1795, Entomol., 4: 30, pl. 4, fig. 48; LeConte, 1852, J. Acad. Nat. Sci. Philadelphia, 2: 164; Fitch, 1858, Nox. Ben. Ins. N.Y. 5th Rept., p. 59; Harris, 1862, Ins. Inj. Veg., ed. 3, p. 111, pl. 11, fig. 13; Packard, 1870, Amer. Nat., 4: 588, figs. 115, 116; Lacordaire, 1872, Genera des coléoptères, 9(2): 834, fn.; LeConte, 1873, Smithson. Misc. Coll., 11(264): 239; Provancher, 1877, Pet. Fauna Entomol. Can., 1: 634; Packard, 1881, U.S. Entomol. Comm. Bull., 7: 58, fig. 17; Harrington,

1883, Can. Entomol., 15: 79; Knobel, 1895; Beetles of New England, p. 34, fig. 126; Leng and Hamilton, 1896, Trans. Amer. Entomol. Soc., 23: 150; Beutenmuller, 1896, J. N.Y. Entomol. Soc., 4: 80; Wickham, 1898, Can.Entomol., 30: 41; Lugger, 1899, Minn. Agr. Exp. Sta. Bull., 66: 215; Harrington, 1899, Ottawa Nat., 13: 67; Felt, 1900, N.Y. St. Mus. Bull., 6: 581; Felt and Joutel, 1904, N.Y. St. Mus. Bull., 74: 44, pl. 3, figs. 1, 3; Chagnon, 1905, Nat. Can., 32: 43; Felt, 1907, Ins. Affect. Trees, p. 67, pl. 3, figs. 1-3; Blatchley, 1910, Coleoptera in Indiana, p. 1087, fig. 471; Smith, 1910, N.J. St. Mus. Rept., 1909: 336; Douglass, 1912, St. Entomol. Indiana. 4th Ann. Rept., p. 108 (habits); Johannsen, 1912, Maine Agr. Exp. Sta. Bull., 207: 464 (habits); Blackman and Ellis, 1916, N.Y. St. Coll. For. Bull., 16: 52, fig. 22 (habits); Kotinsky, 1921, USDA Farm. Bull., 1169: 55, fig. 34 (habits); Mutchler and Weiss, 1923, N.J. Dept. Agr. Circ., 58: 10 (habits); Craighead, 1923, Can. Dept. Agr. Bull., (n.s.) 27: 130; Leonard, 1928, Cornell Agr. Exp. Sta. Mem., 101: 455; Park, 1931, Ecology, 12: 192; Felt and Rankin, 1932, Ins. Dis. Ornam Trees Shrubs, p. 222, fig. 71 (habits); Chagnon, 1933-40, Coleop. Prov. Quebec, p. 278, pl. 19, fig. 5; Britton and Friend, 1935, Conn. Agr. Exp. Sta. Bull., 369: 300, figs. 66-68 (habits); Doane et al., 1936, For. Ins., p. 191; Kaston, 1937, Conn. Agr. Exp. Sta. Bull., 396: 354, 355, 357 (parasites); Fenton, 1939, Okla. Agr. Exp. Sta. Circ., 84: 18 (habits); Pechuman, 1940, Bull. Brooklyn Entomol. Soc., 35: 113 (habits); Hoffmann, 1942, USDA Misc. Pub., 466: 11; Knull, 1946, Ohio Biol. Surv. Bull., 39: 270; Fattig, 1947, Emory Univ. Mus. Bull., 5: 42; Craighead, 1950, USDA Misc. Pub., 657: 267, fig. 52C (habits); Beal, 1952, Duke Univ. Sch. For. Bull., 14: 56 (habits); English, 1958, Ill. Nat. Hist. Surv. Circ., 47: 29, fig. 24 (habits); Breuning, 1960, Frust. Entomol., 3: 9; Dillon and Dillon, 1961, Man. Beetles East. North America, p. 650, pl. D, pl. 65, no. 2; Chagnon and Robert, 1962, Prin. Coleop. Prov. Quebec, p. 278, pl. 19, fig. 5; Abdullah and Abdullah, 1966, Proc. Roy. Entomol. Soc. London, (B) 35: 94; Gardiner, 1969, Can. Dept. Fish. For. Int. Rept., 0-14: 98 (larva); Bayer and Shenefelt, 1969, Univ. Wisc. Res. Bull., 275: 32, fig. 40; Hatch, 1971, Univ. Wash. Pub. Biol., 16: 155; Kirk and Balsbaugh, 1975, South Dakota Agr. Exp. Sta. Tech. Bull., 42: 100; Gosling and Gosling, 1976, Gr. Lakes Entomol., 10: 33, fig. 171; Headstrom, 1977, Beetles of America, p. 381; Laliberte et al., 1977, Fabreries, 3: 98; Solomon, 1977, Can. Entomol., 109: 298; Rice and Enns, 1981, Trans. Mo. Acad. Sci., 15: 104; Drooz, 1985, USDA Misc. Pub., 1426: 298, fig. 133 (habits); Gosling, 1986, Gr. Lakes Entomol., 19: 157.

Saperda (Saperda) tridentata: Breuning, 1952, Entomol. Arb. Mus. Frey, 3: 162, pl. 4, fig. 15; Gilmour, 1965, Cat. Lam. du Monde, 8: 669.

Saperda tridenta: Emmons, 1854, Nat. Hist. N.Y., Agr., 5: 122, pl. 34, fig. 6 (error).

Saperda trilineata: Hubbard, 1877, Psyche, 2: 40 (error).

Compsidea tridentata: Haldeman, 1847, Trans. Amer. Philos. Soc., (2)10: 55; Haldeman, 1847, Proc. Amer. Philos. Soc., 4: 373.

188 University of California Publications in Entomology

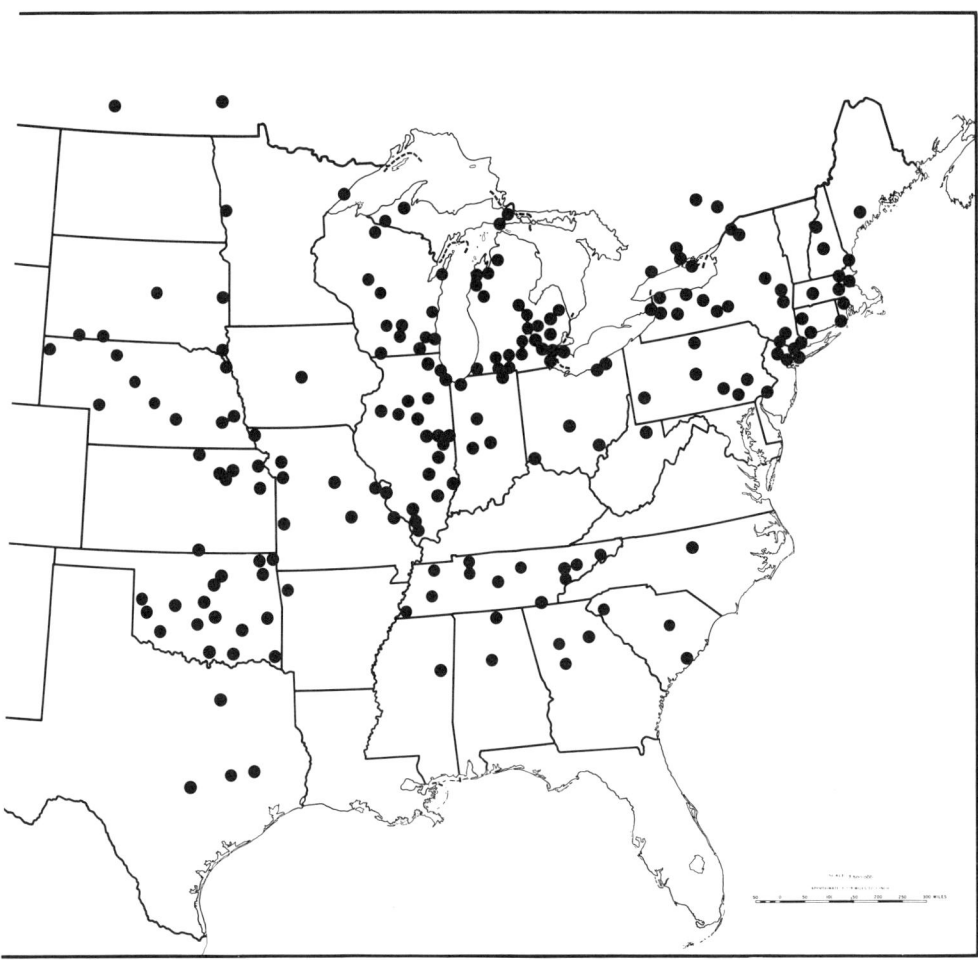

Figure 34. Known geographic range of *Saperda tridentata* Olivier.

Eutetrapha tridentata: Aurivillius, 1923, Coleop. Cat., 74: 489; Baker, 1972, USDA Misc. Pub., 1175: 188, fig. 61 (habits); Stein and Tagestad, 1976, USDA For. Serv. Res. Pap., RM-171: 15.
Compsidea tridentata var. *dubiosa* Haldeman, 1847, Trans. Amer. Philos. Soc., (2)10: 55.
Eutetrapha tridentata ab. *dubiosa*: Aurivillius, 1923, Coleop. Cat., 74: 490.
Saperda tridentata var. *rubronotata* Fitch, 1859, Nox. Ins. N.Y. 5th Rept., p. 840.
Eutetrapha tridentata ab. *rubronotata*: Aurivillius, 1923, Coleop. Cat., 74: 490.

Saperda (Saperda) tridentata m. *rubronotata*: Breuning, 1952, Entomol. Arb. Mus. Frey, 3: 163, pl. 4, fig. 16; Gilmour, 1965, Cat. Lam du Monde, 8: 669.
Saperda tridentata var. *intermedia* Fitch, 1859, Nox. Ins. N.Y. 5th Rept., p. 840.
Eutetrapha tridentata ab. *intermedia*: Aurivillius, 1923, Coleop. Cat., 74: 490.
Saperda (Saperda) tridentata m. *intermedia*: Breuning, 1952, Entomol. Arb. Mus. Frey, 3: 163; Gilmour, 1965, Cat. Lam. du Monde, 8: 669.
Saperda tridentata trifasciata Casey, 1913, Memoirs on the Coleoptera, 4: 359.
Eutetrapha tridentata var. *trifasciata*: Aurivillius, 1923, Coleop. Cat., 74: 490
Saperda (Saperda) tridentata m. *trifasciata*: Breuning, 1952, Entomol. Arb. Mus. Frey, 3: 163; Gilmour, 1965, Cat. Lam. du Monde, 8: 669.
Saperda lateralis v. *disconotata* Pic, 1907, Echange, 23: 152.
Saperda lateralis ab. *disconotata*: Aurivillius, 1923, Coleop. Cat., 74: 486.

Male. Form small, subparallel; integument black to reddish brown; pubescence dense, short, appressed, grayish and orange. Head with front convex, quadrate; interantennal area barely impressed; vertex with orange pubescent vittae around eyes, usually extending onto neck to join pronotal vittae; eyes with lower lobes large, slightly elongate, much longer than genae; antennae slightly longer than body, segments uniformly pubescent, scape slender, cylindrical, third segment longer than first, fourth subequal to first. Pronotum about as long as wide; disk convex with a thin longitudinal, median line; punctures moderately coarse, dense; pubescence fine, dense, sides with rather broad, orange pubescent vittae, apical and basal margins narrowly orange pubescent; prosternum densely gray pubescent; meso- and metasternum uniformly, densely gray pubescent. Elytra about 2-1/2 times as long as broad; each side with a submarginal carina; punctures coarse, contiguous, becoming finer and sparser toward apex; pubescence fine, dense, appressed, vittae orange pubescent, basal one-fourth with a pair extending down from humeri, right-angled across and directed back near suture, middle and apical one-third with oblique vittae, vittae usually joining near lateral margins and rarely, along suture, basal and subapical vittae with vague dark spots behind; apices rounded. Legs finely pubescent; front and middle tarsal claws with distinct processes. Abdomen densely gray pubescent; last sternite shallowly emarginate at apex. Length, 10-14 mm.

Female. Form more robust. Head with front flattened; antennae shorter than body. Abdomen with last sternite elongate, medially, narrowly impressed. Length, 13-17 mm.

Type locality. Of *tridentata*, Canada; *dubiosa*, none; *intermedia*, New York; *rubronotata*, New York; *trifasciata*, Indiana; *disconotata*, none (United States).

Range. Eastern North America to Kansas (Figure 34).

Flight period. April to August.

Host plants. Ulmus.

Parasites. Xorides albopictus Cresson, *Heterospilus* sp., *Atanycolus ulmicola* Viereck, *Capitanius saperdae* Ashmead.

The orange pubescent vittae and submarginal carinae of the elytra distinguish this species. The vittae vary in width and often are not connected laterally.

The variety *disconotata* Pic was placed in synonymy with *S. tridentata* m. *rubronotata* Fitch by Breuning in 1952.

Saperda vestita Say

Saperda vestita Say, 1824, Narr. Exp. source St. Peters River, p. 290; Haldeman, 1847, Trans. Amer. Philos. Soc., (2)10: 55; LeConte, 1852, J. Acad. Nat. Sci. Philadelphia, 2: 163; Emmons, 1854, Nat. Hist. N.Y., Agr., 5: 121, pl. 34, fig. 4; LeConte, 1859, Compl. Writings T. Say, 1: 193; Packard, 1870, Amer. Nat., 4: 591, figs. 117, 118; Lacordaire, 1872, Genera des coléoptères, 9(2): 834, fn.; LeConte, 1873, Smithson. Misc. Coll., 11(264): 238; Provancher, 1877, Pet. Fauna Entomol. Can., 1: 634; Packard, 1881, U.S. Entomol. Comm. Bull., 7: 123, 124, fig. 59; Harrington, 1890, Entomol. Soc. Ontario. 20th Rept., 1889; 52; Packard, 1890, U.S. Entomol. Comm. 5th Rept., p. 226, 474, fig. 171; Knobel, 1895, Beetles of New England, p. 34, fig. 124; Beutenmuller, 1896, J. N.Y. Entomol., Soc., 4: 80; Leng and Hamilton, 1896, Trans. Amer. Entomol. Soc. 23: 149; Wickham, 1898, Can. Entomol., 30:41; Harrington, 1899, Ottawa Nat., 13: 62; Webster, 1900, J. Columbus Hort. Soc., 15: 3, fig. 3; Felt and Joutel, 1904, N.Y. St. Mus. Bull., 74: 54, pl. 5, figs. 1, 5 (habits); Chagnon, 1905, Nat. Can., 32: 43; Felt, 1907, Ins. Affect. Trees, p. 91, pl. 6, figs. 7-14, 16; Smith, 1910, N.J. St. Mus. Rept., 1909: 336; Blatchley, 1910, Coleoptera in Indiana, p. 1089, fig. 473; Britton, 1916, Conn. St. Entomol. 15th Rept., 1915: 186, pl. 13b (habits); Kotinsky, 1921, USDA Farm. Bull., 1169: 57, fig. 35 (habits); Mutchler and Weiss, 1923, N.J. Dept. Agr. Circ., 58: 8 (habits); 1923, Craighead, Can. Dept. Agr. Bull., (n.s.) 27: 130; Britton and Zappe, 1927, Conn. Agr. Exp. Sta. Bull., 292: 148, pl. 10 (habits); Leonard, 1928, Cornell Agr. Exp. Sta. Mem., 101: 456; Felt and Rankin, 1932, Ins. Dis. Ornam. Trees Shrubs, p. 291 (habits); Chagnon, 1933-40, Coleop. Prov. Quebec, p. 276, pl. 19, fig. 3; Loding, 1945, Geol. Surv. Ala. Mon., 11: 125; Knull, 1946, Ohio Biol. Surv. Bull., 39: 271; Fattig, 1947, Emory Univ. Mus. Bull., 5: 42; Craighead, 1950, USDA Misc. Pub., 657: 268 (habits); Dillon and Dillon, 1961, Man. Beetles East. North America, p. 651, pl. 65, no. 8; Chagnon and Robert, 1962, Prin. Coleop. Prov. Quebec, p. 276, pl. 19, fig. 3; Abdullah and Abdullah, 1966, Proc. Roy. Entomol. Soc. London, (B) 35: 94; Bayer and Shenefelt, 1969, Univ. Wisc. Res. Bull., 275: 32, fig. 40; Gardiner, 1969, Can. Dept. Fish. For. Int. Rept., 0-14: 101 (larva); Hatch, 1971, Univ. Wash. Pub. Biol., 16: 155; Baker, 1972, USDA Misc. Pub., 1175: 187 (habits); Stein and Tagestad, 1976, USDA For. Serv. Res. Pap., RM-171: 38; Headstrom,

1977, Beetles of America, p. 382, fig. 533; Gosling and Gosling, 1977, Gr. Lakes Entomol., 10: 33, fig. 172; Laliberte et al., 1977, Fabreries, 3: 98; Rice and Enns, 1981, Trans. Mo. Acad. Sci., 15: 104; Gosling, 1984, Gr. Lakes Entomol., 17: 73; Drooz, 1985, USDA Misc. Pub., 1426: 297 (habits); Gosling, 1986, Gr. Lakes Entomol., 19: 157.*Saperda (Saperda) vestita*: Breuning, 1952, Entomol. Arb. Mus. Frey, 3: 18; Gilmour, 1965, Cat. Lam. du Monde, 8: 669.

Saperda vestida: Lugger, 1899, Minn. Agr. Exp. Sta. Bull., 66: 215 (error).

Saperda vestita morpha *immaculata* Breuning, 1976, Rev. Suisse Zool., 83: 740.

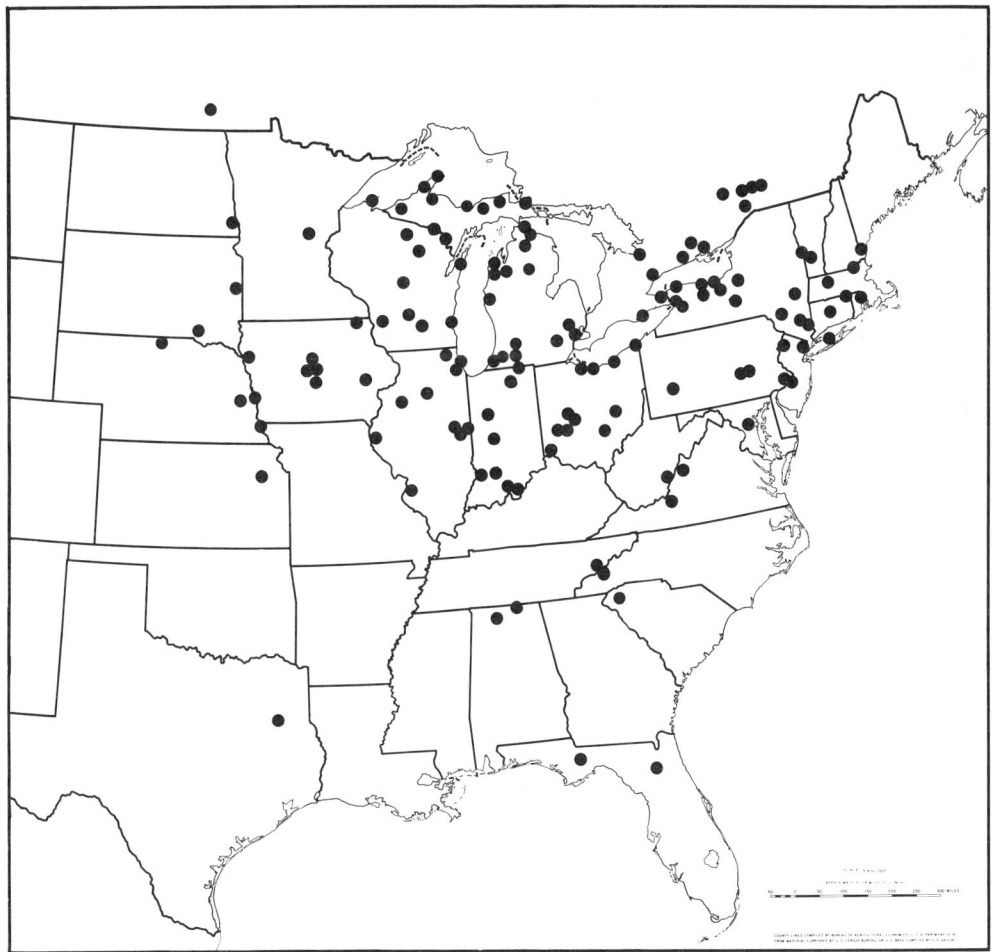

Figure 35. Known geographic range of *Saperda vestita* Say.

Saperda atkinsoni Curtis, 1829, British Entomol., 6: pl. 275; Guerin, 1844, Icon. Regne Anim. Ins., p. 445, pl. 45, fig. 5.

Male. Form moderate-sized, tapering posteriorly; integument black to pale reddish brown, antennae paler; pubescence dense, appressed, olive-brown. Head with front convex, quadrate; interantennal area shallowly impressed; pubescence dense, obscuring surface; eyes with lower lobes slightly longer than broad, genae very small; antennae slightly longer than body, uniformly pubescent, segments from fourth pale brownish, scape slender, conical, third segment longer than first, fourth subequal to first. Pronotum broader than long; disk convex, usually with a median longitudinal line; pubescence dense, obscuring surface; prosternum densely pubescent; meso- and metasternum densely pubescent. Elytra about 2-1/2 times as long as broad; basal punctures coarse, subconfluent, becoming finer and sparser toward apex; pubescence dense, uniform, each elytron with two brown spots near suture before middle, two median spots before sutural pair, and usually two smaller spots at apical one-third; apices rounded. Legs densely pubescent; front and middle tarsal claws with processes. Abdomen densely pubescent; last sternite subtruncate to shallowly emarginate at apex. Length, 12-17 mm.

Female. Form more robust. Head with front plane, genae longer; antennae slightly shorter than body. Abdomen with last sternite longitudinally impressed at middle. Length, 15-23 mm.

Type locality: Of *vestita*, southern Lake Michigan; *atkinsoni, immaculata*, Pennsylvania.

Range. Eastern North America (Figure 35).

Flight period. May to September.

Host plants. *Acer, Populus, Tilia*.

The uniformly olive-brown pubescence and brownish spots of the elytra readily distinguish this species. The apical pair of spots is often lacking.

Saperda discoidea Fabricius
(Figure 36)

Saperda discoidea Fabricius, 1798, Entomol. Syst., Suppl., p. 147; Fabricius, 1801, Syst. Eleuth., 2: 322; LeConte, 1852, J. Acad. Nat. Sci. Philadelphia, 2: 163; Fitch, 1856, Nox. Ben. Ins. N.Y., 3rd Rept., p. 122; Fitch, 1856, Trans. N.Y. St. Agr. Soc., 16: 440; Lacordaire, 1872, Genera des coléoptères, 9(2): 334, fn.; LeConte, 1873, Smithson. Misc. Coll., 11(264): 238; Packard, 1881, U.S. Entomol. Comm. Bull., 7: 70; Harrington, 1884, Can. Entomol., 16: 102; Harrington, 1885, Can. Entomol., 17: 47; Harrington, 1889, Entomol. Soc. Ontario, 20th Rept., p. 52; Hamilton, 1895, Trans. Amer. Entomol. Soc., 22: 369; Beutenmuller, 1896, J. N.Y. Entomol. Soc., 4: 80; Leng and Hamilton, 1896, Trans. Amer. Entomol. Soc., 23: 150; Wickham, 1896, Trans. Amer. Entomol. Soc., 23: 150; Wickham, 1898, Can. Entomol., 30: 41; Harrington, 1899, Ottawa Nat., 13:

67; Lugger, 1899, Minn. Agr. Exp. Sta. Bull., 66: 215; Felt and Joutel, 1904, N.Y. St. Mus. Bull., 74: 52, pl. 3, figs. 2, 5, 6 (habits); Felt, 1907, Ins. Affect. Trees, p. 269; Blatchley, 1910, Coleoptera in Indiana, p. 1089; Smith, 1910, N.J. St. Mus. Rept., 1909: 336; Mutchler and Weiss, 1923, N.J. Dept. Agr. Circ., 58: 9 (habits); Craighead, 1923, Can. Dept. Agr. Bull., (n.s.) 27: 130, pl. 44; Blackman and Stage, 1924, N.Y. St. Coll. For. Tech. Pub., 17: 121; Leonard, 1928, Cornell Agr. Exp. Sta. Mem., 101: 456; Felt and Rankin, 1932, Ins. Dis. Ornam. Trees Shrubs, p. 259; Barrett, 1932, Univ. Calif. Pub. Entomol., 5: 291; Doane et al., 1936 For. Ins., p. 191; Kaston, 1937, Conn. Agr. Exp. Sta. Bull., 396: 357; Procter, 1938, Biol. Surv. Mt. Desert, 6: 154 (habits); Knull, 1946, Ohio Biol. Surv. Bull., 39: 271; Fattig, 1947, Emory Univ. Mus. Bull., 5: 42; Craighead, 1950, USDA Misc. Pub., 657: 267 (habits); Duffy, 1953, Mon. British Timber Beetles, p. 290, fig. 279; Dillon and Dillon, 1961, Man. Beetles East. North America, p. 650, pl. 64, nos. 15, 16; Gardiner, 1969, Can. Dept. Fish. For. Int. Rept., 0-14: 102 (larva); Baker, 1972, USDA Misc. Pub., 1175: 187 (habits); Gosling and Gosling, Gosling, 1977, Gr. Lakes Entomol., 10: 31, fig. 162; Headstrom, 1977, Beetles of America, p. 382; Rice and Enns, 1981, Trans. Mo. Acad. Sci., 15: 103; Gosling, 1984, Gr. Lakes Entomol., 17: 73; Drooz, USDA Mis. Pub., 1426: 298 (habits).

Saperda (Saperda) discoidea: Breuning, 1952, Entomol. Arb. Mus. Frey, 3: 167; Gilmour, 1965, Cat. Lam. du Monde, 8: 669.

Saperda discoides: Abdullah and Abdullah, 1966, Proc. Roy. Entomol. Soc. London, (B) 35: 91 (error).

Saperda fuscipes Say, 1826, J. Acad. Nat. Sci. Philadelphia, 2: 273; Haldeman, 1847, Proc. Amer. Philos. Soc., 4: 373 (synonymy); LeConte, 1852, J. Acad. Nat. Sci. Philadelphia, 2: 163; LeConte, 1859, Compl. Writings T. Say, 2: 331; Lacordaire, 1872, Genera des coléoptères, 9(2): 834, fn.

Stenostola fuscipes: Haldeman, 1847, Trans. Amer. Philos. Soc., (2)10: 56.

Stenostola fuscipes var. *dorsalis* Haldeman, 1847, Trans. Amer. Philos. Soc., (2)10: 56.

Saperda discoidea ab. *dorsalis*: Aurivillius, 1923, Coleop. Cat., 74: 485.

Male. Form small to moderate-sized, subparallel; integument black to dark reddish brown, femora and often tibiae reddish orange; pubescence very short, dense, appressed, grayish. Head with front convex, quadrate; interantennal area not impressed; pubescence short, appressed, denser around eyes; eyes with lower lobes large, much longer than genae; antennae longer than body, segments uniformly pubescent, scape slender, subcylindrical, third segment longer than first and fourth, fourth longer than first. Pronotum slightly longer than broad; disk convex, median line narrow; sides slightly impressed near base; punctures moderately coarse, dense, contiguous; pubescence dense, grayish, each side of middle with a broad brownish vitta, extreme sides with short brownish vittae; prosternum densely whitish pubescent; meso- and metasternum densely whitish pubescent. Elytra almost three times as long as broad; punctures moderately

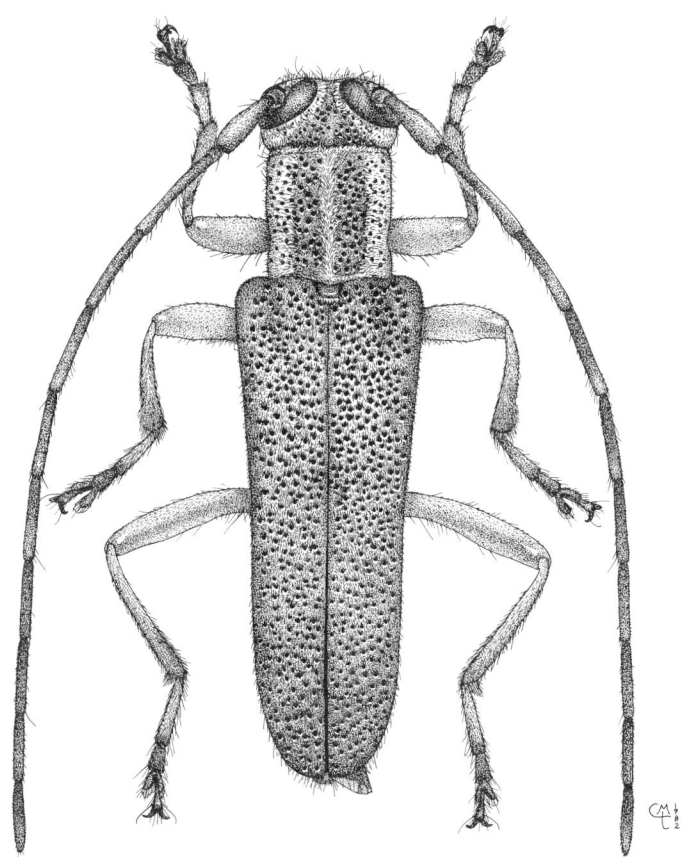

Figure 36. *Saperda discoidea* Fabricius, male.

coarse, dense, contiguous, becoming finer toward apex; pubescence short, dense, uniform, appressed, short, erect hairs numerous; apices broadly rounded. Legs finely pubescent; front and middle tarsal claws with a large process. Abdomen densely whitish pubescent; last sternite truncate at apex. Length, 9-13 mm.

Female. Form more robust; integument dark brownish to reddish brown; pubescence yellow-brown, elytra with two irregular vittae at middle. Head with front plane; antennae shorter than body. Pronotum broader at apex; disk usually with two vague tubercles just before middle; pubescence usually uniformly dense. Elytra about 2-1/2 times as long as broad; each elytron with an irregular, pubescent, transverse vitta near middle, pubescence usually denser along margins and at apex. Abdomen with last sternite elongate, medially impressed. Length, 13-17 mm.

Figure 37. Known geographic range of *Saperda discoidea* Fabricius.

Type locality. Of discoidea, America boreali; *fuscipes*, none; *dorsalis*, none.
Range. Eastern North America (Figure 37).
Flight period. May to September.
Host plants. Carya, Caryocar, Crataegus, Juglans, Ulmus.
Parasites. Xorides albopictus Cresson; *Deuteroxorides caryae* Harrington.

The reddish legs and uniform pubescence of the elytra distinguish males of this species from other *Saperda*. The two irregular pubescent spots of the elytra separate the females.

Saperda cretata Newman

Saperda cretata Newman, 1838, Entomol. Mon. Mag., 5: 396; LeConte, 1852, J. Acad. Nat. Sci. Philadelphia, 2: 164; Lacordaire, 1872, Genera des coléoptères, 9(2): 834, fn; LeConte, 1873, Smithson. Misc. Coll., 11(264):238; Osborn, 1881, Trans. Iowa St. Hort. Soc., 15: 11; Knobel, 1895, Beetles of New England, p. 34, fig. 123; Leng and Hamilton, 1896, Trans. Amer. Entomol. Soc., 23: 149; Wickham, 1898, Can. Entomol., 30: 41; Chittenden, 1898, USDA Div. Entomol. Circ., 32: 8, fig. 2; Lugger, 1899, Minn. Agr. Exp. Sta. Bull., 66: 215, fig. 135; Lugger, 1899, Minn. Exp. Sta. 5th Ann. Rept., p. 131, fig. 135; Felt and Joutel, 1904, N.Y. St. Mus. Bull. 74: 50, pl. 4, figs. 1, 2; Chittenden, 1907, USDA Bur. Entomol. Circ., 32:7, fig. 2 (habits); Blatchley, 1910, Coleoptera in Indiana, p. 1088, fig. 472; Brooks, 1920, USDA Bull., 886: 1 (habits); Wellhouse, 1922, Cornell Agr. Exp. Sta. Mem., 56: 1103; Mutchler and Weiss, 1923, N.J. Dept. Agr. Circ., 58: 13; Leonard, 1928, Cornell Agr. Exp. Sta. Mem., 101: 455; Knull, 1946, Ohio Biol. Surv. Bull., 39: 270, pl. 21, fig. 83; Dillon and Dillon, 1961, Man. Beetles East. North America, p. 650, pl. 65, no. 7; Abdullah and Abdullah, 1966, Proc. Entomol. Soc. London, (B) 35: 91; Baker, 1972, USDA Misc. Pub., 1175: 188 (habits); Gosling and Gosling, 1976, Gr. Lakes Entomol., 10: 31, fig. 161; Headstrom, 1977, Beetles of America, p. 381, fig. 532; Drooz, 1985, USDA Misc. Pub., 1426: 298.

Saperda (*Saperda*) *cretata*: Breuning, 1952, Entomol. Arb. Mus. Frey, 3: 168, pl. 4, fig. 2.

Saperda crenata: Bayer and Shenefelt, 1969, Univ. Wisc. Res. Bull., 275: 31, fig. 39 (error).

Male. Form moderate-sized, subparallel; integument brown; pubescence very fine, brownish and white, dense, appressed. Head with front convex, quadrate; interantennal area shallowly impressed; pubescence fine, dense, appressed, brownish, neck with a white spot on each side not extending up to eyes; eyes with lower lobes slightly elongate, genae very short; antennae slightly longer than body, uniformly pubescent, scape slender, cylindrical, third segment longer than first, fourth subequal to first. Pronotum slightly broader than long; disk convex, usually with a vague median, longitudinal carina; pubescence dense, each side with a broad white pubescent vittae, base at middle with a small white spot; prosternum with white vittae outside of coxal cavities; meso- and metasternum with broad white vittae at sides, expanding toward middle on metasternum. Elytra about 2-1/2 times longer than broad; basal punctures moderately coarse, separated, becoming finer toward apex; pubescence fine, brownish, each elytron with a broad, elongate white pubescent chevron mostly before middle and an apostrophe-like spot before apex, base usually with a small white spot on each side of scutellum; apices rounded. Legs finely, densely pubescent; front and middle tarsal claws with large processes. Abdomen with white pubescent vittae at sides; last sternite truncate at apex. Length, 15-17 mm.

Female. Form more robust. Antennae slightly shorter than body. Abdomen with last sternite longitudinally impressed at middle. Length, 16-21 mm.

Type locality. United States.
Range. Northeastern North America.
Flight period. May to August.
Host plants. Crataegus, Malus.

The shape of the white elytral spots distinguishes *S. cretata* from *S. fayi*. In *S. cretata* the anterior markings are rectangular with irregular anterior and posterior margins, and the subapical spots resemble an apostrophe. If present, the basal spots are greatly reduced.

Saperda fayi Bland

Saperda fayi Bland, 1863, Proc. Entomol. Soc. Philadelphia, 2: 320; LeConte, 1873, Smithson. Misc. Coll., 11(264): 238; Zimmerman, 1878, Can. Entomol., 10: 220; Hamilton, 1888, Can. Entomol., 20: 6 (habits); Hamilton, 1889, Can. Entomol., 21: 104 (habits); Harrington, 1889, Entomol. Soc. Ontario, 20th Rept., p. 52 (habits); Beutenmuller, 1896, J. N.Y. Entomol. Soc., 4: 80; Leng and Hamilton, 1896, Trans. Amer. Entomol. Soc., 23: 149; Wickham, 1898, Can. Entomol., 30: 41; Chittenden, 1898, USDA Div. Entomol. Circ., 32: 8; Felt and Joutel, 1909, N.Y. St. Mus. Bull., 74: 62, pl. 6, figs. 1-4, pl. 13, figs 1, 2 (habits); Felt, 1907, Ins. Affect. Trees, p. 283, pl. 6, figs. 20-23; Smith, 1910, N.J. St. Mus. Rept., 1909: 336; Blatchley, 1910, Coleoptera in Indiana, p. 1088; Wellhouse, 1922, Cornell Agr. Exp. St. Mem., 56: 1103; Mutchler and Weiss, 1923, N.J. Dept. Agr. Circ., 58: 7 (habits); Leonard, 1928, Cornell Agr. Exp. Sta. Mem., 101: 456; Felt and Rankin, 1932, Ins. Dis. Ornam. Trees Shrubs, p. 459, fig. 229 (habits); Knull, 1946, Ohio Biol. Surv. Bull., 39: 272; Craighead, 1950, USDA Misc. Pub., 657: 266; Abdullah and Abdullah, 1966, Proc. Roy Entomol. Soc. Lond., (B) 35: 91; Baker, 1972, USDA Misc. Pub., 1175: 188 (habits); Gosling and Gosling, 1976, Gr. Lakes Entomol., 10: 31, fig. 163; Chamberland, 1976, Fabreries, 2: 89; Laliberte et al., 1977, Fabreries, 3: 98; Headstrom, 1977, Beetles of America, p. 382; Drooz, 1985, USDA Misc. Pub., 1426: 298 (habits).

Saperda (Saperda) fayi: Breuning, 1952, Entomol. Arb. Mus. Frey, 3: 165, pl. 4, fig. 18.

Saperda shoemakeri Davis, 1923, Bull. Brooklyn Entomol. Soc., 18: 97; Chemsak and Linsley, 1982, Checklist of Ceramb., p. 107. New synonymy.

Saperda (Saperda) fayi m. *shoemakeri*: Breuning, 1952, Entomol. Arb. Mus. Frey, 3: 167; Gilmour, 1965, Cat. Lam. du Monde, 8: 669.

Saperda (Saperda) fayi m. *immaculipennis* Breuning, 1952, Entomol. Arb. Mus. Frey, 3: 167; Gilmour, 1965, Cat. Lam. du Monde, 8: 669.

Male. Form moderate-sized, subparallel; integument brownish, appendages darker; pubescence fine, dense, brownish and white. Head with front convex, quadrate; interantennal area shallowly impressed; pubescence fine, dense, grayish, neck often with a small white spot on each side joining pronotal vittae; eyes with lower lobes almost rounded, genae very short; antennae slightly longer than body, uniformly pubescent, scape slightly robust, conical, third segment longer than first, fourth slightly longer than first. Pronotum about as long as broad; disk convex, middle with a narrow longitudinal line; pubescence rather sparse, each side with a densely pubescent white vitta; prosternum with a white vitta at sides; meso- and metasternum with broad white vittae at sides. Elytra about 2-1/2 times longer than broad; basal punctures moderately coarse, contiguous, becoming finer and sparser toward apex; pubescence very fine, brownish, each elytron with two elongate, white, pubescent spots along suture at middle and two smaller spots before apex, base with short vittae extending from pronotal pair; apices rounded. Legs finely pubescent, front and middle tarsal claws with vague processes. Abdomen with broad white pubescent vittae at sides, bases of segments one to four narrowly white pubescent; last sternite broadly truncate at apex. Length, 9-12 mm.

Female. Form more robust. Antennae slightly shorter than body. Elytra with white spots larger. Abdomen with last sternite longitudinally impressed at middle. Length, 11-15 mm.

Type locality. Of *fayi*, Ohio; *shoemakeri*; Slide Mountain, Catskill Mountains, New York; *immaculipennis*, Ridgeway, Ontario.

Range. Northeastern North America.

Flight period. May to July.

Host plants. Crataegus.

This species averages smaller in size than *S. cretata,* and the shape of the white elytral spots is different.

Saperda inornata Say

Saperda inornata Say, 1824, J. Acad. Nat. Sci. Philadelphia, 3: 407; LeConte, 1852, J. Acad. Nat. Sci. Philadelphia, 2: 164; LeConte, 1859, Compl. Writings T. Say, 2: 189; Lacordaire, 1872, Genera des coléoptères, 9(2): 834, fn.; Grimble and Knight, 1970, Ann. Entomol. Soc. Amer., 63: 1309, 2 figs. (habits); Nord and Knight, 1972, Gr. Lakes Entomol., 5: 28; Nord and Knight, 1972, Gr. Lakes Entomol., 5: 87; Nord and Knight, 1972, Gr. Lakes Entomol., 5: 93; Nord, Grimble, and Knight, 1972, Ann. Entomol. Soc. Amer., 65: 127; Gosling and Gosling, 1976, Gr. Lakes Entomol., 10: 32, fig. 165; Melville, 1980, Bull. Zool. Nomen., 37: 89; Drooz, 1985, USDA Misc. pub., 1426: 298 (habits).

Saperda (Mecas) inornata: Packard, 1881, U.S. Entomol. Comm. Bull., 7: 141.

Saperda (Saperda) inornata: Breuning, 1952, Entomol. Arb. Mus. Frey, 3: 169: Gilmour, 1965, Cat. Lam. du Monde, 8: 670.
Mecas inornata: Walsh., 1867, Proc. Entomol. Soc. Philadelphia, 6: 264; Packard, 1890, U.S. Entomol. Comm. 5th Rept., p. 427.
Saperda concolor LeConte, 1852, J. Acad. Nat. Sci. Philadelphia, 2: 163; Lacordaire, 1872, Genera des coléoptères, 9(2): 834, fn.; LeConte, 1873, Smithson. Misc. Coll., 11(264): 239; LeConte, 1873, Smithson. Misc. Coll., 11(265): 346; Provancher, 1877, Pet. Fauna Entomol. Canada, 1: 635; Packard, 1881, U.S. Entomol. Comm. Bull., 7: 118; LeConte and Horn, 1883, Smithson. Misc. Coll., 507: 331; Hamilton, 1885, Can. Entomol., 17: 36; Hamilton, 1888, Can. Entomol., 20: 8 (habits); Hamilton, 1889, Can. Entomol., 21: 105 (habits); Davis, 1891, Ins. Life, 4: 66; Beutenmuller, 1896, J. N.Y. Entomol. Soc., 4: 80; Leng and Hamilton, 1896, Trans. Amer. Entomol. Soc., 23: 151; Wickham, 1898, Can. Entomol., 30: 42; Lugger, 1899, Minn. Agr. Exp. Sta. Bull., 66: 215; Felt and Joutel, 1904, N.Y. St. Mus. Bull., 74: 73, pl. 6, figs. 12-14; Chagnon, 1905, Nat. Can., 32: 43; Smith, 1910, N.J. St. Mus. Rept., 1909: 336; Blatchley, 1910, Coleoptera in Indiana, p. 1086; Craighead, 1923, Can. Dept. Agr. Bull., (n.s.) 27: 129, pl. 23, fig. 10, pl. 41; Leonard, 1928, Cornell. Agr. Exp. Sta. Mem., 101: 456; Chagnon, 1933-40, Coleop. Prov. Quebec, p. 276; Doane et al., 1936, For. Ins., p. 191; Knull, 1946, Ohio Biol. Surv. Bull., 39: 273; Craighead, 1950, USDA Misc. Pub., 657: 264, fig. 57 (habits); Essig, 1958, Ins. West. North America, p. 462; Papp, 1959, Bull. So. Calif. Acad. Sci., 58: 92; Dillon and Dillon, 1961, Man. Beetles East. North America, p. 651, pl. 65, no. 3; Chagnon and Robert, 1962, Prin. Coleop. Prov. Quebec, p. 276; Wong and McLeod, 1965, Can. Dept. For. Bimon. Prog. Rept., 21: 3; Abdullah and Abdullah, 1966, Proc. Roy. Entomol. Soc. London, (B) 35: 91; McLeod and Wong, 1967, Manitoba Entomol., 1: 27 (habits); Bayer and Shenefelt, 1969, Univ. Wisc. Res. Bull., 275: 31, fig. 39; Baker, 1972, USDA Misc. Pub., 1175: 188 (habits); Stern and Tagestad, 1976, USDA For. Serv. Res. Pap., RM-171: 36; Headstrom, 1977, Beetles of America, p. 382; Laliberte et al., 1977, Fabreries, 3: 98; Rice and Enns, 1981, Trans. Mo. Acad. Sci., 15: 103.
Saperda concolor var. *unicolor* Felt and Joutel, 1904, N.Y. St. Mus. Bull., 74: 74, pl. 6, fig. 15; Britton, 1919, Conn. Agr. Exp. Sta. Bull., 211: 347, pl. 14, 15 (habits); Mutchler and Weiss, 1923, N.J. Dept. Agr. Circ., 58: 11; Aurivillius, 1923, Coleop. Cat., 74: 484; Leonard, 1928, Cornell Agr. Exp. Sta. Mem., 101: 456; Procter, 1938, Biol. Surv. Mt. Desert, 6: 154.
Saperda concolor unicolor: Leng, 1920, Cat. Coleop. North Amer., p. 285; Chemsak and Linsley, 1975, Checklist of Beetles, Cerambycidae (red version), p. 205.
Saperda (Saperda) inornata unicolor: Breuning, 1952, Entomol. Arb. Mus. Frey, 3: 170; Gilmour, 1965, Cat. Lam. du Monde, 8: 670.
Saperda mecasoides Casey, 1913, Memoirs on the Coleoptera, 4: 359.
Saperda concolor mecasoides: Leng, 1920, Cat. Coleop. North Amer., p. 285.

Male. Form small to moderate-sized, subparallel; integument black; pubescence dense, short, grayish, appressed, obscuring surface. Head with front convex, quadrate; interantennal area shallowly impressed; pubescence dense, appressed, long, dark, erect hairs numerous; eyes with lower lobes longer than broad, much longer than genae; antennae about as long as body, segments from third usually narrowly dark annulate at apices, scape slender, cylindrical, third segment longer than first and fourth, fourth longer than first. Pronotum slightly broader than long; disk convex, median line vague; sides shallowly impressed near base; punctures rather fine, irregular; pubescence dense, appressed, sides vaguely vittate, middle often with a narrow pubescent vitta; prosternum densely pubescent; meso- and metasternum densely pubescent. Elytra about 2-1/2 times longer than broad; punctures moderately coarse, shallow, contiguous to subconfluent, becoming finer toward apex; pubescence dense, short, appressed, obscuring surface, short, suberect hairs numerous; apices moderately narrowly rounded. Legs densely pubescent; tarsal claws lacking a process. Abdomen densely pubescent; last sternite rounded at apex. Length, 8-11 mm.

Female. Form more robust. Antennae slightly shorter than body. Elytra with apices broadly rounded. Abdomen with last sternite linearly impressed at middle. Length, 9-13 mm.

Type locality. Of *inornata*, Missouri Territory; *concolor*, Santa Fe, New Mexico; *mecasoides*, New York; *unicolor*, eastern United States.

Range. Eastern North America to Idaho and Arizona.

Flight period. March to July.

Host plants. *Populus* spp., *Salix* spp.

The uniformly dense, grayish pubescence, annulate antennae, and dense, shallow punctation of the elytra distinguish this species from the others.

Although most specimens examined are relatively uniform in pubescence, several from Arizona are much more thickly clothed.

There has been considerable confusion between this species and some species of *Mecas*, particularly *M. cana*. Numerous references list "*Mecas inornata*" as the species utilizing *Ambrosia*, *Helianthus*, and *Xanthium* as host plants. These are recorded under *Mecas cana saturnina* (LeConte).

TRIBE PHYTOECIINI FAIRMAIRE

Fairmaire, 1864, in Jacquelin du Val, Gen. Coleop. Europe, 4: 171, 194 (Phytoeciites).
Lacordaire, 1872, Genera des coléoptères, 9(2): 849 (Phytoeciides vraies).
LeConte, 1873, Smithson. Misc. Coll., 11(265): 346 (part).
LeConte and Horn, 1883, Smithson. Misc. Coll., 507: 332 (part).
Leng and Hamilton, 1896, Trans. Amer. Entomol. Soc., 23: 151 (part).
Blatchley, 1910, Coleoptera in Indiana, p. 1089 (part).
Craighead, 1923, Can. Dept. Agr. Bull. (n.s.) 27: 135 (Phytoecides, part).
Bradley, 1930, Man. Genera Beetles, p. 243.
Chagnon, 1933-40, Coleop. Prov. Quebec, p. 278.
Knull, 1946, Ohio Biol. Surv. Bull., 39: 274.
Duffy, 1953, Mon. British Timber Beetles, p. 292.
Dillon and Dillon, 1961, Man. Beetles East. North America, p. 652.
Arnett, 1962, Beetles U.S., 103: 873.
Chagnon and Robert, 1962, Princ. Col. Prov. Quebec, p. 278 (part).
Bayer and Shenefelt, 1969, Univ. Wisc. Res. Bull., 275: 32.
Hatch, 1971, Univ. Wash. Pubs. Biol., 16: 155.
Rice and Enns, 1981, Trans. Mo. Acad. Sci., 15: 104.

Form moderate-sized, cylindrical. Head with front moderately convex, broader than long; antennae usually shorter than body; eyes finely faceted, emarginate, not divided; palpi slender, last segment oval. Pronotum cylindrical; prosternum with intercoxal process narrow, coxal cavities closed behind, externally angulate; mesosternum with coxal cavities open to epimera; metasternum with episternum broad, strongly tapering posteriorly. Legs short; femora linear; anterior coxae conical, protuberant; tarsal claws appendiculate, often bifid. Abdomen normally segmented.

The emarginate, not completely divided eyes separate this tribe from the Tetraopini. The Phytoeciini differ from the Saperdini by the bifid or appendiculate tarsal claws and from the Hemilophini by the broad metepisternum and subequal abdominal sternites.

Two genera occur in our fauna.

KEY TO THE NORTH AMERICAN GENERA OF *PHYTOECIINI*

Tarsal claws bifid (Figure 38) .... *Mecas*
Tarsal claws appendiculate .... *Oberea*

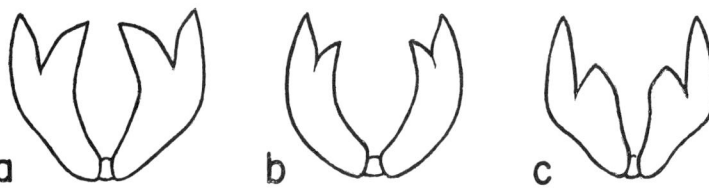

Figure 38. Some variations in the form of the tarsal claws in the genus *Mecas*.

Genus *Mecas* LeConte

Mecas LeConte, 1852, J. Acad. Sci. Philadelphia, 2(2):155; Thomson, 1864, Syst. ceramb., p. 114; LeConte, 1873, Smithson. Misc. Coll., 11(265):347; Horn, 1878, Trans. Amer. Entomol. Soc., 7:44; Bates, 1881, Biol. Centrali-Americana, Coleop., 5:203; LeConte and Horn, 1883, Smithson. Misc. Coll., 26(507):332; Leng and Hamilton, 1896, Trans. Amer. Entomol. Soc., 73:151, 152; Casey, 1913, Memoirs on the Coleoptera, 4:360; Bradley, 1930, Man. Genera Beetles, p. 247; Breuning, 1955, Mem. Soc. Roy. Entomol. Belg., 27:138; Breuning, 1960, Frust. Entomol., 3:4; Dillon and Dillon, 1961, Man. Beetles East. North America, p. 652; Arnett, 1962, Beetles U. S., 103:873; Hatch, 1971, Univ. Wash. Pub. Biol., 16:155; Chemsak and Linsley, 1973, Proc. California Acad. Sci., (4)39:144; Rice and Enns, 1981, Trans. Mo. Acad. Sci., 15:105.

Stenostola: LeConte, 1852, J. Acad. Nat. Sci., Philadelphia, 2:154 (part); Lacordaire, 1872, Genera des coléoptères, 9(2):863 (part).

Form elongate, usually parallel. Head with front convex, interantennal area usually concave; mandibles rather short, apices curved, acute; palpi slender, maxillary pair longer than labial; eyes rather small, finely faceted, deeply emarginate, upper lobe small; antennae usually slender, sparsely or densely fringed with long hairs beneath, particularly on basal segments, third segment usually longer than first, fourth subequal to or shorter than third, outer segments short or long. Pronotum wider than long, sides usually rounded; disk variably pubescent, often with glabrous calluses; prosternum short, intercoxal process narrow, expanded at apex, coxal cavities closed behind; mesosternum with coxal cavities open; metasternum with episternum broad in front, narrowing behind. Legs short; intermediate tibiae with a dorsal sinus; tarsal claws bifid. Abdomen normally segmented; last sternite deeply impressed in the male, linearly impressed in the female.

Type species. Phytoecia femoralis Haldeman (monobasic).

This genus is distinctive from others in its tribe by the proportions of the antennal segments, the frequent presence of dorsal calluses on the pronotum, the shape of the metepisternum, and by the nature of the impressions of the last abdominal sternite, which are concave in the male and linear in the female. Many of the species resemble members of the genus *Saperda,* but the bifid claws readily separate them (Figure 38).

There are 10 species of *Mecas* known from America north of Mexico.

KEY TO THE NORTH AMERICAN SPECIES OF *MECAS*

1   Pronotum not densely fringed with short, erect golden pubescence, erect hairs moderately long; elytral apices broadly rounded or rotundate-truncate; abdomen with pubescence of sternites unicolorous. *Mecas,* s. str 2

Pronotum densely fringed with short, erect, golden pubescence with scattered long setae intermixed, dorsal surface with a pair of longitudinal vittae on each side of middle composed of short, appressed, golden pubescence; integument usually concolorous golden-yellow, less commonly mottled or vittate with black, rarely wholly black; elytral apices angulate, obliquely truncate or emarginate; abdomen often with last three sternites margined laterally with longitudinal bands of yellowish-white pubescence suggesting luminescent organs of a lampyrid. *Mecas (Dylobolus).* Length, 9-19 mm. Southwestern United States to Guatemala ...... *rotundicollis*

2(1)   Pronotum with pubescence intact, not interrupted by polished black callosities 3

Pronotum with at least two polished black callosities and usually a median elongate impunctate area on disk 6

3(2)   Pronotum and elytra with concolorous pubescence which obscures the surface; sternum uniformly densely pubescent 4

Pronotum and elytra with longitudinal bands of dense, appressed, often yellowish pubescence at middle

	and sides, remaining pubescence not completely obscuring surface; sternum margined with a row of dense, appressed, yellowish pubescence. Length, 6.5-8 mm. Southeastern United States to New Mexico..................................*marginella*
4(3)	Tarsal claws with inner tooth much smaller than outer one (Fig. 38b). Smaller species, 6-11 mm in length 5
	Tarsal claws with inner tooth almost as long as outer one (Fig. 38a). Length, 10-14 mm. Kansas to Texas ....*confusa*
5(4)	Femora always reddish; pubescence finer, not completely obscuring surface. Length, 6-8 mm. Southeastern United States ....*femoralis*
	Femora always black; pubescence thick, obscuring surface. Length, 6-11 mm. Southeastern United States to Arizona, Colorado, and northeastern Mexico....................................*cineracea*
6(2)	Pronotum with four rounded glabrous calluses in addition to median impunctate area....................7
	Pronotum with two rounded glabrous calluses in addition to elongate median impunctate area 8
7(6)	Pronotum broader at base than at apex; antennae extending to apices of elytra in males, slightly shorter in females, outer segments not annulate. Length, 11-12 mm. Texas, northern Mexico ....*linsleyi*
	Pronotum with base as broad as apex; antennae not extending to elytral apices in males, to about second abdominal segment in females, outer segments annulate. Length, 6-12 mm. Great Plains to southeastern United States, New Mexico and northeastern Mexico ....*pergrata*
8(6)	Antennae at least as long as body, outer segments elongate 9
	Antennae much shorter than body, outer segments short; elytra lacking long erect dark hairs over apical half. Length, 10-15 mm. Washington to northern Baja California and Colorado ....*bicallosa*
9(8)	Elytra sparsely, separately punctate, lacking long erect hairs; pronotal calluses small, median impunctate area vague 10
	Elytra coarsely, contiguously punctate, densely clothed with long erect hairs; pronotal calluses large, median impunctate area distinct. Length, 8-13 mm. Arizona to Nayarit and Distrito Federal, Mexico....................................*menthae*

10(9) Elytra with distinct longitudinal pubescent bands along margins and suture. Length, 10-16 mm.
Florida .... *cana cana*
Elytra uniformly grayish or yellowish pubescent, suture and margins without pubescent bands. Length, 10-16 mm. Eastern United States to South Dakota and north-eastern Mexico ..... *cana saturnina*

Subgenus *Dylobolus* Thomson

Dylobolus Thomson, 1868, Physis, 2:195; Lacordaire, 1872, Genera des coléoptères, 9:897, 900.
Mecas (*Dylobolus*): Chemsak and Linsley, 1973, Proc. Calif. Acad. Sci., (4)39:153.

Form slender, elongate. Antennae slender, third segment slightly curved. Pronotum with sides rounded, disk densely fringed with short, erect golden pubescence. Elytra with apices angulate, usually obliquely emarginate. Legs with tarsal claws with inner tooth almost as long as outer one. Abdomen frequently with yellowish appressed pubescence at sides of apical sternites.

Type species. Dylobolus rotundicollis Thomson (monobasic).

This subgenus differs from the others by the densely fringed discal pubescence of the pronotum and the emarginate or truncate elytral apices. The single known species is a lampyrid mimic.

Mecas (*Dylobolus*) *rotundicollis* (Thomson)
(Figure 39)

Dylobolus rotundicollis Thomson, 1868, Physis, 2:196.
Mecas rotundicollis: Bates, 1881, Biol. Centrali-Americana, Coleop., 5:205; Breuning, 1955, Mem. Soc. Roy. Entomol. Belg., 27:148.
Mecas (*Dylobolus*) *rotundicollis*: Chemsak and Linsley, 1973, Proc. Calif. Acad. Sci., (4)39:153, figs. 4, 5, 6; Hovore et al., 1987, Proc. Calif. Acad. Sci., 44:320, fig. 20.
Mecas ruficollis Horn, 1878, Trans. Amer. Entomol. Soc., 7:44; Bates, 1881, Biol. Centrali-Americana, Coleop., 5:205; Leng and Hamilton, 1896, Trans. Amer. Entomol. Soc., 23:152, 153; Casey, 1913, Memoirs on the Coleoptera, 4:362; Linsley et al., 1961, Amer. Mus. Nov., 2050:31.
Mecas ruficollis morpha *mediomaculata* Breuning, 1955, Mem. Soc. Roy. Entomol. Belg., 27:149.
Mecas rotundicollis morpha *ruficollis*: Breuning, 1955, Mem. Soc. Roy. Entomol. Belg., 27:149.
Mecas laticeps Bates, 1881, Biol. Centrali-Americana, Coleop., 5:204; Breuning, 1955, Mem. Soc. Roy. Entomol. Belg., 27:151.

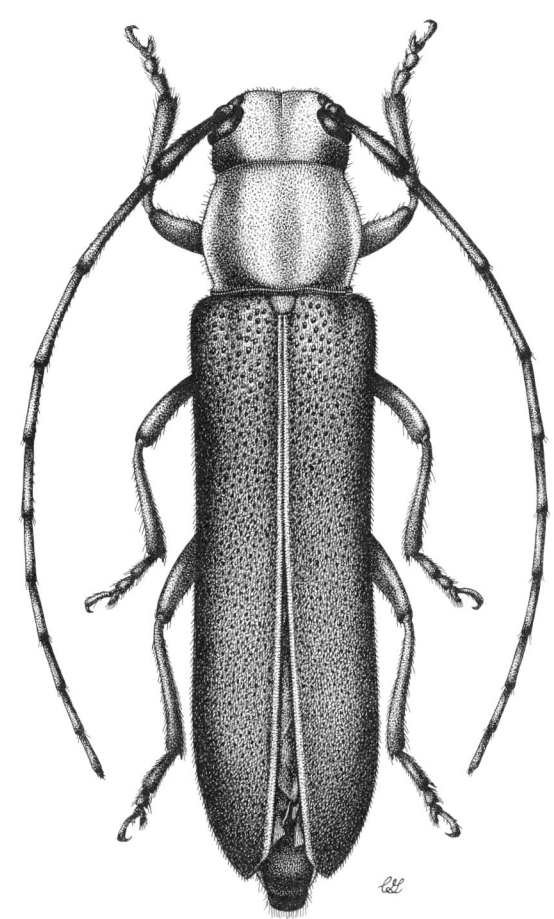

Figure 39. *Mecas (Dylobolus) rotundicollis* (Thomson), male.

Mecas laticeps morpha *sutureflava* Breuning, 1955, Mem. Soc. Roy. Entomol. Belg., 27:151.
Mecas laticeps morpha *mediopunctata* Breuning, 1955, Mem. Soc. Roy. Entomol. Belg., 27:151.
Mecas mexicana Bates, 1881, Biol. Centrali-Americana, Coleop., 5:204.
Mecas rotundicollis morpha *mexicana*: Breuning, 1955, Mem. Soc. Roy. Entomol. Belg., 27:149.
Mecas vitticollis Casey, 1913, Memoirs on the Coleoptera, 4:362; Knull, 1934, Entomol. News, 45:211.

Mecas laticeps morpha *vitticollis*: Breuning, 1955, Mem. Soc. Roy. Entomol. Belg., 27:152.

Male. Form moderate-sized, elongate, sides parallel; color black, head and pronotum orange, usually with dark spots or bands, legs often orange, thoracic sterna often orange, abdomen usually with broad bands of yellow appressed pubescence at sides of last three sternites, elytra frequently with narrow bands of appressed yellowish pubescence down suture and epipleurae. Head rather small; front convex, median line extending from clypeus to neck; interantennal area concave; vertex moderately coarsely, densely punctate; pubescence dense, yellowish, appressed with short, dark, erect hairs numerously interspersed; antennae shorter than body, scape finely, very densely punctate, third segment longer than first, fourth shorter than third, fifth subequal to first, segments from sixth gradually decreasing in length, scape rather densely clothed with short subpressed hairs, underside of segments densely clothed with short, pale, appressed pubescence, basal segments with a few long erect hairs beneath. Pronotum usually broader than long, sides rounded; disk convex, sparsely to rather densely punctate; pubescence usually dense, consisting of short, dense, subappressed, longitudinal bands, one on each side of middle and at lateral margins, longer erect hairs numerously interspersed; prosternum densely pubescent; meso- and metasternum finely densely punctate at middle, coarsely at sides, pubescence dense. Elytra over three times longer than broad; suture and epipleurae usually with narrow bands of appressed yellow pubescence; punctures rather coarse, dense, becoming finer and sparser toward apex; surface clothed with fine appressed pubescence, short, recurved hairs numerously interspersed, apices obliquely truncate. Legs finely, densely pubescent; tarsal claws with inner tooth almost as long as outer. Abdomen elongate, densely pubescent; last three sternites usually clothed with broad yellow bands at sides; last sternite deeply impressed for most of its length. Length, 9-16 mm.

Female. Form similar, more robust. Antennae slightly shorter than in male. Abdomen with last sternite linearly impressed for its entire length; last tergite strongly, obtusely conical at apex. Length, 10-19 mm.

Type locality. Of *rotundicollis*, Mexico; *ruficollis*, Texas; *laticeps*, Guanajuato, Mexico; *mexicana*, Izucar, Mexico; *vitticollis*, Durango City; *mediomaculata*, Guerrerro; *sutureflava*, Temax, Yucatan; *mediopunctata*, Mexico.

Range. Oklahoma to Arizona, Texas, and south to Guatemala.

Flight period. May to December.

Host plants. Adults have been collected on flowers of *Verbesina* and *Guardiola tulocarpa* (Compositae) and on *Eysenhardtia polystachya* (Leguminosae).

This mimetic species resembles different lampyrid models in different parts of its range, as indicated by the polychromatism exhibited in the specimens at hand. Differences are expressed in size, coloration of the head, pronotum, sternum and legs and in the presence or absence of yellowish longitudinal sutural and epipleural elytral bands and in the yellowish lampyrid-like apical sternites of the abdomen. There are varying combinations of these characters, but we have been unable to correlate them geographically, although this may yet be possible with larger series of specimens and model-mimic associations throughout the entire range of the species.

Subgenus *Mecas* s. str.

Mecas LeConte, 1852, J. Acad. Nat. Sci. Philadelphia, (2)2:155; Leconte, 1873, Smithson. Misc. Coll., 11(265):347; Blatchey, 1910, Coleoptera in Indiana, p. 1990; Knull, 1946, Ohio Biol. Survey Bull., 39:274.
Mecas (*Mecas*): Chemsak and Linsley, 1973, Proc. Calif. Acad. Sci., (4)39:159.

Form moderate-sized, parallel; body usually densely clothed with appressed pubescence. Pronotum with or without dorsal calluses, sides broadly to narrowly rounded. Elytra parallel, apices rounded, disk not costate. Legs with tarsal claws variable, inner tooth long or short.

Type species. Phytoecia femoralis Haldeman (monobasic).

The members of this subgenus are easily recognizable by the densely pubescent body, subcylindrical and usually densely pubescent pronotum, which frequently has glabrous dorsal calluses.

Mecas (*Mecas*) *marginella* LeConte

Mecas marginella LeConte, 1873, Smithson. Misc. Coll., 11(264):239; Horn, 1878, Trans. Amer. Entomol. Soc., 23:152; Blatchley, 1910, Coleoptera Indiana, p. 1090; Casey, 1913, Memoirs on the Coleoptera, 4:361; Fattig, 1947, Emory Univ. Mus. Bull. 5:43; Breuning, 1955, Mem. Soc. Roy. Entomol. Belg., 27:147; Linsley, Knull and Slatham, 1961, Amer. Mus. Nov., 2050:32; Kirk, 1969, S. C. Agr. Exp. Sta. Tech. Bull. 1033:87; Rice et al., 1985, Coleop Bull., 39:23.
Mecas (*Mecas*) *marginella*: Chemsak and Linsley, 1973, Proc. Calif. Acad. Sci., (4)39:162; Hovore et al., 1987, Proc. Calif. Acad. Sci., 44:320, fig. 20.

Male. Form small, subparallel; color black, pronotum with three longitudinal bands of yellowish to whitish appressed pubescence, elytra with narrow bands of pale pubescence down suture and lateral margins. Head with front convex, deeply punctate, densely clothed with appressed pale pubescence, long erect hairs numerous; interantennal area broadly concave; vertex coarsely, densely punctate; antennae a little longer than body, very sparsely gray pubescent beneath, long erect hairs numerous on basal segments, third segment longer than scape, fourth subequal to third, fifth shorter than fourth. Pronotum broader than long, sides subparallel; punctures moderately coarse, dense, calluses absent; pubescence dense, appressed, lateral bands broad, yellowish, median band narrower, usually whitish, remainder of surface finely pubescent, long erect hairs numerous; prosternum densely pubescent; meso- and metasternum densely pubescent, rather coarsely punctate at sides, metepisternum yellow pubescent over posterior half. Elytra over twice as long as broad; punctures coarse, close, becoming finer toward apex; pubescence between longitudinal bands fine, appressed, with longer erect hairs numerously interspersed; apices rounded. Legs finely, densely pubescent; tarsal claws with teeth subequal in length. Abdomen densely pubescent, narrowly yellow at sides of apical sternites; last sternite deeply impressed for its entire length. Length, 6.5-8 mm.

Female. Form and size similar. Antennae about as long as body. Abdomen with last sternite shallowly impressed near apex. Length, 6.5-8 mm.

Type locality. Western states and Texas.

Range. Southeastern United States to New Mexico.

Flight period. March to July.

Host plants. Unknown: one specimen was collected on *Colubrina texensis* (Rhamnaceae) in Texas, but it is unlikely that this shrub is a host. Also taken in *Thelesperma*.

The absence of pronotal calluses and the distinctive pubescent bands make this species easily recognizable.

Mecas (Mecas) confusa Chemsak and Linsley
(Figure 40)

Mecas (Mecas) confusa Chemsak and Linsley, 1973, Proc. Calif. Acad. Sci., (4)39:163, fig. 8; Hovore et al., 1987, Proc. Calif. Acad. Sci., 44:320.

Mecas inornata Blanchard, 1887 (not Say, 1824), Entomol. Amer., 3:86; Horn, 1886, Trans. Amer. Entomol. Soc., 15:301; Leng and Hamilton, 1896, Trans. Amer. Entomol. Soc., 23:152; Blatchley, 1910, Coleoptera in Indiana, p. 1090.

Male. Form moderate-sized, subparallel; color black, body densely clothed with thick, grayish, recumbent pubescence which obscures the surface. Head with front convex, finely densely punctate, darker suberect hairs short, about half as long as second antennal segment; interantennal area very shallowly concave; vertex sparsely punctate, large punctures well separated; antennae about as long as body, scape finely gray pubescent, remaining segments to ninth gray pubescent beneath, third segment longer than scape, fourth shorter than third, fifth shorter than first, remaining segments gradually decreasing in length. Pronotum broader than long, sides rounded, base impressed; disk convex, calluses absent; large deep punctures irregular, well separated, each puncture bearing a long erect hair; pro-, meso- and metasterna densely clothed with recumbent pubescence which obscures the surface. Elytra less than 2-1/2 times as long as broad; punctures coarse, close, linearly arranged, becoming obsolete at apex; recumbent pubescence completely obscuring surface, base with numerous rather short suberect hairs,

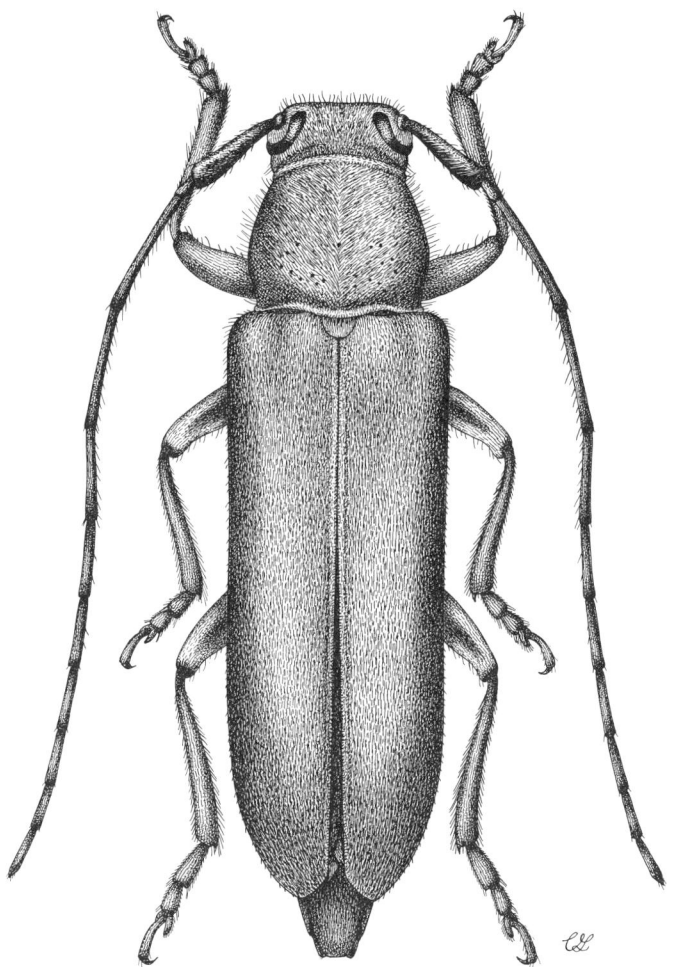

Figure 40. *Mecas (Mecas) confusa* Chemsak and Linsley, male.

these becoming shorter and recurved toward apex; apices obliquely subtruncate. Legs very densely pubescent; tarsal claws with inner tooth almost as long as outer. Abdomen very densely pubescent; last sternite impressed for its entire length. Length, 10-13 mm.

Female. Similar in form and size. Abdomen with last sternite linearly impressed, apex broadly v-shaped. Length, 10-14 mm.

Type locality. Luling, Gonzales Co., Texas.
Range. Kansas to Texas.
Flight period. April to July.

This species closely resembles *M. cineracea*, but may be separated by its larger size, denser overall pubescence, shorter erect hairs on the front of the head, and by the structure of the tarsal claws. In *M. confusa* the inner tooth of the claws is elongate and almost as long as the outer one; in *cineracea* the tooth is very small. The two species are sympatric, at least in parts of Texas, but it is not now known whether they infest the same or different host plants. *M. confusa* has been collected on *Heterotheca*.

Mecas (Mecas) femoralis (Haldeman)

Phytoecia femoralis Haldeman, 1847, Trans. Amer. Philos. Soc., (2)10:59.
Mecas femoralis: LeConte, 1852, J. Acad. Nat. Sci. Philadelphia, (2)2:155; Horn, 1878, Trans. Amer. Entomol. Soc., 7:44; Leng and Hamilton, 1896, Trans. Amer. Entomol. Soc., 23:152, 153; Casey, 1913, Memoirs on the Coleoptera, 4:360 (fn.); Fattig, 1947, Emory Univ. Mus. Bull. 5:43; Breuning, 1955, Mem. Soc. Roy. Entomol. Belg., 27:143; Turnbow and Hovore, 1979, Entomol. News, 90:227.
Mecas (Mecas) femoralis: Chemsak and Linsley, 1973, Proc. Calif. Acad. Sci., (4)39:165.

Male. Size small, subparallel; color black, femora reddish; pubescence grayish. Head with front convex, densely punctate, densely clothed with appressed pubescence and numerous suberect, dark hairs; vertex rather coarsely, closely punctate, densely pubescent; antennae about as long as body, basal segments sparsely gray pubescent beneath, long, erect hairs sparse, third segment longer than first, fourth shorter than third, fifth shorter than fourth. Pronotum about as long as broad; sides almost subparallel; disk convex, rather coarsely, closely punctate; pubescence dense, short, appressed, long, erect hairs numerous; prosternum densely pubescent; meso- and metasternum densely clothed with recumbent pubescence, sides more coarsely punctate. Elytra about 2-1/2 times as long as broad; punctures coarse, dense, becoming finer toward apex; pubescence dense, appressed, with longer, erect hairs numerously interspersed; apices rounded. Legs moderately densely pubescent; tarsal claws with inner tooth very short. Abdomen densely pubescent; last sternite shallowly impressed over most of its length. Length, 6-8 mm.

Female. Very similar in size and shape. Abdomen with last sternite impressed over apical half. Length, 6-8 mm.

Type locality. Not given.
Range. Southeastern United States.
Flight period. May to July.
Host plants. Unknown.

Mecas femoralis may be recognized by its small size, rather uniform pubescence, lack of pronotal calluses, and reddish femora. Examples are rare in collections. The eleven specimens we have seen show little variation.

Mecas (Mecas) cineracea Casey

Mecas cineracea Casey, 1913, Memoirs on the Coleoptera, 4:360; Vogt, 1949, Pan- Pac. Entomol., 25:184 (record); Turnbow and Franklin, 1980, J. Ga. Entomol. Soc., 15:346; Rice and Enns, 1981, Trans. Mo. Acad. Sci., 15:105.
Saperda cineracea: Breuning, 1955, Mem. Soc. Roy. Entomol. Belg., 27: 139.
Mecas (Mecas) cineracea: Chemsak and Linsley, 1973, Proc. Calif. Acad. Sci., (4)39:168; Hovore et al., 1987, Proc. Calif. Acad. Sci., 44:320.
Saperda bicallosa Breuning, 1955 (not Martin, 1924), Mem. Soc. Roy. Entomol. Belg., 27:140.

Male. Form rather small, parallel; color black, body densely clothed with gray recumbent pubescence, Head with front convex, appressed pubescence obscuring punctures, longer erect hairs numerous; interantennal area almost plane; vertex moderately coarsely, separately punctate; antennae about as long as body, basal segments finely gray pubescent beneath, long erect hairs decreasing in number toward apex, third segment longer than first, remaining segments gradually decreasing in length. Pronotum broader than long, sides almost parallel, shallowly impressed at base; disk convex, without calluses, punctures coarse, separated; appressed pubescence obscuring surface, long erect hairs numerous; prosternum densely pubescent; meso- and metasternum densely pubescent, coarsely punctured at sides. Elytra about 2-1/2 times as long as broad; punctures coarse, contiguous, becoming finer toward apex; pubescence obscuring surface, long suberect hairs numerous near base; apices obliquely truncate. Legs very densely pubescent; tarsal claws with inner tooth small. Abdomen densely pubescent; last sternite shallowly impressed for its entire length. Length, 6-10 mm.

Female. Similar in form and size. Antennae shorter than body. Abdomen with last sternite linearly impressed. Length, 7-11 mm.

Type locality. Harris Co., Texas.

Range. Southeastern United States to Arizona and Colorado and northeastern Mexico.

Flight period. April to August.

Host plants. Helenium microcephalum, Baileya multiradiata.

This species may be recognized by its small size, lack of pronotal calluses, uniform coloration and pubescence, and by the small inner tooth of the tarsal claws. It was incorrectly transferred to *Saperda* by Breuning (1955).

A series of specimens from western New Mexico and Arizona have denser pubescence than Texas examples. However, in our material there appears to be a gradient in this character from east to west.

Mecas (Mecas) linsleyi Knull

Mecas linsleyi Knull, 1975, Ohio J. Sci., 75:130, fig. 1.

Mecas (Mecas) linsleyi: Hovore et al., 1987, Proc. Calif. Acad. Sci., 44: 321.

"*Male.* Form elongate, nearly 3-1/2 times as long as wide. Color: Head, antennae, base of pronotum, ventral surface, tibiae, tarsi, tips of femora, and five shining callosities on pronotum black; pronotum and femora orange; elytra black with orange tint at base and along outer margin, rimmed with dense yellow appressed pubescence, including scutellum. Head with front convex, densely punctured, interior lobes of eyes margined by a row of big punctures; densely clothed with appressed short gray pubescence, pubescence lacking in a narrow lone which extends from occiput part way down front, an upright black seta arising from each puncture; antennae extending approximately to apices of elytra, not annulate, ratio of length of segments 1 to 11, 12:3:18:16:12:12:10:10:10:8:10, scape deeply punctured beneath, first eight segments clothed with minute white recumbent pubescence and long black setae beneath, more evident on basal six segments. Pronotum slightly wider than long, wider at base than at apex, convex, one smooth, round shining black callosity in front of middle on each side of a median elongate callosity extending from near base to about middle, and a round callosity behind middle on each side, a transverse depression in front of scutellum and a narrow one at apex; surface densely clothed with short white recumbent pubescence, not obscuring the densely well-separated punctured surface, punctures much larger on basal half, short white upright setae arising from small punctures and longer black upright setae arising from the larger punctures; base with a narrow transverse line of yellow appressed pubescence. Elytra over 2-1/2 times as long as wide, much wider at base than base of thorax; sides slightly converging to well-rounded apices; surface densely punctured, punctures larger at base, clothed with short recumbent white pubescence nearly obscuring punctures, an upright black seta arising from each puncture. Ventral surface obscured by recumbent white pubescence; abdomen with intermixed longer white setae; last sternite impressed over most of its length. Tarsal claws with a short tooth near apex. Length, 11 mm.

Female. Form similar, more robust. Antennae not reaching apices of elytra. Last tergite convex. Last sternite impressed near apex; margin broadly emarginate. Length 12 mm." (original description.)

Type locality. Bentsen Rio Grande State Park, Hidalgo Co., Texas.
Range. Texas, northern Mexico.
Flight period. March to June.
Host plants. The type series was collected on foliage of *Aster spinosus* which is probably the larval host.

The shape of the pronotum and longer, unannulated antennae separate this species from *M. pergrata*.

Mecas (Mecas) pergrata (Say)

Saperda pergrata Say, 1824, J. Acad. Nat. Sci. Philadelphia, 3:407; Haldeman, 1847, Trans. Amer. Philos. Soc., (2)10:55; LeConte, 1859, Compl. Writings T. Say, 2:190.

Stenostola pergrata: Haldeman, 1847, Proc. Amer. Philos. Soc., 4:373; LeConte, 1852, J. Acad. Nat. Sci. Philadelphia, (2)2:154; Lacordaire, 1872, Genera des coléoptères, 9(2):864, fn.

Mecas pergrata: Horn, 1878, Trans. Amer. Entomol. Soc., 7:44; Leng and Hamilton, 1896, Trans. Amer. Entomol. Soc., 23:152, 153; Townsend, 1902, Trans. Tex. Acad. Sci., 5:80; Blatchley, 1910, Coleoptera in Indiana, p. 1090, 1091; Casey, 1913, Memoirs on the Coleoptera, 4:361; Craighead, 1923, Can. Dept. Agr. Bull. 27:138; Linsley and Martin, 1933, Entomol. News, 44:183; Knull, 1946, Ohio Biol. Surv. Bull. 39:274, 275, pl. 22, fig. 86; Fattig, 1947, Emory Univ. Mus. Bull. 5:43; Breuning, 1955, Mem. Soc. Roy. Entomol. Belg., 27:140, 144, fig. 1; Kirk, 1969, S. C. Agr. Exp. Sta. Tech. Bull. 1033:87; Gosling and Gosling, 1976, Gr. Lakes Entomol., 10:35, fig. 179; Rice and Enns, 1981, Trans. Mo. Acad. Sci., 15:105.

Mecas (Mecas) pergrata: Chemsak and Linsley, 1973, Proc. Calif. Acad. Sci., (4)39:169; Hovore et al., 1987, Proc. Calif. Acad. Sci., 44:321.

Mecas pergrata morpha semiruficollis Breuning, 1955, Mem. Soc. Roy. Entomol. Belg., 27:140, 145.

Stenostola gentilis LeConte, 1852, J. Acad. Nat. Sci. Philadelphia, (2)2:154.

Mecas discovittata Breuning, 1955, Mem. Soc. Roy. Entomol. Belg., 17:140, 143.

Male. Form moderate-sized, parallel; color black, femora pale reddish, elytra occasionally partly reddish; pubescence dense, short, recumbent, grayish. Head with front convex, punctures rather fine, well separated; pubescence dense, appressed, long, dark, erect hairs numerously interspersed; appressed pubescence thicker around eyes; vertex rather densely punctate; antennae shorter than body, segments gray pubescent beneath, outer segments annulate, third segment longer than first, fourth subequal to first, remaining segments gradually decreasing in length. Pronotum broader than long, sides slightly rounded; disk convex, four glabrous calluses present in addition to median callus; punctures rather sparse, scattered; apex and base usually with a narrow band of dense yellowish pubescence, remaining surface partially obscured, long, erect hairs numerously interspersed; prosternum densely pubescent; meso- and metasternum densely pubescent, densely punctate at sides. Scutellum densely clothed with yellowish recumbent pubescence. Elytra about 2-1/2 times as long as broad; punctures rather coarse, contiguous at base, becoming finer toward apex; pubescence short, recumbent, partially obscuring surface, longer suberect hairs numerous, suture and lateral margins narrowly clothed with dense, yellowish, appressed pubescence; apices rounded. Legs finely, densely pubescent; last sternite shallowly impressed over most of its length. Length, 6-11 mm.

Female. Similar in form and size. Antennae extending to about second abdominal segment. Abdomen with last sternite linearly impressed. Length, 6- 12 mm.

Type locality. Of *pergrata*, Platte River, Nebraska; *gentilis*, Missouri Territory; *semiruficollis*, Texas; *discovittata*, Colorado.

Range. Great Plains to southeastern United States, New Mexico.

Flight period. April to July.

Host plants. *Aster* (roots), *Helianthus*. Adults also taken on *Heterothoca*.

The five glabrous spots of the pronotum, reddish femora and densely pubescent lines on the suture and lateral margins of the elytra readily distinguish this species. In certain parts of the range, the elytra tend to be reddish down the disk and frequently the pronotum is also partially reddish.

Habits. According to Craighead (1923), larvae feed in the stems of *Aster* and down into the roots, completely hollowing the latter. Subsequently, that portion of the stem of the plant breaks off at the surface of the ground. Small heaps of frass are exuded about the base of the plant. Only one larva is found in each stem.

Mecas (Mecas) bicallosa Martin
(Figure 41)

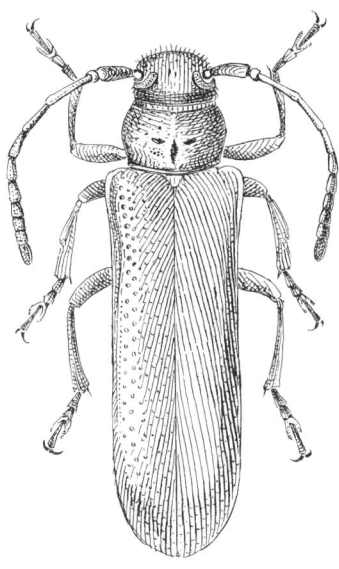

Figure 41. *Mecas (Mecas) bicallosa* Martin, female.

Mecas bicallosa Martin, 1924, Entomol. News, 35:244; Barr and Penrose, 1969, Great Basin Nat., 29:90; Hatch, 1971, Univ. Wash. Publs. Biol., 16:155.
Saperda bicallosa: Breuning, 1955, Mem. Soc. Roy. Entomol. Belg., 27:139-140.
Mecas (Mecas) bicallosa: Chemsak and Linsley, 1973, Proc. Calif. Acad. Sci., (4)39:172; Hovore, 1988, Wasmann J. Biol., 46: 25.

Male. Form moderate-sized, parallel, rather robust; color black, body densely clothed with short, appressed, grayish pubescence. Head with front convex, pubescence obscuring surface, long erect hairs very numerous; interantennal area plane, vertex deeply punctate; antennae extending to about third abdominal segment, segments through fourth gray pubescent, dark at apices, basal segments with numerous long, suberect hairs, segments from third with long hairs beneath, these decreasing in number toward apex, third segment longer than first, fourth subequal to first, remaining segments short, subequal in length. Pronotum broader than long, sides rounded, disk convex, with two glabrous calluses at middle and a smaller median one behind middle; punctures rather fine, deep, separated; pubescence obscuring surface, very long, erect hairs numerous; prosternum densely pubescent, front coxal cavities narrowly open behind; meso- and metasternum densely clothed with recumbent and subdepressed pubescence. Elytra more than twice as long as broad; punctures at base coarse, dense, becoming finer toward apex; pubescence obscuring surface, long, suberect hairs abundant over basal half; apices rounded, often vaguely, obtusely toothed. Legs very densely pubescent; tarsal claws with inner tooth very small, short. Abdomen densely pubescent; last sternite shallowly, rather broadly impressed. Length, 10-13 mm.

Female. Form similar. Antennae slightly shorter. Abdomen with last sternite narrowly linearly impressed, apex shallowly concave. Length, 10-15 mm.

Type locality. Martins Springs, Lassen Co., California.
Range. Washington to northern Baja California, to Colorado.
Flight period. April to August.
Host plants. Artemisia tridentata.

The bicallused pronotum and abbreviated distal antennal segments characterize this species. Breuning (1955) incorrectly synonymized *M. bicallosa* with *M. cineracea* Casey and transferred both to *Saperda*.

Mecas (Mecas) menthae Chemsak and Linsley
(Figure 42)

Mecas (Mecas) menthae Chemsak and Linsley, 1973, Proc. Calif. Acad. Sci., (4)39:174, figs, 10, 11.
Mecas marginella Linsley, Knull, and Statham, 1961 (not LeConte, 1873), Amer. Mus. Nov., 2050:32.

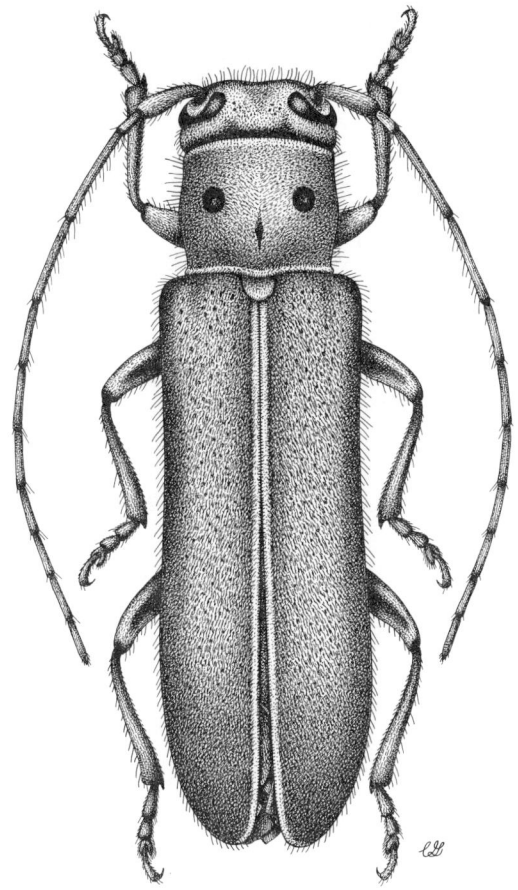

Figure 42. *Mecas (Mecas) menthae* Chemsak and Linsley, female.

Male. Form moderate-sized, subparallel; color black; pubescence dense, short, appressed, grayish to gray-brown, long erect, dark hairs numerous. Head with front convex, rather finely, separately punctate, vertex deeply, separately punctate; pubescence dense, short, appressed, antennal tubercles dark pubescent above, long erect hairs numerous on front and vertex; antennae slightly longer than elytra, segments to tenth gray pubescent beneath, segments from fifth narrowly pale annulate at base, long erect hairs fairly numerous beneath on basal segments, third segment longer than first, fourth shorter than third but longer than first, fifth equal to first, remaining segments gradually decreasing in length. Pronotum broader than long, sides broadly rounded; disk convex, each side of middle with a rather large, glabrous callus, middle with an elongate callus toward base; punctures rather fine, separated; pubescence short, appressed, obscuring surface, long erect hairs numerous; prosternum densely pubescent; meso- and metasternum densely pubescent, finely, densely punctate at sides. Scutellum densely clothed with appressed pubescence. Elytra more than 2-1/2 times as long as broad; punctures coarse, contiguous to about apical one-third, very fine at apex; pubescence dense, short appressed, lateral margins and suture with a narrow band of appressed pubescence, long erect hairs numerous basally, shorter toward apex; apices rounded. Legs densely pubescent; tarsal claws with inner tooth slightly shorter than outer. Abdomen densely pubescent; last sternite deeply impressed for about 3/4 of its length. Length, 8-12 mm.

Female. Similar in form, slightly more robust. Antennae about as long as elytra. Abdomen with last sternite linearly impressed. Length, 9-13 mm.

Type locality. 8 miles W El Palmito, Sinaloa, Mexico.

Range. Southern Arizona to Nayarit and Distrito Federal, Mexico.

Flight period. June to October.

Host plants. Agastache.

This species is distinctive in the rather uniform grayish pubescence and the narrow bands of dense appressed pubescence along the lateral and sutural margins.

Adults may be found in the curves of smaller top leaves of the host plant. Larvae bore in the roots of the living plants. This is the first record of a non-composite host for this genus.

Mecas (Mecas) cana (Newman)

Saperda cana Newman, 1840, Entomol., 1:12; LeConte, 1852, J. Acad. Nat. Sci. Philadelphia, (2)2:164; Lacordaire, 1872, Genera des coléoptères, 9(2):834, fn.

Mecas cana: Gahan, 1888, Trans. Amer. Entomol. Soc., 15:300; Leng and Hamilton, 1896, Trans. Amer. Entomol. Soc., 23:152; Casey, 1913, Memoirs on the Coleoptera, 4:360; Breuning, 1955, Mem. Soc. Roy. Entomol. Belg., 27:148.

Mecas (Mecas) cana: Chemsak and Linsley, 1973, Proc. Calif. Acad. Sci., (4)39:177.

Male. Form moderate-sized, subparallel; color black, body densely clothed with gray recumbent pubescence. Head with front convex, pubescence obscuring punctures, longer, dark, suberect hairs numerous; interantennal area almost plane; vertex finely separately punctate; antennae slightly shorter than body, segments to sixth rather sparsely gray pubescent beneath, long erect hairs decreasing in number toward apex, third segment longer than scape, fourth shorter than third but longer than first, remaining segments gradually decreasing in length. Pronotum broader than long, sides rounded to subparallel; basal and apical margins narrowly margined; disk convex, each side with a flat glabrous callus before middle, middle usually with a vague linear callus near base; punctures moderately coarse, rather sparse, partially obscured by pubescence; longer erect hairs pale with dark setae interspersed mostly at sides; prosternum densely pubescent; meso- and metasternum densely pubescent; metasternum deeply, rather densely punctate at sides. Elytra over 2-1/2 times longer than broad; punctures moderately coarse, well separated, becoming finer and sparser toward apex; pubescence obscuring surface, longer suberect hairs numerous; apices obliquely subtruncate. Legs densely pubescent; tarsal claws with inner tooth small. Abdomen occasionally reddish, densely pubescent; last sternite moderately impressed for its entire length. Length, 9-15 mm.

Female. Form similar. Antennae shorter than body. Abdomen with last sternite linearly impressed. Length, 10-16 mm.

Type locality. St. John's Bluff, Florida.

Range. Southeastern United States to Texas, northeastern Mexico, and South Dakota.

This species may be separated from *M. confusa* by the short inner tooth of the tarsal claws. The pronotal calluses readily distinguish it from *M. (M.) cineracea* and the elongate antennal segments from *M. (M.) bicallosa*.

Two allopatric subspecies can be recognized.

Mecas (Mecas) cana cana (Newman)

Saperda cana Newman, 1840, Entomol., 1:12; LeConte, 1852, J. Acad. Nat. Sci. Philadelphia, (2)2:164.

Mecas cana: Gahan, 1888, Trans. Amer. Entomol. Soc., 15:300; Leng and Hamilton, 1896, Trans. Amer. Entomol. Soc., 23:152; Casey, 1913, Memoirs on the Coleoptera, 4:360; Breuning, 1955, Mem. Soc. Roy. Entomol. Belg., 27:148.

Mecas (Mecas) cana cana: Chemsak and Linsley, 1973, Proc. Calif. Acad. Sci., (4)39:178.

Body densely grayish pubescent; elytra narrowly clothed at suture and lateral margins with bands of appressed pubescence. Length, 10-12.5 mm.

Type locality. St. John's Bluff, Florida.

Range. Florida.

Flight period. April to October.

Host plants. Ambrosia, Flaveria linearis.

This subspecies appears to be restricted to Florida, primarily the southern portion. Although Breuning (1955) states that the type of *cana* appears to be lost, it is present in the collection of the British Museum (Natural History).

Mecas (Mecas) cana saturnina (LeConte)

Stenostola saturnina LeConte, 1859, Smithson. Contr. Knowledge, 11:21.

Mecas saturnina: Gahan, 1888, Trans. Amer. Entomol. Soc., 15:300; Horn, 1888, Trans. Amer. Entomol. Soc., 15:301; Knowlton and Thatcher, 1936, Utah Acad. Sci. Arts Letters, 13:281; Breuning, 1955, Mem. Soc. Roy.

Entomol. Belg., 27:146; Wilson, 1960, Comm. Inst. Biol. Control. Tech. Comm. 1:62; Stride and Warwick, 1962, Animal Behaviour, 10:112 (habits); Stride and Straatman, 1963, Austral. J. Zool., 11:446, figs. 1, 4-7 (habits, larva, pupa); Harris and Piper, 1970, Comm. Inst. Biol. Contr. Tech. Bull. 13:128; Phillips et al., 1973, Texas Agr. Exp. Sta. MP-1116:3, figs. 1,2; Goeden, 1978, USDA Handb. 483:384; Carter, 1978, Sunflower Sci. Tech. Agron. 19, p. 210; Hilgendorf and Goeden, 1981, Bull. Entomol. Soc. Amer., 27: 103.

Mecas (Mecas) cana saturnina: Chemsak and Linsley, 1973, Proc. Calif. Acad. Sci., (4)39:179; Hovore et al., 1987, Proc. Calif. Acad. Sci., 44:321.

Mecas cana saturnina: Rice and Enns, 1981, Trans. Mo. Acad. Sci., 15:105.

Mecas inornata: Horn, 1878 (not Say, 1824), Trans. Amer. Entomol. Soc., 7: 44; Blanchard, 1887, Entomol. Amer., 3: 86; Gahan, 1888, Trans. Amer. Entomol. Soc., 15: 300; Horn, 1888, Trans. Amer. Entomol. Soc., 15: 301; Beutenmuller, 1896, J. N.Y. Entomol. Soc., 4: 81; Leng and Hamilton, 1896, Trans. Amer. Entomol. Soc., 23: 152; Townsend, 1902, Trans. Tex. Acad. Sci., 5: 80; Garnett, 1918, Can. Entomol., 50: 283; Baerg, 1921, J. Econ. Entomol., 14: 99 (habits); Linsley and Martin, 1933, Entomol. News, 44: 183; Knull, 1946, Ohio Biol. Surv. Bull., 39: 274; Fattig, 1947, Emory Univ. Mus. Bull., 5: 43; Dillon and Dillon, 1961, Man. Beetles East. North America, p. 652, p. 65, no. 17; Stein and Tagestad, 1976, USDA For. Serv. Res. Pap., RM-171: 17 (part); Headstrom, 1977, Beetles of America, p. 382; Rogers, 1977, Envir. Entomol., 6: 833; Rogers and Serda, 1979, J. Kansas Entomol. Soc., 52: 546; Turnbow and Franklin, 1980, J. Ga. Entomol. Soc., 15: 348.

Mecas brevicollis Casey, 1913, Memoirs on the Coleoptera, 4:362.

Saperda concolor: Kirk and Balsbaugh, 1975 (not LeConte, 1852), S.D. Agr. Exp. Sta. Tech. bull., 42: 100.

Similar in form and size to typical subspecies. Pubescence uniformly gray or yellowish, elytra without pubescent bands on margins and suture. Length, 9-16 mm.

Type locality. Of *saturnina*, Kansas; *brevicollis*, Kansas.

Range. Alabama to northeastern Mexico to South Dakota.

Flight period. April to August.

Host plants. Ambrosia, Xanthium, Helianthus, Iva, Gaillarda, Parthenium. Adults have also been taken on *Prosopis* and *Salvia* in Texas.

Genus *Oberea* Mulsant

Oberea Mulsant, 1839, Hist. Nat. Coleop. France, Longicornes, p. 194; Haldeman, 1847, Trans. Amer. Philos. Soc., (2)10: 56; LeConte, 1852, J. Acad. Nat. Sci. Philadelphia, 2: 151; Thomson, 1860, Class. ceram., p. 62; Thomson, 1864, Syst. ceram., p. 121; Lacordaire, 1872, Genera des coléoptères, 9(2): 864; LeConte, 1873, Smithson. Misc. Coll., 11 (265): 347; Provancher, 1877, Pet. Fauna Entomol. Can., 1: 635; Horn, 1878, Trans.

Amer. Entomol. Soc., 7: 45; LeConte and Horn, 1883, Smithson. Misc. Coll., 507: 332; Leng and Hamilton, 1896, Trans. Amer. Entomol. Soc., 23: 153; Wickham, 1897, Can. Entomol., 29: 204; Wickham, 1898, Can. Entomol., 30: 43; Blatchley, 1910, Coleoptera in Indiana, p. 1091; Casey, 1913, Memoirs on the Coleoptera, 4: 364; Craighead, 1923, Can. Dept.Agr. Bull., (n.s.) 27: 135; Mutchler and Weiss, 1923, N.J. Dept. Agr. Cir., 58: 22; Champlain and Knull, 1925, Entomol. News, 36: 141; Bradley, 1930, Man. Genera Beetles, p. 247; Chagnon, 1933-40, Coleop. Prov. Quebec, p. 278; Knull, 1946, Ohio Biol. Surv. Bull., 39: 275; Craighead, 1950, USDA Misc. Pub., 657: 255; Breuning, 1960-62, Frust. Entomol., 3, 4, 5: 16; Dillon and Dillon, 1961, Man. Beetles East. North America, p. 652; Arnett, 1962, Beetles U.S., 103: 873; Chagnon and Robert, 1962, Prin. Coleop. Prov. Quebec, p. 278; Hicks, 1962, Coleop. Bull., 16: 5; Bayer and Shenefelt, 1969, Univ. Wisc. Res. Bull., 275: 32; Hatch, 1971, Univ. Wash. Publ. Biol., 16: 156; Baker, 1972, USDA Misc. Pub., 1175: 189; Headstrom, 1977, Beetles of America, p. 383; Rice and Enns, 1981, Trans. Mo. Acad. Sci., 15: 104; Drooz, 1985, USDA For. Serv. Misc. Pub., 1426: 299.

Isosceles Newman, 1842, Entomol., 1: 318; Thomson, 1864, Syst. ceram., p. 122.

(Type species: *Isosceles macilenta* Newman, Thomson designation, 1864.)

Form elongate, slender, usually parallel. Head with front quadrate, convex, interantennal area shallowly impressed, mandibles curved near apices, acute; palpi unequal, slender; eyes moderate, finely faceted, deeply emarginate, upper lobes widely separated; genae shorter than lower eye lobes; antennae slender, usually about as long as body in males, slightly shorter in females, basal segments usually lightly fringed beneath, scape short, subconical, third segment longer than scape, fourth shorter than third, longer than first, outer segments gradually decreasing in length. Pronotum usually broader than long, cylindrical, sides usually broadly rounded; disk usually with rounded, glabrous calluses; prosternum short, intercoxal process narrow, broadly expanded at apex, coxal cavities closed behind; mesosternum with coxal cavities open to epimeron; metasternum with episternum broad in front, strongly narrowing posteriorly. Elytra often with sides broadly impressed near middle; apices rounded to emarginate-truncate. Legs short; intermediate tibiae with a dorsal sinus; tarsal claws appendiculate. Abdomen normally segmented; last sternite deeply impressed in males, linearly impressed in females.

Type species. Cerambyx oculatus Linnaeus (Thomson designation, 1864).

The more slender, elongate body form and appendiculate tarsal claws separate this genus from *Mecas*.

Oberea is a very large, almost worldwide genus. Thus far, the group is not known to occur in the Neotropical region, but is abundant in the Nearctic. In North America, most of the species occur in the eastern part and a single species is found on the Pacific Coast.

We are listing the Breuning (1962) infrasubspecific names in synonymy with the appropriate species. These were clearly described as varieties and are unavailable names according to Article 16 of the International Code of Zoological Nomenclature.

KEY TO THE NORTH AMERICAS SPECIES OF *OBEREA*

1	Pronotum with at least two rounded glabrous calluses	2
	Pronotum lacking discal calluses, densely punctate; integument yellowish with two dark longitudinal vittae on elytra. Length, 10.5-14 mm. Eastern United States	*gracilis*
2(2)	Pronotum with four rounded calluses on disk	3
	Pronotum with two calluses on disk	5
3(2)	Pronotum with dorsal calluses black, glabrous	4
	Pronotum with dorsal calluses pale, punctate. Integument orange, antennae and elytra black. Elytra densely gray pubescent. Length, 15-21 mm. Eastern North America	*ruficollis*
4(3)	Elytra almost always black, densely grayish pubescent. Legs with femora almost impunctate, very sparsely pubescent. Length, 11-14 mm. Pacific Coast to Colorado	*quadricallosa*
	Elytra usually partially testaceous, pubescence not partially obscuring surface. Legs with femora densely punctate and pubescent. Length, 11-17 mm. Eastern North America	*schaumi*
5(2)	Integument black to brownish, legs often yellowish	6
	Integument at least partially testaceous. Appendages variably colored	7
6(5)	Legs yellow. Elytra with pubescence fine, not obscuring surface. Form small. Length, 7-11 mm. Northeastern North America to Manitoba	*flavipes*
	Legs dark. Elytra with pubescence dense, partially obscuring surface. Form moderate-sized. Length, 11-13 mm. Midwestern North America	*oculaticollis*
7(5)	Elytra uniformly dark, not longitudinally vittate	8
	Elytra testaceous to lightly infuscated, usually with dark longitudinal vittae	11
8(7)	Underside at least partially infuscated. Elytra with pubescence very fine, usually darker, not partially obscuring surface. Legs usually unicolorous	9
	Underside orange. Elytra with pubescence dense, grayish, at least partially obscuring surface. Femora pale, tibiae and tarsi dark. Length, 10.5-17 mm. Eastern United States to Texas and North Dakota	*ocellata*

9(8)	Pronotum testaceous with black spots and often narrowly black along base and/or apex. Legs usually dark	10
	Pronotum infuscated at least behind dorsal black calluses. Legs testaceous. Length, 10-13 mm. Northeastern North America	*delongi*
10(9)	Pronotum broadly impressed at apex and base; disk with median calluses often pale, middle with a linear longitudinal callus. Form larger, more robust. Length, 11-16 mm. Eastern North America	*affinis*
	Pronotum barely impressed, cylindrical; disk shining, base usually transversely black. Form smaller, less robust. Length, 8-15 mm. Eastern North America to Utah	*perspicillata*
11(7)	Antennae concolorous, not annulate	12
	Antennae with outer segments narrowly dark, annulate at apices. Pronotum shining, finely punctate, apex and base moderately impressed transversely. Abdomen of males with last sternite shallowly impressed. Length, 8-13 mm. Eastern North America	*tripunctata*
12(11)	Pronotum with disk subopaque, conversely, confluently punctate, apex and base deeply impressed transversely. Abdomen of males with last sternite deeply, circularly impressed. Form usually broader. Length, 13-18 mm. Eastern North America	*myops*
	Pronotum with disk shining, punctures moderately coarse, apex shallowly impressed transversely. Abdomen of males with last sternite shallowly impressed. Form slender, elongate. Length, 10-16 mm. Northeastern North America to Manitoba	*praelonga*

Oberea gracilis (Fabricius)

Saperda gracilis Fabricius, 1801, Syst. Eleuth., 2: 324.
Oberea gracilis: Haldeman, 1847, Trans. Amer. Philos. Soc., (2)10: 57; LeConte, 1852, J. Acad. Nat. Sci. Philadelphia, 2: 152; Lacordaire, 1872, Genera des coléoptères, 9(2): 866, fn.; Horn, 1878, Trans. Amer. Entomol. Soc., 7: 47; Leng and Hamilton, 1896, Trans. Amer. Entomol. Soc., 23: 156; Blatchley, 1910, Coleoptera in Indiana, p. 1093; Smith, 1910, N.J. St. Mus. Rept., 1909: 337; Casey, 1913, Memoirs on the Coleoptera, 4: 372; Mutchler and Weiss, 1923, N.J. Dept. Agr. Cir., 58: 19; Knull, 1946, Ohio Biol. Surv. Bull., 39: 281; Fattig, 1947, Emory Univ. Mus. Bull., 5: 45;

Hicks, 1962, Coleop. Bull., 16: 6; Breuning, 1962, Frust. Entomol., 5: 225; Turnbow and Hovore, 1979, Entomol. News, 90: 227 (habits).

Male. Form moderate-sized, elongate; integument yellow to yellowish orange, tips of mandibles, eyes, antennae, tarsi, apices of tibiae, and longitudinal vittae from humeri almost to apices black: pubescence moderately dense, very short and appressed and long and erect. Head with front slightly broader than long, punctures coarse, dense, subconfluent, pubescence suberect; genae about half as long as lower eye lobes; antennae about as long as body, erect hairs sparse beneath, segments beneath clothed with very fine, grayish, appressed pubescence, third segment longer than fourth, fourth longer than first. Pronotum broader than long, sides broadly rounded; apex and base broadly, shallowly impressed; disk convex, coarsely, confluently punctate; pubescence very fine, appressed, long, erect, dark hairs numerous; prosternum finely pubescent; meso- and metasternum densely micropunctate, sides with numerous coarse punctures, pubescence dense, appressed. Elytra about three times longer than broad, sides broadly impressed over middle half; punctures coarse, subserial on disk, finer toward apex, area between suture and costa with four rows of punctures, two inner rows somewhat irregular; pubescence sparse-appearing, appressed pubescence very fine, long, erect hairs numerous; apices obliquely truncate. Legs short; femora very finely, densely punctate, finely pubescent. Abdomen densely micropunctate, sides of basal segments with coarser punctures; pubescence fine, dense; last sternite moderately deeply excavated for most of its length, apical angles rounded. Length, 10.5-15 mm.

Female. Form similar. Antennae shorter than body. Abdomen with last sternite with a longitudinal, impressed line, apex vaguely triangularly impressed. Length, 11-14 mm.

Type locality. Carolina.
Range. Eastern United States to Florida.
Flight period. April to July.
Host plants. Quercus.

O. gracilis is easily recognized by the absence of dorsal callosities on the pronotum and the yellowish body color with a dark vitta on each elytron.

In the material available for study there is little variation in coloration and the size range is not great.

Oberea ruficollis (Fabricius)

Saperda ruficollis Fabricius, 1792, Entomol. Syst., 1(2): 311; Fabricius, 1801, Syst. Eleuth., 2: 322.

Oberea ruficollis: Haldeman, 1847, Trans. Amer. Philos. Soc., (2)10: 56; LeConte, 1852, J. Acad. Nat. Sci. Philadelphia, 2: 152; Lacordaire, 1872, Genera des coléoptères, 9(2): 865, fn.; Horn, 1878, Trans. Amer. Entomol. Soc., 7: 47; Leng and Hamilton, 1896, Trans. Amer. Entomol. Soc., 23: 156; Wickham, 1898, Can. Entomol., 30: 43; Blatchley, 1910, Coleoptera in

Indiana, p. 1093; Smith, 1910, N.J. St. Mus. Rept. 1909: 337; Casey, 1913, Memoirs on the Coleoptera, 4: 372; Aurivillius, 1923, Coleop. Cat., 74: 537; Craighead, 1923, Can. Dept. Agr. Bull., (n.s.) 27: 136, pls. 2, 13, 24, 32; Mutchler and Weiss, 1923, N.J. Dept. Agr. Cir., 58: 19; Champlain and Knull, 1925, Entomol. News, 36: 142; Leonard, 1928, Cornell Agr. Exp. Sta. Mem., 101: 457; Knull, 1946, Ohio Biol. Surv. Bull., 39: 280; Fattig, 1947, Emory Univ. Mus. Bull., 5: 44; Craighead, 1950, USDA Misc. Pub., 657: 256, fig. 560 (habits); Duffy, 1953, Mon. British Timber Beetles, p. 295, figs. 283, 284; Dillon and Dillon, 1961, Man. Beetles East. North America, p. 656, pl. 65, no. 11; Breuning, 1962, Frust. Entomol., 5: 226; Hicks, 1962, Coleop. Bull., 16: 6; Baker, 1972, USDA Misc. Pub., 1175: 190 (habits); Gosling and Gosling, 1976, Gr. Lakes Entomol., 10: 34, fig. 176; Headstrom, 1977, Beetles of America, p. 384; Soloman, 1977, Can. Entomol., 109: 297; Rice and Enns, 1981, Trans. Mo. Acad. Sci., 15: 105; Drooz, 1985, USDA Misc. Pub., 1426: 300 (habits).

Oberia ruficollis: Beal, 1952, Duke Univ. Sch. For. Bull., 14: 45 (habits) (error).

Saperda plumbea Olivier, 1795, Entomol., 4(68): 21, pl. 4, fig. 42; Harris, 1938, Rept. Comm. Zool. Surv. St. (Mass.), p. 91.

Phytoecia tibialis Haldeman, 1847, Trans. Amer. Philos. Soc., (2)10: 57.

Oberea tibialis: Horn, 1878, Trans. Amer. Entomol. Soc., 7: 46; Aurivillius, 1923, Coleop. Cat., 74: 537.

Oberea ruficollis v. *tibialis*: Breuning, 1962, Frust. Entomol., 5: 227.

Oberea ruficollis m. *tibialis*: Breuning, 1966, Cat. Lam. du Monde, 9: 826.

Oberea ruficollis v. *rufolineata* Breuning, 1962, Frust. Entomol., 5: 227

Oberea ruficollis m.*rufolineata*: Breuning, 1966, Cat. Lam. du Monde, 9: 826.

Male. Form moderately large, elongate; integument orange, tips of mouth-parts, eyes, antennae, elytra, tibiae, and tarsi black; pubescence dense, short, recumbent, short, erect hairs moderately dense. Head with front slightly broader than long, densely punctate, interspaces micropunctate; pubescence fine, appressed, denser around eyes; antennae about as long as body, erect hairs sparse beneath, segments densely clothed with very short, appressed, grayish pubescence, third segment longer than fourth, fourth longer than first. Pronotum broader than long, sides rounded; apex and base broadly impressed transversely; disk with four punctate calluses, center with a raised line behind middle; punctures dense, coarse, subconfluent; pubescence very fine, dense, long, erect hairs sparsely interspersed; prosternum finely, densely pubescent; meso- and metasternum moderately coarsely, irregularly punctate at sides, pubescence fine, dense, longer, suberect hairs numerous. Elytra slightly less than 3-1/2 times longer than broad, sides tapering from behind humeri ,then slightly expanding near apex; punctures coarse, contiguous, becoming finer toward apex, subserially arranged on disk; disk with vague costae extending from base inside of humeri to about apical one-fifth, area between costa and suture with four rows of punctures; pubescence dense, grayish, appressed, longer, erect hairs denser on basal half; apices rounded to subtruncate. Legs short; femora densely micropunctate, finely pubescent; tibiae finely, densely pubescent. Abdomen densely micropunctate with large punctures sparsely interspersed at sides; pubescence fine, dense; last sternite shallowly, triangularly impressed, apex shallowly notched. Length, 15-19 mm.

Female. Form similar. Antennae shorter than body. Abdomen with last sternite shallowly impressed at apex, medially, linearly impressed. Length, 15-21 mm.

Type locality. Of *ruficollis*, Virginia; *plumbea*, America septentrionale; *tibialis*, Pennsylvania; *rufolineata*, Stanford, Florida.

Range. Eastern North America from Ontario to Florida west to Missouri.

Flight period. May to August.

Host plants. Sassafras.

This is one of the largest species of *Oberea* in our fauna. It may be easily recognized by the orange head and pronotum and the black, densely gray-pubescent elytra. The pronotum has the vague dorsal calluses punctate.

According to Craighead (1923), larvae occur in the stems and roots of young, living sassafras.

Oberea quadricallosa LeConte
(Figure 43)

Oberea quadricallosa LeConte, 1874, Trans. Amer. Entomol. Soc., 3: 68; Horn, 1878, Trans. Amer. Entomol. Soc., 7: 46; Leng and Hamilton, 1896, Trans. Amer. Entomol. Soc., 23: 157; Beutenmuller, 1896, J. N.Y. Entomol. Soc., 4:81; Casey, 1913, Memoirs on the Coleoptera, 4: 365; Hardy, 1927, Rept. Prov. Mus., 1926: C37; Knowlton and Thatcher, 1936, Utah Acad. Sci. Arts Letters, 13: 281; Keen, 1952, USDA Misc. Pub., 273: 47; Essig, 1958, Ins. West. North America, p. 462 (habits); Hicks, 1962, Coleop. Bull., 16: 11; Hatch, 1971, Univ. Wash. Pub. Biol., 16: 156; Stern and Tagestad, 1976, USDA For. Serv. Res. Garnett (not LeConte, 1874), 1918, Can. Entomol., 50: 284.Pap., RM-171: 26; Hovore, 1988, Wasmann J. Biol., 46: 24.

Oberea schaumii: Garnett, 1918 (not LeConte, 1874), Can. Entomol., 50:284.

Oberea schaumi v. *quadricallosa*: Hardy, 1926, Rept. Prov. Mus., 1925: 10; Breuning, 1962, Frust. Entomol., 5: 223.

Oberea schaumi m.*quadricallosa*: Breuning, 1966, Cat. Lam. du Monde, 9: 826.

Oberea quadricallosa m. *infrarufa* Breuning, 1962, Frust. Entomol., 5: 223.

Oberea ferruginea: Keen, 1938, USDA Misc. Pub., 273: 37; Keen, 1952, USDA Misc. Pub., 273: 47 (misidentified).

Male. Form moderate-sized, elongate, parallel; integument yellowish, head, antennae, elytra, scutellum, spots on pronotum, parts of meso- and metasternum, and abdomen black; pubescence dense, short, grayish, appressed. Head with front slightly broader than long, punctures obscured by dense, grayish, appressed pubescence, long, dark, erect hairs numerous; antennae shorter than body, erect hairs sparse beneath, segments densely clothed with short appressed pubescence, dark above, grayish beneath, third segment longer than fourth, fourth longer than first. Pronotum broader than long, sides broadly rounded; apex broadly, shallowly impressed, base broadly, more deeply impressed; disk with four glabrous, black calluses, each side at angle of coxal cavity with a glabrous black spot, middle with a longitudinal carina; punctures shallow, irregular; pubescence fine, pale, not obscuring surface, long, erect hairs numerous; prosternum densely pubescent; meso- and metasternum densely micropunctate, larger punctures sparse, pubescence dense, short and appressed and long and suberect. Elytra a little more than three times longer than broad, sides subparallel, vaguely narrowing near middle; punctures coarse, contiguous, becoming finer toward apex; disk vaguely bicostate on each side, two or three rows of punctures present between suture and first costa and three rows between costae; pubescence appressed, obscuring surface, longer, suberect hairs numerous on basal one-third; apices subtruncate. Legs short; femora almost impunctate, very sparsely pubescent. Abdomen densely micropunctate with larger punctures at sides; pubescence dense, fine; last sternite deeply impressed, sides at apex moderately keeled. Length, 12-16 mm.

Female. Form similar, slightly more robust. Antennae extending to about third abdominal segment. Abdomen with last sternite longitudinally impressed at middle, apex shallowly impressed triangularly; last tergite with a broad tubercle near apex. Length, 11-17 mm.

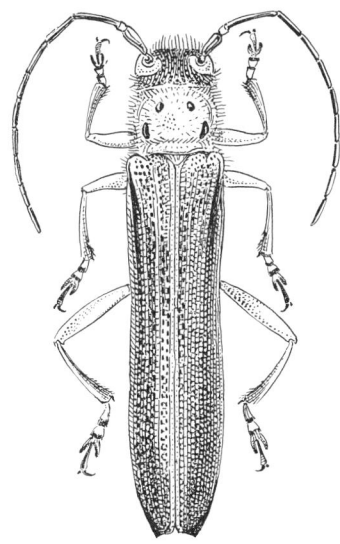

Figure 43. *Oberea quadricallosa* LeConte, male.

Type locality. Of quadricallosa, Western California and Nevada; *infrarufa*, Mariposa, California.
Range. British Columbia to southern California, Colorado, and Nevada.
Flight period. May to August.
Host plants. Salix, Populus.

This species may be recognized by the fairly large size, black head, antennae, and elytra and the four black spots on the disk of the pronotum. Additionally the glabrous femora are distinctive.

The coloration of *quadricallosa* appears to be relatively constant throughout its range. Some variation is expressed in the amount of infuscation of the underside. There is a tendency for denser-appearing, grayish pubescence in individuals from the eastern side of the Sierra Nevada.

Oberea schaumi LeConte

Oberea schaumii LeConte, 1852, J. Acad. Nat. Sci. Philadelphia, 2: 153; Lacordaire, 1872, Genera des coléoptères, 9(2): 866, fn.; LeConte, 1873, Smithson. Misc. Coll., 11(265): 346; Horn, 1878, Trans. Amer. Entomol. Soc., 7: 46; Packard, 1881, U.S. Entomol. Comm. Bull., 7: 115; Knobel, 1895, Beetles New England, p. 34, fig. 132; Leng and Hamilton, 1896, Trans. Amer. Entomol. Soc., 23: 154; Beutenmuller, 1896, J. N.Y. Entomol. Soc., 4: 81; Wickham, 1898, Can. Entomol., 30: 43; Chagnon, 1905, Nat. Can., 32: 44; Blatchley, 1910, Coleoptera in Indiana, p. 1091; Craighead, 1923, Can. Dept. Agr. Bull., (n.s.) 27: 137; Mutchler and Weiss, 1923, N.J. Dept. Agr. Cir., 58: 16; Craighead and Middleton, 1930, USDA Misc. Pub., 74: 8; Hicks, 1945, Can. Entomol., 77: 214; Knull, 1946, Ohio Biol. Surv. Bull., 39: 276; Fattig, 1947, Emory Univ. Mis. Bull., 5: 43; Dillon and Dillon, 1961, Man. Beetles East. North America, p. 653, pl. 65, no. 9; Hicks, 1962, Coleop. Bull., 16: 11, fig. 3; Knight, 1963, Proc. No. Cent. Br. Entomol. Soc. Amer., 18: 65; Myers, Knight and Grimble, 1968, Ann. Entomol. Soc. Amer., 61: 1418; Gardiner, 1969, Can. Dept. Fish. For. Int. Rept., 0-14: 104; Grimble and Knight, 1971, Ann. Entomol. Soc. Amer., 64: 1417 (fig. 1); Nord and Knight, 1972, Gr. Lakes Entomol., 5: 28; Nord and Knight, 1972, Gr. Lakes Entomol., 5: 87; Nord and Knight, 1972, Gr. Lakes Entomol., 5: 93; Nord, Grimble and Knight, 1972, Ann. Entomol. Soc. Amer. 65: 144 (habits); Headstrom, 1977, Beetles of America, p. 383; Solomon, 1977, Can. Entomol., 109: 298; Laliberte et al., 1977, Fabreries, 3: 95; Rice and Enns, 1981, Trans. Mo. Acad. Sci., 15: 105; Drooz, 1985, USDA Misc. Pub., 1426: 300 (habits).

Oberea schaumii var. *schaumii*: Horn, 1878, Trans. Amer. Entomol. Soc., 7: 46.

Oberea schaumi: Casey, 1913, Memoirs on the Coleoptera, 4: 365; Leonard, 1928, Cornell Agr. Exp. Sta. Mem., 101: 456; Park, 1931, Ecology, 12 189; Knull, 1932, Entomol. News, 43: 64 (habits); Chagnon, 1933-40, Coleop. Prov. Quebec., p. 279; Craighead, 1950, USDA Misc. Pub., 657: 256

(habits); Chagnon and Robert, 1962, Prin. Coleop. Prov. Quebec., p. 279; Breuning, 1962, Frust. Entomol., 5: 222; Baker, 1972, USDA Misc. Pub., 1175: 190 (habits); Gosling and Gosling, 1976, Gr. Lakes Entomol., 10: 33, fig. 173.
Oberea wapleri Chevrolat, 1852, Rev. Zoo. (2)4:420.
Oberea pruinosa Casey, 1913, Memoirs on the Coleoptera, 4: 365; Knull, 1946, Ohio Biol. Surv. Bull., 39: 277; Hicks, 1962, Coleop. Bull., 16: 11.
Oberea schaumi v. *pruinosa*: Breuning, 1962, Frust. Entomol., 5: 224.
Oberea schaumi m. *pruinosa*: Breuning, 1966, Cat. Lam. du Monde, 9: 826.
Oberea ferruginea Casey, 1913, Memoirs on the Coleoptera, 4: 366; Craighead, 1923, Can. Dept. Agr. Bull., (n.s.) 27: 27: 137; Craighead, 1950, USDA Misc. Pub., 657: 255 (habits); Baker, 1972, USDA Misc. Pub., 1175: 190; Headstrom, 1977, Beetles of Amer., p. 384; Drooz, 1985, USDA Misc. Pub., 1426: 300.
Oberea caseyi Plavilstshikov, 1926, Encycl. Entomol., (B)2, Coleop., 1: 64 (new name for *ferruginea* Casey.)
Oberea schaumi v. *caseyi*: Breuning, 1962, Frust. Entomol., 5: 244.
Oberea schaumi m *caseyi*: Breuning, 1966, Cat. Lam. du Monde, 9: 826.
Oberea quadricallosa cylindricollis Casey, 1924, Memoirs on the Coleoptera, 11: 295.
Oberea cylindricollis: Hicks, 1962, Coleop. Bull., 16: 11.
Oberea schaumi v. *subcylindricollis* Breuning, 1962, Frust. Entomol., 5: 223.
Oberea schaumi m. *subcylindricollis*: Breuning, 1966, Cat. Lam. du Monde, 9: 826.
Oberea schaumi v. *vittipennis* Breuning, 1962, Frust. Entomol., 5: 224.
Oberea schaumi m. *vittipennis*: Breuning, 1966, Cat. Lam. du Monde, 9: 826.

Male. Form moderate-sized, elongate; integument testaceous, head partially dark, pronotum with four dark spots, elytra usually at least partially infuscated, underside and legs at least partially dark; pubescence very fine, appressed, erect hairs very sparse. Head with front broader than long, moderately coarsely, irregularly punctate, densely clothed with pale appressed pubescence; genae about one-third shorter than lower eye lobes; antennae extending to about fourth abdominal segment, basal segments with several erect hairs beneath, fine grayish pubescence moderately dense, appressed, third segment longer than fourth, fourth longer than first. Pronotum broader than long, sides broadly rounded; apex shallowly impressed, base more deeply impressed; disk with two rounded, glabrous, black calluses at sides of middle before center and two larger black calluses at sides before base, center with an elongate, longitudinal callus usually extending between apical and basal impressions; punctures irregular, moderately coarse, areas between micropunctate; pubescence dense, pale, appressed, long, erect hairs interspersed; prosternum finely, densely pubescent; meso- and metasternum densely micropunctate with coarser punctures interspersed, pubescence dense, appressed. Elytra a little more than three times longer than broad, sides slightly impressed at middle; each elytron longitudinally carinate down middle; punctures coarse, subserial, becoming finer and obsolete toward apex; pubescence fine, dense, very short, appressed, base with a few long, suberect hairs; apices narrowly rounded. Legs short; femora stout, rather broad, finely, densely punctate and pubescent; hind tibiae arcuate, broadened at apices. Abdomen densely micropunctate with large punctures moderately sparsely interspersed; pubescence dense, appressed; last sternite deeply excavated for its entire length, apical margin deeply impresssed; last tergite not produced. Length, 11-16 mm.

Female. Form more robust. Antennae extending to about second abdominal segment. Abdomen with last sternite linearly impressed medially, apex triangularly impressed, apical margin broadly v-shaped. Length, 11-17 mm.

Type locality. Of *schaumi*, Louisiana; *wapleri*, New Orleans; *pruinosa*, near St. Louis, Missouri; *ferruginea*, Kansas; *cylindricollis*, northern Illinois.

Range. Eastern North America to Texas and Alberta.

Flight period. June and July.

Host plants. Populus deltoides, Salix.

This species varies considerably in coloration. The elytra range from pale to dark, with all intermediates evident. The underside is variably infuscated and the legs are usually pale except for the tarsi. The four glabrous black spots of the pronotum are present in all examined specimens.

The broader, stouter, and densely punctate and pubescent femora separate *O. schaumi* from *O. quadricallosa*, its western counterpart.

Oberea flavipes Haldeman

Oberea flavipes Haldeman, 1847, Trans. Amer. Philos. Soc., (2)10: 57; LeConte, 1852, J. Acad. Nat. Sci. Philadelphia, 2: 153; Lacordaire, 1872, Genera des coléoptères, 9(2): 866, fn.; Horn, 1878, Trans. Amer. Entomol. Soc., 7: 46; Leng and Hamilton, 1896, Trans. Amer. Entomol. Soc., 23: 155; Casey, 1913, Memoirs on the Coleoptera, 4: 364; Casey, 1914, Memoirs on the Coleoptera, 5: 369; Craighead, 1923, Can. Dept. Agr. Bull., (n.s.) 27: 141; Fattig, 1947, Emory Univ. Mus. Bull., 5: 44; Hicks, 1962, Coleop. Bull., 16: 9.

Oberea bimaculata form *flavipes*: Leng and Hamilton, 1896, Trans. Amer. Entomol. Soc., 23: 156.

Oberea tripunctata var. *flavipes*: Blatchley, 1910, Coleoptera in Indiana, p. 1092.

Oberea bimaculata v. *flavipes*: Breuning, 1962, Frust. Entomol., 5: 232.

Oberea bimaculata m. *flavipes*: Breuning, 1966, Cat. Lam. du Monde, 9: 827.

Male. Form small, slender, subparallel; integument black to brownish, legs yellowish; pubescence moderately dense, short, appressed, short, erect hairs fairly numerous. Head with front about as broad as long, densely punctate and pubescent, long, erect hairs numerous; genae very short; antennae about as long as body, segments not annulate, basal segments with few erect hairs beneath, segments with one or two long setae at apices, scape slender, third segment longer than fourth, fourth slightly longer than first. Pronotum about as long as broad; apex and base narrowly, shallowly impressed transversely; disk with two glabrous calluses before middle, middle toward base with a short linear callus; punctures moderately coarse, denser at sides; pubescence moderately dense, fine, appressed, long, erect hairs abundant; prosternum finely pubescent; meso- and metasternum finely, densely punctate, punctures coarser at sides, pubescence dense, appressed. Elytra at least four times longer than broad, sides slightly, broadly impressed toward middle; punctures moderately coarse, subserial, disk with vague costae at middle and at sides back from humeri; pubescence moderately dense, grayish, short, appressed, short, erect hairs more numerous near base; apices obliquely truncate. Legs short; femora robust, very finely punctate. Abdomen densely micropunctate with larger punctures interspersed; pubescence dense, appressed; last sternite moderately deeply, semicircularly impressed, apex shallowly emarginate. Length, 7-10 mm.

Female. Form similar. Antennae slightly shorter than body. Abdomen with last sternite triangularly impressed at apical half. Length, 7-11 mm.

Type locality. Pennsylvania.
Range. Northeastern North America to Manitoba.
Flight period. May to July.
Host plants. Phlox.

The slender body form and dark integument with yellowish legs make this species easily recognizable.

Oberea oculaticollis (Say)

Saperda oculaticollis Say, 1824, J. Acad. Nat. Sci. Philadelphia, 3: 406; LeConte, 1852, J. Acad. Nat. Sci. Philadelphia, 2: 164; LeConte, 1859, Compl. Writings T. Say, 2: 189.

Oberea oculaticollis: LeConte, 1859, Compl. Writings T. Say, 2: 189; Lacordaire, 1872, Genera des coléoptères, 9(2): 866, fn.; Horn, 1878, Trans. Amer. Entomol. Soc., 7: 46; Leng and Hamilton, 1896, Trans. Amer. Entomol. Soc., 23: 157; Hicks, 1944, Can. Entomol., 76: 163.

Oberea tripunctata var. *bimaculata* form *oculaticollis*: Leng and Hamilton, 1896, Trans. Amer. Entomol. Soc., 23: 155.

Oberea bimaculata v. *oculaticollis*: Breuning, 1962, Frust. Entomol., 5: 232.

Oberea bimaculata m. *oculaticollis*: Breuning, 1966, Cat. Lam. du Monde, 9: 827.

Oberea brooksi Wallis, 1926, Can. Entomol., 58: 94.

Male. Form moderate-sized, slender, subparallel; integument black, appendages paler, usually dark reddish brown; pubescence dense, grayish, short, appressed, obscuring surface. Head with front about as long as broad, punctures rather fine, dense, pubescence dense, appressed, denser around eyes, dark suberect hairs numerous; genae about as long as second antennal segment; antennae about as long as body, segments unicolorous, outer segments with two long, erect hairs at apices, scape moderate, third segment longer than first, fourth subequal to first. Pronotum about as long as broad; base and apex barely impressed; disk convex, with two black, glabrous calluses before middle, middle with a narrow longitudinal line, usually more prominent toward base; punctures moderately coarse, moderately dense; pubescence dense, long erect hairs sparse; prosternum densely pubescent; meso- and metasternum coarsely, densely punctate at sides, pubescence dense. Elytra about 3 3/4 times longer than broad, sides broadly impressed near middle; punctures moderately coarse, subserial, disk with four ill-defined rows of punctures between suture and vague subhumeral costae; pubescence very fine, dense, partially obscuring surface, short erect hairs more numerous basally; apices shallowly emarginate truncate, angles lightly dentate. Legs short; femora finely punctate and pubescent. Abdomen moderately finely punctate, punctures denser at sides; pubescence dense, appressed; last sternite moderately deeply impressed, apex deeply notched. Length, 11-13 mm.

Female. Form more robust. Antennae shorter than body. Abdomen with last sternite triangularly impressed toward apex. Length, 11-13 mm.

Type locality. Of *oculaticollis*, Missouri; *brooksi*, Transcona, Manitoba.

Range. Midwestern North America from Manitoba to Texas.

Flight period. May to July.

Host plants. Unknown.

The dark integument and dense, grayish pubescence readily distinguish this species.

Oberea ocellata Haldeman

Oberea ocellata Haldeman, 1847, Trans. Amer. Philos. Soc., (2)10: 56; LeConte, 1852, J. Acad. Nat. Sci. Philadelphia, 2: 152; Lacordaire, 1872, Genera des coléoptères, 9(2): 866, fn.; Horn, 1878, Trans. Amer. Entomol. Soc., 7: 46; Knobel, 1895, Beetles New England, p. 34, fig. 133; Leng and Hamilton, 1896, Trans. Amer. Entomol. Soc., 23: 156; Chittenden, 1899, USDA Div. Entomol. Bull., (n.s.) 19: 98; Blatchley, 1910, Coleoptera in Indiana, p. 1092; Casey, 1913, Memoirs on the Coleoptera, 4: 371; Craighead, 1923, Can. Dept. Agr. Bull., (n.s.) 27: 137, pl. 44; Mutchler and Weiss, 1923, N.J. Dept. Agr. Cir., 58: 17; Leonard, 1928, Cornell Agr. Exp. Sta. Mem., 101: 457; Hicks, 1945, Can. Entomol., 77: 214; Knull, 1946, Ohio Biol. Surv. Bull., 39: 280; Fattig, 1947, Emory Univ. Mus. Bull., 5: 44; Craighead, 1950, USDA Misc. Pub., 657: 255 (habits); Duffy, 1953, Mon. British Timber beetles, p. 295, figs. 285, 286; Dillon and Dillon, 1961, Man. Beetles East. North America, p. 653, pl. 65, no. 10; Breuning, 1962, Frust. Entomol., 5: 224; Hicks, 1962, Coleop. Bull., 16: 10; Bayer and Shenefelt, 1969, Univ. Wisc. Res. Bull., 275: 32, fig. 40; Baker, 1972, USDA Misc. Publ., 1175: 190 (habits); Headstrom, 1977, Beetles of

America, p. 383; Gosling and Gosling, 1977, Gr. Lakes Entomol., 10: 34, fig. 175; Rice and Enns, 1981, Trans. Mo. Acad. Sci., 15: 105; Gosling, 1984, Gr. Lakes Entomol., 17: 72; Drooz, 1985, USDA Misc. Pub., 1426: 300 (habits).
Oberea ocellata var. *ocellata*: Horn, 1878, Trans. Amer. Entomol. Soc. 7: 46.
Oberea ocellata var. *discoidea* Horn, 1878, Trans. Amer. Entomol. Soc., 7: 47.
Oberea ocellata v. *discoidea*: Smith, 1910, N.J. St. Mus. Rept., 1909: 336; Leonard, 1928, Cornell Agr. Exp. Sta. mem., 101: 457; Breuning, 1962, Frust. Entomol., 5: 225.
Oberea ocellata ab. *discoidea*: Aurivillius, 1923, Coleop. Cat., 74: 536.
Oberea discoidea: Casey, 1924, Memoirs on the Coleoptera, 11: 295.
Oberea ocellata discoidea: Hicks, 1962, Coleop. Bull., 16: 10; Kirk, 1970, S.C. Agr. Exp. Sta. Tech. Bull., 1038: 84.
Oberea ocellata m. *discoidea*: Breuning, 1966, Cat. Lam. du Monde, 9: 826.
Oberea ulmicola Chittenden, 1904, in Webster, Bull. Ill. St. Lab. Nat. Hist., 7: 4, pl. 1, fig. 1; 1904, Webster, 1904, Bull. Ill. St. Lab. Nat. Hist., 7: 1 (habits); Forbes, 1908, 24th Rept. St. Entomol. Nox. Ben. Ins. Illinois, p. 118, pl. 8 (habits); Hoffmann, 1942, USDA Misc. Pub., 466: 11; Knull, 1946, Ohio Biol. Surv. Bull., 39: 278; Hicks, 1962, Coleop. Bull., 16: 9; Bayer and Shenefelt, 1969, Univ. Wisc. Res. Bull., 275: 33, fig. 40; Baker, 1972, USDA Misc. Pub., 1175: 190; Drooz, 1985, USDA Misc. Pub., 1426: 300.
Oberea bimaculata v. *ulmicola*: Breuning, 1962, Frust. Ent., 5: 231.
Oberea bimaculata m. *ulmicola*: Breuning, 1966, Cat. Lam. du Monde, 9: 827.
Oberea ocellata plagiata Casey, 1913, Memoirs on the Coleoptera, 4: 371.
Oberea ocellata ab. *plagiata*: Aurivillius, 1923, Coleop. Cat., 74: 536.
Oberea plagiata: Casey, 1924, Memoirs on the Coleoptera, 11: 295.
Oberea ocellata v. *plagiata*: Breuning, 1962, Frust. Entomol., 5: 225.
Oberea ocellata m. *plagiata*: Breuning, 1966, Cat. Lam. du Monde, 9: 826.

Male. Form moderate-sized to moderately large, elongate; integument testaceous, tips of mouth parts, eyes, antennae, two spots on pronotum, elytra, tibiae and tarsi black, vertex of head and disk of pronotum occasionally infuscated; pubescence dense, short, recumbent, short, erect hairs moderately dense. Head with front broader than long, densely punctate, interspaces micropunctate; pubescence fine, appressed, longer, erect hairs numerous; antennae slightly shorter than body, erect hairs sparse beneath segments densely clothed with very short, dark, appressed pubescence dorsally, pubescence grayish beneath, third segment longer than fourth, fourth longer than first. Pronotum broader than long, sides rounded, apex and base broadly impressed, base more deeply; disk shining with two dark, glabrous calluses a little before middle, middle with a longitudinal callus extending to basal impression; punctures moderately coarse, irregular; pubescence fine, pale, short and appressed and long and erect; prosternum finely pubescent; meso- and metasternum micropunctate with coarser punctures at sides, pubescence moderately dense, pale. Elytra about 3-1/2 times as long as broad, sides slightly impressed at middle half; punctures coarse, subserially arranged behind basal one-sixth, two sutural rows of punctures irregular, each elytron with a vague median costa; pubescence dense, pale, appressed, longer erect hairs more numerous basally; apices rounded to subtruncate. Legs short; femora densely micropunctate, finely pubescent; tibiae densely pubescent. Abdomen densely micropunctate with larger punctures rather sparsely interspersed; pubescence fine, dense; last sternite shallowly, subtriangularly impressed. Length, 10.5-15.5 mm.

Female. Form similar, slightly more robust. Antennae shorter than body. Abdomen with last sternite linearly impressed longitudinally at middle. Length, 11-17 mm.

Type locality. Of *ocellata*, Pennsylvania; *discoidea*, Florida; *plagiata*, Southern Pines, N. C.; *ulmicola*, Decatur, Illinois.

Range. Eastern United States from New England to Florida west to Texas and North Dakota.

Flight period. April to September.

Host plants. Carya, Cornus, Malus, Morus, Prunus, Pyrus, Quercus, Rhus, Ulmus.

Parasites. Euderus lividus (Ashmead) (Chalcidae).

In the material available for study, most specimens from Florida have the vertex of the head and disk of the pronotum infuscated. This characteristic seems to occur less frequently northward to Pennsylvania and we can find no geographical basis for the separation of the two color forms.

The shining, two-spotted or infuscated pronotum and fine, dense, appressed, grayish pubescence of the pronotum readily separate this species. Additionally the reddish pronotum and underside and dark tibiae and tarsi are distinctive.

Oberea delongi Knull

Oberea delongi Knull, 1928, Entomol. News, 39: 12; Hicks, 1945, Can. Entomol., 77: 214; Knull, 1946, Ohio Biol. Surv. Bull., 39: 281, pl. 26, fig. 113; Breuning, 1962, Frust. Entomol., 5: 224; Hicks, 1962, Coleop. Bull., 16: 10; Solomon, 1969, Ann. Entomol. Soc. Amer., 62: 1214; Gosling and Gosling, 1976, Gr. Lakes Entomol., 10: 34, fig. 174; Solomon, 1977, Can. Entomol., 109: 298; Rice and Enns, 1981, Trans. Mo. Acad. Sci., 15: 105.

Oberea delongi v. *anteruficollis* Breuning, 1962, Frust. Entomol., 5: 224.

Oberea delongi m. *anteruficollis*: Breuning, 1966, Cat. Lam. du Monde, 9: 826.

Male. Form small to moderate-sized, elongate; integument shining black, part of head and prothorax, legs and part of last sternite yellowish; pubescence moderately dense, short, recumbent and erect. Head with front broader than long, finely, densely punctate; pubescence dense, short and appressed, long, erect hairs numerous, pubescence around eyes more dense; genae about half as long as lower eye lobes; antennae extending to fourth abdominal sternite, outer segments often paler brown, basal segments with very few erect hairs beneath, segments densely clothed with pale, appressed pubescence, scape rather broad, densely punctate, third segment longer than fourth, fourth longer than first. Pronotum slightly broader than long, sides rounded, apex and base broadly impressed; disk with two glabrous calluses before middle and an often vague, longitudinal callus at center behind rounded calluses; punctures moderately coarse, dense, subconfluent; pubescence moderately dense, short and long; prosternum finely, densely pubescent; meso- and metasternum finely, very densely punctate, densely clothed with pale, recumbent pubescence. Elytra a little over three times as long as wide, sides slightly, broadly impressed at middle half; punctures moderately coarse, subserially arranged, rows slightly irregular, disk with a vague median costa on each side; pubescence moderately dense, appressed pale hairs very short, longer, suberect hairs moderate; apices subtruncate. Legs short; femora stout, very sparsely punctate; middle and hind tibiae with bristling, suberect hairs. Abdomen very finely, densely punctate, densely clothed with pale recumbent pubescence; last sternite moderately deeply, semicircularly impressed, apex broadly v-shaped. Length, 8-11 mm.

Female. Form similar. Antennae extending to third abdominal segment. Abdomen with last sternite triangularly impressed at apical one-half. Length, 8.5-13 mm.

Type locality. Of *delongi*, Cedar Point, Ohio; *anteruficollis*, Hartford, Connecticut.

Range. Northeastern North America.

Flight period. April to August.

Host plants. Populus deltoides.

This species may be easily recognized by the shining, black callosities of the pronotum, yellowish legs, very fine pubescence of the elytra, and by the relatively short antennae. The infuscation of the head and pronotum varies in degree but the pronotum always has at least the basal half black.

Oberea affinis Leng and Hamilton

Oberea affinis Horn, 1878, Trans. Amer. Entomol. Soc., 7: 48 (nomen nudum).
Oberea tripunctata var. *bimaculata* form *affinis* Leng and Hamilton, 1896, Trans. Amer. Entomol. Soc., 23: 155.
Oberea affinis: Casey, 1913, Memoirs on the Coleoptera, 4: 364; Hicks, 1962, Coleop. Bull., 16: 6, fig. 2.
Oberea bimaculata ab. *affinis*: Aurivillius, 1923, Coleop. Cat., 74: 535.
Oberea bimaculata var. *affinis*: Leonard, 1928, Cornell Agr. Exp. Sta. Mem., 101: 457; Procter, 1938, Biol. Surv. Mt. Desert, 6: 154; Knull, 1946, Ohio Biol. Surv. Bull., 39: 280.
Oberea bimaculata v. *affinis*: Breuning, 1962, Frust. Entomol., 5: 230.
Oberea bimaculata m. *affinis*: Breuning, 1966, Cat. Lam. du Monde, 9: 827.

Male. Form moderate-sized, slender, elongate; integument black, prothorax except for lateral vittae and median spot at basal margin yellowish, disk occasionally with two black spots; pubescence dense, short, dark, appressed with numerous, longer suberect hairs. Head with front about as broad as long, densely punctate and pubescent, long, suberect hairs numerous; antennae extending to about fourth abdominal segment, long, erect hairs abundant beneath on basal segments, two very long setae present at apices of segments three to ten, segments grayish pubescent beneath, third segment longer than first, fourth subequal to first. Pronotum slightly broader than long, sides broadly rounded; apex and base broadly impressed transversely; disk with two glabrous, occasionally infuscated, calluses, basal median dark callus transversely plicate, middle behind base usually with a linear longitudinal callus; punctures moderately coarse, scattered, interspaces micropunctate; pubescence fine, golden, appressed, very long, erect hairs sparsely interspersed; prosternum finely punctate; meso- and metasternum moderately coarsely punctate at sides; pubescence dark, appressed. Elytra more than 3-1/2 times longer than broad, sides slightly narrowing at middle half; punctures coarse, somewhat serially arranged, with three rows between suture and discal costa, somewhat irregular at extreme base; pubescence moderately dense, short, dark, appressed, longer suberect hairs sparse; apices obliquely truncate, angles often dentate. Legs short; femora minutely punctate, finely pubescent. Abdomen with sternites moderately coarsely punctate at sides, densely micropunctate medially; pubescence appressed; last sternite deeply excavated for its entire length, apex deeply emarginate. Length, 11-14 mm.

Female. Form similar. Antennae slightly shorter. Abdomen with last sternite triangularly impressed at apex, middle with an impressed longitudinal line. Length, 11-16 mm.

Type locality. None.
Range. Eastern North America.
Flight period. June to July.
Host plants. Rubus.

The shape and coloration of the pronotum distinguishes this species from *O. perspicillata*. The apex and base are broadly, transversely impressed, the two glabrous calluses are usually yellowish, and the basal median dark callus is transversely plicate. Additionally, the sides have broad, dark, longitudinal vittae. Occasionally the pronotal calluses are dark, but the shape of the pronotum separate *affinis* from *perspicillata*.

Oberea perspicillata Haldeman
(Figure 44)

Oberea perspicillata Haldeman, 1847, Trans. Amer. Philos. Soc., (2)10: 57; LeConte, 1852, J. Acad. Nat. Sci. Philadelphia, 2: 153; Lacordaire, 1872, Genera des coléoptères, 9(2): 866, fn.; Saunders, 1873, Rept. Entomol. Soc.

Ontario, p. 9 (habits); Riley, 1874, 6th Ann. Rept. Nox. Ben. Ins. Mo., p. 111; Casey, 1913, Memoirs on the Coleoptera, 4: 369; Fattig, 1947, Emory Univ. Mus. Bull., 5: 44.

Oberea bimaculata v. *perspicillata*: Breuning, 1962, Frust. Entomol., 5: 231.

Oberea bimaculata m. *perspicillata*: Breuning, 1966, Cat. Lam. du Monde, 9: 827.

Oberea bimaculata Horn, 1878 (not Olivier, 1795), Trans. Amer. Entomol. Soc., 7: 46; Webster, 1892, Ohio Agr. Exp. Sta. Bull., 45: 199, fig. 24 (habits); Knobel, 1895, Beetles of New England, p. 34, fig. 131; Beutenmuller, 1896, J. N.Y. Entomol. Soc., 4: 81; Webster and Mally, 1897, USDA Div. Entomol. Bull., (n.s.) 9: 43; Harrington, 1899, Ottawa Nat., 13: 62; Lugger, 1899, 5th. Ann. Rept. Minn. St. Exp. Sta., p. 131, fig. 136 (habits); Lugger, 1899, Minn. Agr. Exp. Sta. Bull., 66: 215, pl. 4, fig. 136 (habits); Webster, 1900, J. Columbus Hort. Soc., 15: 6; Chagnon, 1905, Nat. Can., 32: 44; Patch, 1908, Maine Agr. Exp. Sta. Bull., 162: 359 (habits); Smith, 1910, N.J. St. Mus. Rept., 1909: 336; Blatchley, 1910, Coleoptera in Indiana, p. 1092, fig. 475; Johannsen, 1912, Maine Agr. Exp. Sta. Bull., 207: 463; Casey, 1913, Memoirs on the Coleoptera, 4: 370; Mutchler and Weiss, 1923, N.J. Dept. Agr. Cir., 58: 16 (habits); Craighead, 1923, Can. Dept. Agr. Bull., (n.s.) 27: 138, pl. 16; Craighead and Middleton, 1930, USDA Misc. Pub., 74: 8; Chagnon, 1933-40, Coleop. Prov. Quebec, p. 279, pl. 19, fig. 11; Gibson, 1934, Can. Dept. Agr. Bull., (n.s.) 99: 56, fig. 79 (habits); Slate et al., 1942, N.Y. St. Agr. Exp. Sta. Cir., 153: 54, fig. 14 (habits); Knull, 1946, Ohio Biol. Surv. Bull., 39: 279; Fattig, 1947, Emory Univ. Mus. Bull., 5: 44; Weigel and Baumhofer, 1948, USDA Misc. Pub., 626: 78, fig. 131 (habits); Dillon and Dillon, 1961, Man. Beetles East. North America, p. 653, pl. 65, no. 12; Breuning, 1962, Frust. Entomol., 5: 230; Chagnon and Robert, 1962, Prin. Coleop. Prov. Quebec, p. 279, pl. 19, fig. 11; Hatch, 1971, Univ. Wash. Pub. Biol., 16: 156, pl. 19, fig. 7; Stein and Tagestad, 1976, USDA For. Serv. Res. Pap., RM-171: 25; Chamberland, 1976, Fabreries, 2: 89; Headstrom, 1977, Beetles of America, p. 383; Gosling and Gosling, 1977, Gr. Lakes Entomol., 10: 34, fig. 178; Laliberte et al., 1977, Fabreries, 3: 95; Solomon, 1977, Can. Entomol., 109: 297; Rice and Enns, 1981, Trans. Mo. Acad. Sci., 15: 104; Gosling, 1984, Gr. Lakes Entomol., 17: 72.

Oberea bimaculata var. *tripunctata*: Horn, 1878, Trans. Amer. Entomol. Soc., 7: 46; Leonard, 1928, Cornell Agr. Exp. Sta. mem., 101: 457.

Oberea tripunctata var. *bimaculata*: Leng and Hamilton, 1896, Trans. Amer. Entomol. Soc., 23: 154; Wickham, 1898, Can. Entomol., 30: 43.

Oberia bimaculata: Pettit, 1904, Mich. St. Agr. Coll. Exp. Sta. Spec. Bull., 24: 29 (error).

Oberea basalis LeConte, 1852, J. Acad. Nat. Sci. Philadelphia, 2: 153; Lacordaire, 1872, Genera des coléoptères, 9(2): 866, fn.; Casey, 1913, Memoirs on the Coleoptera, 4: 370; Leonard, 1928, Cornell Agr. Exp. Sta. Mem., 101: 457; Dillon and Dillon, 1961, Man. Beetles East. North

America, p. 653, pl. 65, no. 5; Stein and Tagestad, 1976, USDA For. Serv. Res. Pap., RM-171: 24.

Oberea bimaculata var. *basalis*: Horn, 1878, Trans. Amer. Entomol. Soc., 7: 46; Blatchley, 1910, Coleoptera in Indiana, p. 1092; Smith, 1910, N.J. St. Mus. Rept., 1909: 336; Mutchler and Weiss, 1923, N.J. Dept. Agr. Cir., 58: 16; Knull, 1946, Ohio Biol. Surv. Bull., 39: 279; Fattig, 1947, Emory Univ. Mus. Bull., 5: 44; Bayer and Shenefelt, 1969, Univ. Wisc. Res. Bull., 275: 32, fig. 40.

Oberea tripunctata var. *bimaculata* form *basalis*: Leng and Hamilton, 1896, Trans. Amer. Entomol. Soc., 23: 154.

Oberea bimaculata v. *basalis*: Breuning, 1962, Frust. Entomol., 5: 231.

Oberea bimaculata m. *basalis*: Breuning, 1966, Cat. Lam. du Monde, 9: 827.

Oberea texana Horn, 1878, Trans. Amer. Entomol. Soc., 7: 47; Leng and Hamilton, 1896, Trans. Amer. Entomol. Soc., 23: 157; Casey, 1913, Memoirs on the Coleoptera, 4: 364.

Oberea tripunctata v. *texana*: Breuning, 1962, Frust. Entomol., 5: 229.

Oberea tripunctata m. *texana*: Breuning, 1966, Cat. Lam. du Monde, 9: 827.

Oberea montana Casey, 1913, Memoirs on the Coleoptera, 4: 369; Hatch, 1971, Univ. Wash. Pub. Biol., 16: 156.

Oberea perspicillata var. *montana*: Aurivillius, 1923, Coleop. Cat., 74: 536.

Oberea bimaculata var. *montanus*: Knowlton and Wood, 1950, Bull. Brooklyn Entomol. Soc., 95: 4.

Oberea filum Casey, 1913, Memoirs on the Coleoptera, 4: 369.

Oberea basalis var. *filum*: Aurivillius, 1923, Coleop. Cat., 74: 535.

Oberea bimaculata v. *filum*: Breuning, 1962, Frust. Entomol., 5: 231.

Oberea bimaculata m. *filum*: Breuning, 1966, Cat. Lam. du Monde, 9: 827.

Oberea delicatula Casey, 1913, Memoirs on the Coleoptera, 4: 368; Fattig, 1947, Emory Univ. Mus. Bull., 5: 43.

Oberea exilis Casey, 1913, Memoirs on the Coleoptera, 4: 368; Hicks, 1962, Coleop. Bull., 16: 9.

Oberea bimaculata v. *exilis*: Breuning, 1962, Frust. Entomol., 5: 230.

Oberea bimaculata m. *exilis*: Breuning, 1966, Cat. Lam. du Monde, 9: 827.

Oberea iowensis Casey, 1913, Memoirs on the Coleoptera, 4: 370.

Oberea basalis var. *iowensis*: Aurivillius, 1923, Coleop. Cat., 74: 535.

Oberea bimaculata v. *iowensis*: Breuning, 1962, Frust. Entomol., 5: 230.

Oberea bimaculata m. *iowensis*: Breuning, 1966, Cat. Lam. du Monde, 9: 827.

Oberea insignis Casey, 1913, Memoirs on the Coleoptera, 4: 370; Fattig, 1947, Emory Univ. Mus. Bull., 5: 44.

Oberea bimaculata v. *insignior* Breuning, 1962, Frust. Entomol., 5: 231. (new name for *insignis* Casey).

Oberea bimaculata m. *insignior*: Breuning, 1966, Cat. Lam. du Monde, 9: 827.

Oberea umbra Casey, 1913, Memoirs on the Coleoptera, 4: 370; Fattig, 1947, Emory Univ. Mus. Bull., 5: 44; Kirk, 1969, SC Agr. Exp. Sta. Tech. Bull., 1033: 87.

Oberea bimaculata v. *umbra*: Breuning, 1962, Frust. Entomol. 5: 232.

Oberea bimaculata m. *umbra*: Breuning, 1966, Cat. Lam. du Monde, 9: 827.
Oberea dolosa Casey, 1913, Memoirs on the Coleoptera, 4: 371.
Oberea bimaculata ab.? *dolsa*: Aurivillius, 1923, Coleop. Cat., 74: 535.
Oberea bimaculata v. *dolosa*: Breuning, 1962, Frust. Entomol., 5: 231.
Oberea bimaculata m. *dolosa*: Breuning, 1966, Cat. Lam. du Monde, 9: 827.
Oberea flavocephala Blatchley, 1922, Can. Entomol., 54: 32.
Oberea bimaculata v. *flavocephala*: Breuning, 1962, Frust. Entomol., 5: 231.
Oberea bimaculata m. *flavocephala*: Breuning, 1966, Cat. Lam. du Monde, 9: 827.
Oberea tripunctata v. *submandarina* Breuning, 1962, Frust. Entomol., 5: 229.
Oberea tripunctata m. *submandarina*: Breuning, 1966, Cat. Lam. du Monde, 9: 826.
Oberea bimaculata v. *basicollis* Breuning, 1962, Frust. Entomol., 5: 230.
Oberea bimaculata m. *basicollis*: Breuning, 1966, Cat. Lam. du Monde, 9: 827.
Oberea bimaculata v. *bicallosicollis* Breuning, 1962, Frust. Entomol., 5: 232.
Oberea bimaculata m. *bacallosicollis*: Breuning, 1966, Cat. Lam. du Monde, 9: 827 (error).

Male. Form small, slender; integument black, pronotum except three spots testaceous; pubescence dense, fine, appressed, erect hairs short. Head with front slightly broader than long, punctures moderately coarse, dense, pubescence dense, appressed, eye margins with very long, erect hairs; genae about as long as second antennal segment; antennae unicolorous, slightly shorter than body, basal segments with short erect hairs beneath, segments from third with two very long erect hairs at apices, third segment longer than fourth, fourth slightly longer than first. Pronotum cylindrical, shining, usually slightly broader than long; apex barely impressed, base shallowly impressed; disk convex, two black, glabrous calluses located slightly before middle, basal median black spot transversely rugose; middle usually with a vague linear callus; punctures coarse, shallow, dense, subconfluent; pubescence fine, long erect hairs numerous; prosternum finely pubescent, usually with a dark patch at sides of coxal cavities; meso- and metasternum coarsely punctate at sides, pubescence dense, appressed, long, suberect hairs numerous. Elytra four times or slightly less as long as broad, sides broadly impressed toward middle; punctures moderately coarse, subserial, three rows present between suture and median costae; pubescence dense, fine, appressed, short erect hairs more numerous basally; apices truncate to very shallowly emarginate, angles often vaguely dentate. Legs short; femora very finely punctate. Abdomen densely micro-punctate, coarse punctures denser at sides; pubescence dense, appressed, long, suberect hairs numerous; last sternite shallowly excavated, apex usually broadly u-shaped. Length, 8-14 mm.

Female. Form similar, slightly more robust. Antennae extending to about fourth abdominal segment. Abdomen with last sternite triangularly impressed at apex, middle linearly impressed. Length, 8.5-15 mm.

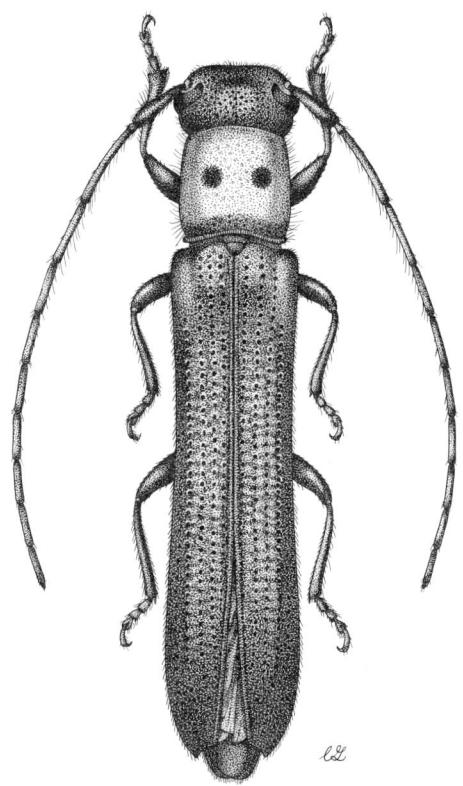

Figure 44. *Oberea perspicillata* Haldeman, female.

Type locality. Of *perspicillata*, none (United States); *basalis*, Tobula, Georgia; *texana*, Texas; *montana*, Nephi, Utah; *filum*, Harrisburg, Pennsylvania; *delicatula*, District of Columbia; *exilis*, Pennsylvania; *iowensis*, Keokuk, Iowa; *insignis*, Southern Pines, North Carolina; *umbra*, Black Mts., North Carolina; *dolosa*, Southern Pines, North Carolina; *flavocephala*, Dunedin, Florida; *submandarina*, Missouri; *basicollis*, Smoky Mts., Tennessee; *bicallosicollis*, "Wilmading," Pennsylvania.

Range. Eastern North America to Utah and Texas.

Flight period. April to August.

Host plants. Rubus, Rosa.

Horn in 1878 considered this species to be the *bimaculatus* of Olivier, originally described from France. Subsequent authors have followed this usage for the black species with an orange pronotum and black discal pronotal calluses. We do not accept the premise that the species described by Olivier in 1795 was an import from the United States, and choose to apply the earliest available name, *perspicillata* Haldeman, to the North American species.

Since *O. perspicillata* is of economic importance to raspberry and blackberry production, numerous biological references are available, mostly as *O. bimaculata*. Past confusion as to the identity of this species makes exact interpretation of the literature difficult. Determining which references apply to what species can only be done by data acquired from future rearing. It is also possible that a complex of species is involved.

Oberea tripunctata (Swederus)

Cerambyx (Saperda) tripunctatus Swederus, 1787, Kongl. Vet. Acad. Nya Handl., 8: 197.

Saperda tripunctata: Fabricius, 1792, Entomol. Syst., 2: 310; Fabricius, 1801, Syst. Eleuth., 2: 321.

Saperda tripunctata: Harris, 1838, Rept. Comm. Zool. Surv. St. (Mass.), p. 91.

Saperda (Oberea) tripunctata: Harris, 1841, Rept. Ins. Mass. Inj. Veg., p. 91 (habits); Emmons, 1854, Nat. Hist. N.Y., Ins., 5: 122, pl. 16, fig. 7.

Oberea tripunctata: Haldeman, 1847, Trans. Amer. Philos. Soc., (2)10: 57; LeConte, 1852, J. Acad. Nat. Sci. Philadelphia, 2: 153; Provancher, 1877, Pet. Fauna Canada, 1: 636; Horn, 1878, Trans. Amer. Entomol. Soc., 7: 47; Kingsley, 1884, Riverside Nat. Hist., 2: 326 (habits); Knobel, 1895, Beetles New England, p. 34, fig. 134; Leng and Hamilton, 1896, Trans. Amer. Entomol. Soc., 23: 154; Wickham, 1898; Can. Entomol., 30: 43; Webster, 1900, J. Columbus Hort. Soc., 15: 6, fig. 6; Chagnon, 1905, Nat. Can., 32: 44; Blatchley, 1910, Coleoptera in Indiana, p. 1092; Smith, 1910, N.J. St. Mus. Rept., 1909: 336; Forbes, 1911, 26th Rept. St. Entomol. Nox. Ben. Ins. Illinois, p. 44, figs. 48-53 (habits); Casey, 1913, Memoirs on the Coleoptera, 4: 367; Ruggles, 1915, J. Econ. Entomol., 7: 79, figs. 1-4, 6 (habits); Felt, 1921, N.Y. St. Mus. Bull., 247-248: 85; Craighead, 1923, Can. Dept. Agr. Bull. (n.s.) 27: 137, pl. 44; Mutchler and Weiss, 1923, NJ Dept. Agr. Cir., 58: 17; Leonard, 1928, N.J. St. Mus. Rept., 1909: 336; Craighead and Middleton, 1930, USDA Misc. Pub., 74: 8; Felt and Rankin, 1932, Ins. Diseases Ornam. Trees Shrubs, p. 203, 218 (habits); Knull, 1932, Entomol. News, 43: 64; Chagnon, 1933-40, Coleop. Prov. Quebec, p. 279, p. 19, fig. 12; Langford and Cory, 1939, Univ. Md. Ext. Serv. Bull., 84: 47 (habits); Soraci, 1941, N.J. Dept. Agr. Cir., 326: 49 (habits); Hoffmann, 1942, USDA Misc. Pub., 466: 11; Knull, 1946, Ohio Biol. Surv. Bull., 39: 277; Fattig, 1947, Emory Univ. Mus. Bull., 5: 43; Weigel and Baumhofer, 1948, USDA Misc. Pub., 626: 40, fig. 65 (habits); Craighead, 1950, USDA Misc. Pub., 657: 256 (habits); Dillon and Dillon, 1961, Man. Beetles East. North America, p. 653, pl. 65, no. 13; Breuning, 1962, Frust. Entomol., 5: 227; Hicks, 1962, Coleop. Bull., 16: 7, fig. 4; Chagnon and Robert, 1962, Prin. Coleop. Prov. Quebec, p. 279, pl. 19, fig. 12; Bayer and Shenefelt,

1969, Univ. Wisc. Res. Bull., 275: 32, fig. 40; Gardiner, 1969, Can. Dept. Fish. For Inst. Rept., 0-14: 106; Kirk, 1970, S.C. Agr.Exp. Sta. Tech. Bull., 1038: 84; Baker 1972, USDA Misc. Pub., 1175: 189 (habits); Stein and Tagestad, 1976, USDA For. Serv. Res. Pap., RM-171: 26; Gosling and Gosling, 1976, Gr. Lakes Entomol., 10: 34, fig. 177; Chamberland, 1976, Fabreries, 2: 89; Headstrom, 1977, Beetles of America, p. 383, fig. 534; Laliberte et al., 1977, Fabreries, 3: 95; Rice and Enns, 1981, Trans. Mo. Acad. Sci., 15: 105; Drooz, 1985, USDA Misc. Pub., 1426: 299 (habits); Gosling and Gosling, 1986, Gr. Lakes Entomol., 19: 157.

Oberea tripunctata var. *tripunctata*: Leng and Hamilton, 1896, Trans. Amer. Entomol. Soc., 23: 155; Wickham, 1898, Can. Entomol., 30: 43.

Oberea bimaculata var. *tripunctata*: Smith, 1909, N.J. St. Mus. Rept., 1909: 336.

Saperda mandarina Fabricius, 1801, Syst. Eleuth., 2: 321.

Oberea mandarina: LeConte, 1852, J. Acad. Nat. Sci. Philadelphia, 2: 152; Lacordaire, 1872, Genera des coléoptères, 9(2): 866, fn.; Packard, 1887, U.S. Entomol. Comm. Bull., 7: 115; Beutenmuller, 1896, J. N.Y. Entomol. Soc., 4: 81; Knull, 1946, Ohio Biol. Surv. Bull., 39: 278; Fattig, 1947, Emory Univ. Mus. Bull., 5: 43.

Oberea tripunctata var. *mandarina*: Horn, 1878, Trans. Amer. Entomol. Soc., 7: 47; Wickham, 1898, Can. Entomol., 30: 43; Smith, 1910, N.J. St. Mus. Rept., 1909: 336; Leonard, 1928, Cornell Agr. Exp. Sta. Mem., 101: 456; Knull, 1932, Entomol. News, 43: 64; Breuning, 1962, Frust. Entomol., 5: 229.

Oberea tripunctata var. *tripunctata* form *mandarina*: Leng and Hamilton, 1896, Trans. Amer. Entomol. Soc., 23: 155.

Oberea tripunctata m. *mandarina*: Breuning, 1966, Cat. Lam. du Monde, 9: 826.

Oberea amabilis Haldeman, 1847, Trans. Amer. Philos. Soc., (2)10: 57; LeConte, 1852, J. Acad. Nat. Sci. Philadelphia, 2: 152; Lacordaire, 1872, Genera des coléoptères, 9(2): 866, fn.; Provancher, 1877, Pet. Fauna Entomol. Can., 1: 636.

Oberea tripunctata v. *amabilis*: Breuning, 1962, Frust. Entomol., 5: 229.

Oberea tripunctata m. *amabilis*: Breuning, 1966, Cat. Lam. du Monde, 9: 826.

Oberea tripunctata intermedia Casey, 1913, Memoirs on the Coleoptera, 4: 367.

Oberea tripunctata var. *intermedia*: Aurivillius, 1923, Coleop. Cat., 74: 537.

Oberea tripunctata appalachiana Casey, 1913, Memoirs on the Coleoptera, 4: 367.

Oberea tripunctata var. *appalachiana*: Aurivillius, 1923, Coleop. Cat., 74: 537; Breuning, 1962, Frust. Entomol., 5: 228.

Oberea appalachiana: Hicks, 1962, Coleop. Bull., 16: 7.

Oberea tripunctata m. *appalachiana*: Breuning, 1966, Cat. Lam. du Monde, 9: 826.

Oberea praelonga deficiens Casey, 1924, Memoirs on the Coleoptera, 11: 296

Oberea tripunctata v. *deficiens*: Breuning, 1962, Frust. Entomol., 5: 229.
Oberea tripunctata m. *deficiens*: Breuning, 1966, Cat. Lam. du Monde, 9: 826.
Oberea tripunctata v. *subexilis* Breuning, 1962, Frust. Entomol., 5: 228.
Oberea tripunctata m. *subexilis*: Breuning, 1966, Cat. Lam. du Monde, 9: 826.
Oberea tripunctata v. *subdeficiens* Breuning, 1962, Frust. Entomol., 5: 229.
Oberea tripunctata m. *subdeficiens*: Breuning, 1966, Cat. Lam. du Monde, 9: 826.

Male. Form moderate-sized, slender, subparallel; integument testaceous, antennae dark annulate, pronotum with two black calluses and usually a dark basal spot at middle, elytra usually dark vittate laterally and more narrowly along suture, head frequently partially black, underside variably black; pubescence fine, short, appressed, short erect hairs numerous. Head with front about as broad as long, densely, moderately coarse punctate, densely clothed with pale appressed pubescence, long, suberect hairs numerous; genae about as long as second antennal segment; antennae about as long as body, at least outer segments dark annulate at apices, basal segments with few erect hairs beneath, segments usually with two long hairs at apices, scape moderate, third segment longer than fourth, fourth slightly longer than first. Pronotum broader than long; base and apex moderately impressed transversely; disk shining, with two black, glabrous calluses and a median black spot at base; punctures rather fine, irregular, denser at sides, pubescence fairly dense, appressed, long, erect hairs numerous; prosternum finely pubescent; meso- and metasternum finely, densely punctate, punctures coarser at sides, pubescence dense, appressed, with longer suberect hairs numerous. Elytra slightly more than 3-1/2 times as long as broad, sides shallowly impressed toward middle; punctures coarse, subserial, disk with four rows of punctures between suture and vague subhumeral costae; pubescence very fine, short, appressed, short, erect hairs numerous; apices obliquely truncate, angles often minutely dentate. Legs short; femora very finely punctate. Abdomen densely micropunctate with larger punctures interspersed; pubescence dense, short, appressed with longer suberect hairs interspersed; last sternite shallowly impressed, apex very broadly v-shaped. Length, 8-12 mm.

Female. Form similar. Antennae slightly shorter than body. Abdomen with last sternite triangularly impressed at apical half. Length, 8.5-13 mm.

Type locality. Of *tripunctata*, New York; *mandarina*, Carolina; *amabilis*, none; *intermedia*, Indiana; *appalachiana*, Asheville, North Carolina; *deficiens*, New Jersey; *subexilis*, Canada; *subdeficiens*, New York.

Range. Eastern North America to Manitoba and Texas.

Flight period. May to August.

Host plants. Amygdalus, Azalea, Carya, Cornus, Cydonia, Hamamelis, Kalmia, Malus, Oxydendrum, Populus, Prunus, Rhus, Ribes, Ulmus, Viburnum.

This species may be recognized by the following combination of characters: pronotum with two discal calluses, antennae with at least outer segments annulate, and elytra testaceous to lightly infuscated, usually with dark vittae.

Oberea myops Haldeman
(Figure 45)

Oberea myops Haldeman, 1847, Trans. Amer. Philos. Soc., (2)10: 57; Haldeman, 1847, Proc. Amer. Philos. Soc., 4: 373; LeConte, 1852, J. Acad. Nat. Sci. Philadelphia, 2: 152; Lacordaire, 1872, Genera des coléoptères, 9(2): 866, fn.; Casey, 1913, Memoirs on the Coleoptera 4: 364; Casey, 1914, Memoirs on the Coleoptera, 5: 369; Craighead, 1923, Can. Dept. Agr. Bull., (n.s.) 27: 141; Champlain and Knull, 1925, Entomol. News, 36: 142; Driggers, 1929, J. N.Y. Entomol. Soc., 37: 67 (habits); Craighead and Middleton, 1930, USDA Misc. Publ., 74: 8; Phipps, 1930, Maine Agr. Exp. Sta. Bull., 356: 191; Felt and Rankin, 1932, Ins. Diseases Ornam. Trees Shrubs, p. 157 (habits); Procter, 1938, Biol. Surv. Mt. Desert, 6: 154; Knull, 1946, Ohio Biol. Surv. Bull., 39: 279; Fattig, 1947, Emory Univ. Mus. Bull., 5: 43; Craighead, 1950, USDA Misc. Pub., 657: 255 (habits); Hicks, 1962, Coleop. Bull., 16: 10; Baker, 1972, USDA Misc. Pub., 1175: 190 (habits); Headstrom, 1977, Beetles of America, p. 384; Drooz, 1985, USDA Misc. Publ., 1426: 300 (habits).
Oberea tripunctata var. *myops*: Horn, 1978, Trans. Amer. Entomol. Soc., 7: 47; Blatchley, 1910, Coleoptera in Indiana, p. 1092; Smith, 1910, N.J. St. Mus. Rept., 1969: 336; Mutchler and Weiss, 1923, N.J. Dept. Agr. Cir., 58: 17; Leonard, 1928, Cornell Agr. Exp. Sta. Mem., 101: 456; Langford and Cory, 1939, Univ. Md. Ext. Serv. Bull., 84: 46 (habits); Breuning, 1962, Frust. Entomol., 5: 229.
Oberea tripunctata var. *tripunctata* form *myops*: Leng and Hamilton, 1896, Trans. Amer. Entomol. Soc., 23: 155.
Oberea tripunctata m. *myops*: Breuning, 1966, Cat. Lam. du Monde, 9: 826.
Oberea schaumii: Hicks, 1944, Can. Entomol., 76: 163 (misidentified).
Oberea canadensis Fisher, 1945, Can. Entomol., 77: 56; Hicks, 1945, Can. Entomol., 77: 214; Hicks, 1962, Coleop. Bull., 16: 12; Breuning, 1962, Frust. Entomol., 5: 226. New synonymy.

Male. Form moderate-sized, elongate, subparallel; integument testaceous, antennae usually dark, not annulate, pronotum with two black, glabrous spots, head occasionally and underside often partially infuscated, elytra often longitudinally dark vittate along suture and lateral margins. Head with front about as long as broad, densely micropunctate with larger punctures interspersed; pubescence dense, fine, appressed, longer suberect hairs numerous; genae short, about as long as second antennal segment; antennae dark, extending to about fourth abdominal segment, outer segments with two fairly short, erect hairs at apices, scape moderate, third segment longer than fourth, fourth longer than first. Pronotum broader than

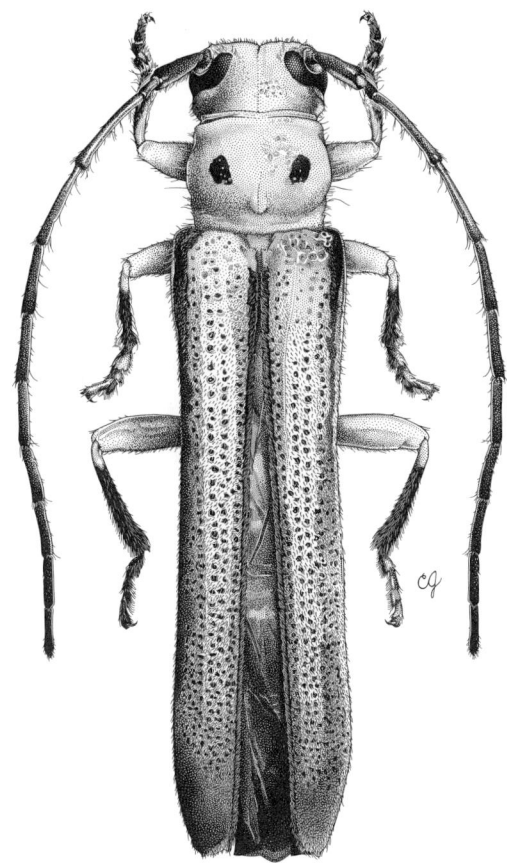

Figure 45. *Oberea myops* Haldeman, female.

long, sides broadly rounded; base and apex deeply impressed transversely; disk subopaque, with two black glabrous calluses; punctures coarse, confluent; pubesence very fine, very short, longer erect hairs numerous; prosternum very finely pubescent; meso- and metasternum micropunctate, sides with larger punctures, pubescence very fine, appressed. Elytra at least 3.8 times as long as broad, sides broadly impressed toward middle; punctures moderately coarse, serial, with three rows between suture and vague median costae, humeral costae vague; pubescence very fine, short, erect hairs more abundant toward base; apices obliquely shallowly emarginate. Legs short; femora very finely punctate. Abdomen densely micropunctate with larger punctures interspersed at sides; pubescence golden, very short, appressed, erect hairs absent; last sternite deeply, circularly impressed, apex deeply emarginate. Length, 13-15 mm.

Female. Form similar. Antennae usually extending to third abdominal segment. Abdomen with last sternite triangularly impressed. Length, 13.5-18 mm.

Type locality. Of *myops*, none (United States); *canadensis*, Ojibway, Ontario, Canada.

Range. Eastern North America to Georgia and Manitoba.

Flight period. May to July.

Host plants. Azalea, Cornus, Cydonia, Kalmia, Malus, Oxydendrum, Prunus, Rhododenron, Salix, Ulmus, Vaccinium.

The larger size, non-annulate antennae, dense punctation of the pronotum and deeply impressed last abdominal sternite of males separate this species from *tripunctata*.

Oberea praelonga Casey

Oberea praelonga Casey, 1913, Memoirs on the Coleoptera, 4: 368; Hicks, 1962, Coleop. Bull., 16: 7, fig. 5.

Oberea tripunctata v. *praelonga*: Breuning, 1962, Frust. Entomol., 5: 228.

Oberea tripunctata m. *praelonga*: Breuning, 1966, Cat. Lam. du Monde, 9: 826.

Oberea pallida Casey, 1913, Memoirs on the Coleoptera, 4: 366; Frost, 1916, Can. Entomol., 48: 389; Knull, 1946, Ohio Biol. Surv. Bull., 39: 277; Hicks, 1962, Coleop. Bull., 16: 9, fig. 1; Baker, 1972, USDA Misc. Pub., 1175: 190 (host); Drooz, 1985, USDA Misc. Pub., 1426: 300.

Oberea tripunctata v. *pallida*: Breuning, 1962, Frust. Entomol., 5: 228.

Oberea tripunctata m. *pallida*: Breuning, 1966, Cat. Lam. du Monde, 9: 826.

Male. Form moderate-sized, slender, elongate; integument testaceous, head often black, antennae with basal segments often dark, elytra usually with sutural and lateral dark vittae, underside variously dark; pubescence very fine, moderately dense, appressed. Head with front about as long as broad, punctures rather fine, separated, pubescence dense, long, suberect hairs numerous; genae about as long as second antennal segment; antennae slightly shorter than body, segments not annulate, segments three to ten with two long, suberect hairs at apices, third segment longer than fourth, fourth slightly longer than first. Pronotum broader than long; apex shallowly, base more deeply impressed; disk shining, convex with two, usually black, glabrous calluses before middle, middle at base usually with a black spot; punctures moderately coarse, subconfluent, pubescence fine, golden, appressed, long, erect hairs numerous; prosternum sparsely pubescent; meso- and metasternum densely micropunctate, coarse punctures present at sides, pubescence dense, appressed, with long, suberect hairs interspersed. Elytra usually at least 3.8 times longer than broad, sides shallowly impressed toward middle; punctures moderately coarse, serial, separated; disk with usually four rows of punctures between suture and vague subhumeral costae; pubescence fine, short, moderately dense, short erect hairs denser basally; apices shallowly emarginate truncate, angles lightly dentate. Legs short; femora minutely punctate. Abdomen densely micropunctate with a few larger punctures interspersed at sides; pubescence very fine, appressed, longer, suberect hairs numerous; last sternite shallowly impressed, apex emarginate. Length, 11-16 mm.

Female. Form similar. Antennae extending to about fourth abdominal segment. Abdomen with last sternite triangularly impressed. Length, 10-16 mm.

Type locality. Of praelonga, Bluff Point, Lake Champlain, New York; *pallida*, Maine.

Range. Northeastern North America to North Carolina westward to Manitoba.

Flight period. May to July.

Host plants. Alnus spp. (including *incana*); *Cornus* spp. (including *baylei*, *stolonifera*); *Ulmus, Malus, Pyrus, Prunus*.

This species may be separated from *O. myops* by the longer elytra, the more finely punctate and shining pronotum, the median basal black spot of the pronotum, and by the much shallower impression of the last abdominal sternite of the males. *O. tripunctata* differs by the shorter elytra, smaller average size, and darkly annulate antennae. Additionally, *O. praelonga* has the pronotum broader-appearing and usually more deeply impressed basally.

A great deal of variation is evident in our material. Some specimens are almost wholly testaceous with hints of infuscation on the pronotal calluses, basal antennal segments, and underside (*pallida* Casey). These pale forms seem to appear randomly throughout the range of the species and are structurally identical to the darker forms. Some specimens have the antennae all dark, while others have the outer segments pale. The elytra and underside are variably infuscated. There appears to be no geographic significance to the color variability. Both forms utilize the same host plants.

Although the name *pallida* Casey has page priority, we prefer the name *praelonga*, which is more descriptive of this species.

Incertae sedis
Oberea bimaculata (Olivier)

Saperda bimaculata Olivier, 1795. Entomol. Hist. Nat. Ins., 4(68): 21, pl. 4, fig. 42.

This species was characterized by Olivier as being black with the head, underside of the prothorax, and exterior border of the elytra reddish. Two black spots are present on the pronotum.

Since this species was described from the Department of Var in France, we consider it highly unlikely that it represents our North American species. Horn (1878) was apparently the first to apply the name *bimaculata* Olivier to the common eastern North American species, previously described as *basalis* by Haldeman in 1847.

Oberea mairei Chevrolat

Oberea mairei Chevrolat, 1856, Rev. Mag. Zool., (2)8: 435.

Chevrolat described this species from Meung, near Orleans, France. Breuning (1962) considered *mairei* a synonym of *gracilis* Fabricius. We regard it as improbable that the two are conspecific.

TRIBE TETRAOPINI THOMSON

Thomson, 1860, Class. ceramb., p. 66 (Tetraopesitae).
Bates, 1866, Ann. Mag. Nat. Hist., (3) 17: 368 (Astatheinae).
Lacordaire, 1872, Genera des coléoptères, 9(2): 849, 871 (Tetraopides).
LeConte, 1873, Smithson. Misc. Coll., 11(265): 347 (Phytoeciini, part).
Bates, 1881, Biol. Centr.-Amer., Coleop, 5: 195.
LeConte and Horn, 1883, Smithson. Misc. Coll., 507: 332 (Phytoeciini, part).
Casey, 1913, Memoirs on the Coleoptera, 4: 373.
Bradley, 1930, Man. Genera Beetles, p. 247.
Saalas, 1936, Ann. Zool.-Soc. Zool.-Bot. Fenn. Vanamo, 4: 164.
Knull, 1946, Ohio Biol. Surv. Bull., 39: 281.
Duffy, 1960, Mon. Neotropical Timber Beetles, p. 280.
Dillon and Dillon, 1961, Man. Beetles East. North America, p. 656.
Arnett, 1962, Beetles U.S., 103: 873 (part).
Bayer and Shenefelt, 1969, Univ. Wisc. Res. Bull. 275: 33.
Hatch, 1971, Univ. Wash. Pub. Biol., 16: 156.
Rice and Enns, 1981, Trans. Mo. Acad. Sci., 15: 106.

Form small to moderate-sized, cylindrical. Head with front convex; eyes finely faceted, completely divided; antennae eleven-segmented, about as long as body in males, shorter in females, basal segments with long, suberect hairs beneath. Pronotum transverse, sides rounded or tuberculate; disk usually with an elevated umbone (Figure 46); prosternum narrow, intercoxal process narrow, expanded apically, coxal cavities closed behind; mesosternum with coxal cavities open to epimeron; metasternum with episternum slightly broader anteriorly. Elytra convex, sides often slightly expanding near apex; apices rounded. Legs short, stout, femora linear; middle tibiae with an external sinus; tarsal claws appendiculate or bifid. Abdomen normally segmented, somewhat inflated with last sternite medially impressed in females.

The completely divided eyes, elevated thoracic umbone and appendiculate or bifid tarsal claws readily separate this tribe from others in our fauna.

Two genera, both of Mexican derivation, occur in North America.

KEY TO THE NORTH AMERICAN GENERA OF TETRAOPINI.

Tarsal claws bifid. Body usually broad, robust, elytra always with dark spots or chevrons . . . . *Tetraopes*

Tarsal claws appendiculate. Body usually more
narrow, elongate, elytra unicolorous .... *Phaea*

Genus *Tetraopes* Schönherr

Tetraopes Schönherr, 1817, Synonymia Insectorum, 1(3):401; Germar, 1824, Insectorum Species, 1:487:Latreille, 1829, *in* Cuvier, Le Regne Animal (ed. 2) 5:124: Latreille, 1830, Dict. Class. d'Hist. Nat., 16:190: Stephens, 1831, Illus. British Entomol., Mandibulata, 4:237; Serville, 1835, Ann. Soc. Entomol. France, 4:68; Castelnau, 1840, Hist. Nat. Insectes, Coleop., 2:486; Spry & Shuckard, 1840, The British Coleoptera delineated, 2(supp.); 76; Westwood, 1849, *in* Cuvier, The Animal Kingdom, new ed. p. 548; Thomson, 1857, Arch. Entomol., 1:60: pl. 8, fig. 4; Thomson, 1860, Class. ceramb., p. 66; Chevrolat, 1861, J. Entomol., 1:190; Thomson, 1864, Syst. ceramb., p. 125; Lacordaire, 1872, Genera des coléoptères, 9:879; LeConte, 1873, Smithson. Misc. Coll., 11(265): 347 (habits); Hamilton, 1896, Trans. Amer. Entomol. Soc., 23:158; Casey, 1913, Memoirs on the Coleoptera, 4:373; Chapman, Mickel, Parker, Miller, and Kelly, 1926, Ecology, 7:421; Bradley, 1930, Man. Genera Beetles, p. 247; Chagnon, 1933-40, Coleop. Prov. Quebec, p. 279; Chagnon, 1938, Nat. Canadian, 5:279 (Habits); Knull, 1946, Ohio Biol. Surv. Bull. 39:282; Dillon and Dillon, 1961, Man. Beetles East. North America, p. 656; Arnett, 1962, Beetles U. S., 103:873; Chemsak, 1963, Univ. Calif. Pub. Entomol., 30:17; Bayer and Shenefelt, 1969, Univ. Wis. Res. Bull. 275:33; Hatch, 1971, Univ. Wash. Pub. Biol., 16:156; Headstrom, 1977, Beetles of America, p. 384; Isman et al., 1977, Can. J. Zool., 55:1024; Marinoni, 1977, Dusenia, 10:49; Rice and Enns, 1981, Trans. Mo. Acad. Sci., 15:106.

Body stout, robust, occasionally slightly elongate, underside densely, cinereously pubescent. Head a little narrower than prothorax, feebly concave between antennal tubercles; front rather convex, transverse; antennae eleven segmented, stout, as long as body or slightly shorter in male, shorter in female, internal bristling hairs moderately abundant basally, decreasing in number apically, eleventh segment usually slightly tapering; eyes finely granulated, completely separated, upper lobe smaller than lower; mandibles stout, strongly excavated dorsally near base. Pronotum transverse, discal umbone strongly elevated and elongate or only slightly evident and broader; four black spots usually present at corners of umbone; sides usually distinctly tuberculate, impressed basally and apically; prosternum narrow, prosternal process narrow but broadly expanded apically, front coxal cavities closed behind; metasternum convex, episternum of metathorax broad in front, narrowing behind. Elytra with surface convex; pubescence variable but always present, usually short, pale pubescence evident; color pattern variable but always with at least an apical chevron or spots present; apices rounded; scutellum black, rounded apically, densely pubescent. Legs short, stout, femora not clavate; middle and hind tibiae with external brush of bristle-like hairs; tarsal claws bifed.

Type species. Lamia tornator Fabricius (Chemsak designation, 1963).

This genus is easily recognizable by the completely divided eyes and bifid tarsal claws.

Thirteen species are known in North America.

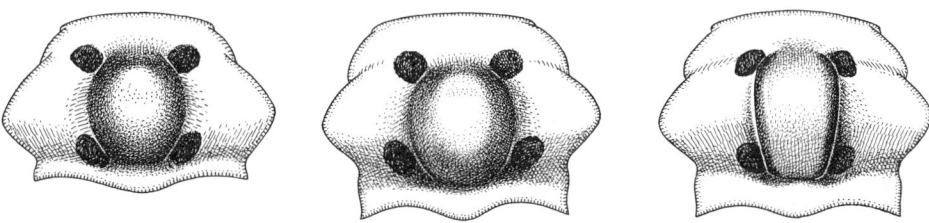

Figure 46. Thoracic umbones of: left, *Tetraopes tetrophthalmus* (Forster) (gradually convex, not abruptly elevated nor distinctly delimited); middle, *T. basalis* LeConte (not abruptly elevated nor distinctly delimited); right, *T. femoratus* LeConte (abruptly elevated and distinctly delimited).

KEY TO THE NORTH AMERICAN SPECIES OF *TETRAOPES*

1	Elytra with an apical dark chevron bordered anteriorly by two dark round or transverse spots; anterior elytral chevron may be present or not	2
	Elytra without apical or anterior chevrons, reddish, usually with black spots; sides of pronotum usually prominently inflated, almost tuberculate	4
2(1)	Thoracic umbone abruptly elevated, distinctly delimited sides of pronotum not appearing tuberculate (Fig. 46)	3
	Thoracic umbone not abruptly elevated, sides of pronotum tuberculate; elytra with anterior dark chevron usually heart-shaped; head and prothorax reddish. Length, 7.5-12 mm. Eastern United States to Michigan	*melanurus*
3(2)	Head and prothorax red; apical segment of maxillary palpi inflated, not tapering to a sharp point; elytra often with two small, dark anterior spots, anterior chevron always lacking. Length, 5-6 mm. Chisos Mts., Texas, to Santa Catalina Mts., Arizona	*linsleyi*
	Head red or black, prothorax usually black or with small anterior red areas; anterior elytral spots lacking; apical segment of maxillary palpi elongate, narrow, sharply pointed at apex; anterior dark chevron usually present. Length, 6-10 mm. Colorado to Guatemala	*discoideus*
4(1)	Thoracic umbone very abruptly, definitely elevated, lateral margins distinctly delimited; umbone definitely longer than wide (Fig. 46)	10
	Thoracic umbone not abruptly, highly elevated, rather flattened, lateral margins delimited or not; umbone not definitely oblong in shape (Fig. 46)	5
5(4)	Body densely clothed with short appressed pubescence which obscures the surface at least partially; long, erect hairs absent from elytra	6
	Pubescence moderate, body not distinctly, densely clothed with short appressed pubescence; long, erect hairs present on elytra	7
6(5)	Pubescence yellowish, very dense, thick and appressed, surface completely obscured, short suberect hairs absent from elytra. Larger species, 11-17 mm. Kansas, Nebraska, Colorado to Texas	*pilosus*
	Pubescence grayish, dense but not thick, not completely obscuring surface, short suberect hairs often present on elytra; umbone usually shining;	

	elytra usually tipped with black. Smaller species, 8-16 mm. Alberta to Texas, Utah, and Arizona .... *annulatus*
7(5)	Antennae distinctly annulate with cinereous pubescence, at least narrowly; if not annulate, umbone convex, not subhexagonal 8
	Antennae black, not annulate, faint traces of annulations present at most; umbone usually subhexagonal in shape; elytra almost always with post-humeral spot present and large; appendages black. Length, 8-15 mm. Eastern North America to Rocky Mountains .... *tetrophthalmus*
8(7)	Last antennal segment gradually tapering apically, antennal segments without fringes of very long hairs at apices...................................... 9
	Last antennal segment abruptly tapering at middle, then very narrowly extending apically; antennal segments with fringe of very long hairs at apices. Length, 11-16, 5 mm. Oklahoma to Texas .... *texanus*
9(8)	Mandibles of male with very large, prominent dorsal tubercle extending for about half their lengths; elytra with sutural spot always present and occasionally a post-humeral spot. Length, 12-15 mm. Northern Texas to western Oklahoma ......*mandibularis*
	Mandibles of male with dorsal tubercle moderate, not extending for one-half the length of mandibles; elytra with black spots at humeri and apical one-third only. Length, 9-12 mm. Ontario, Midwestern states to Texas .... *quinquemaculatus*
10(4)	Punctures of elytra at base distinct, coarse to moderately coarse, fairly abundant 11
	Punctures of elytra at base very fine and sparse; lateral pronotal tubercles prominent, extending to edges of umbone to form a transverse ridge; apical segment of maxillary palpi rather slender, not swollen; scape, femora, and some tibiae reddish. Robust species. Length, 11.5-18 mm. Southern California .... *sublaevis*
11(10)	Antennae black, not annulated, or occasionally at most with very small cinereous bands, scape black; pronotum rather prominently tuberculate laterally; umbone not highly elevated; short pale pubescence dense on elytra and pronotum. Length, 8-14 mm. Southern Texas to El Salvador .... *thermophilus*
	Antennae distinctly annulated with cinereous pubescence 12

12(11) Thoracic umbone always distinctly delimited laterally, usually distinctly longer than wide (Fig. 46), apical segment of axillary palpi inflated, not elongate. Length 8-19 mm. Great Basin, Great Plains to Ohio, Arizona to Texas to central Mexico .... *femoratus*
Thoracic umbone often vaguely defined and not distinctly delimited at sides (Fig. 46), usually but little longer than wide, apical segment of maxillary palpi slender, not swollen, only slightly broader than apical segment of labial palpi. Length, 8-17 mm. Southwest Oregon to southern California .... *basalis*

Tetraopes melanurus Schönherr

Tetraopes melanura Schönherr, 1817, Synonymia Insectorum, 1(3):401.
Tetraopes melanurus: Knull, 1946, Ohio Biol. Surv. Bull., 39:283; Dillon and Dillon, 1961, Man. Beetles East. North America, p. 656, pl. 65, no. 16; Chemsak, 1963, Univ. Calif. Publ. Entomol., 30:21, figs. 3,4; Kirk, 1969, S. C. Agr. Exp. Sta. Tech. Bull., 1033:88; Gosling and Gosling, 1976, Gr. Lakes Entomol., 10:36, fig. 180; Headstrom, 1977, Beetles of America p. 384, pl. 535; Nishio et al., 1983, Kyoto Univ. Coll. Agr. Mem., 122:43.
Lamia canteriator Drapiez, 1819, Ann. Gen. Sci. Phys. Brux., 2:47, pl. 16, fig. 6.
Tetraopes canteriator: Say, 1835, Boston J. Nat. Hist., 1(2):666; Haldeman, 1847, Trans. Amer. Phil. Soc., (2)10:53; LeConte, 1852, J. Acad. Nat. Sci. Philadelphia, (2)2:156; Thomson, 1857, Arch. Entomol., 1:62; Cresson, 1861, Proc. Entomol. Soc. Philadelphia, 1(2):32 (habits); Evett, 1861, Proc. Entomol. Soc. Philadelphia, 1(2):32 (habits); Horn, 1878, Trans. Amer. Entomol. Soc., 7:49; Hamilton, 1896, Trans. Amer. Entomol. Soc., 23:159; Blatchley, 1910, Coleoptera in Indiana, p. 1094; Casey, 1913, Memoirs on the Coleoptera, 4:376.
Tetraopes melanurus ab. *canteriator*: Saalas, 1936, Ann. Zool. Soc. Zool.-Bot. Fenn. Vanamo, 4(1):164.
Lamia (Tetraopes) arator Germar, 1824, Insectorum Species Novae, Coleoptera 1:486.
Tetraopes tornator Stephens, 1829 (not Fabricius 1775), System. Cat. British Insects, p. 413; Stephens, 1831, Illus. British Entomol., Mandibulata, 4:237.

Male. Form stout, robust, size moderate; color reddish orange, underside-black; pubescence dense. Head reddish, narrower than prothorax, moderately coarsely, rather densely punctate, area immediately behind eyes much more coarsely punctate; rather densely clothed with long, dark, suberect hairs and very fine, short, subdepressed, pale pubescence; labrum sclerotized, finely punctate; mandibles robust, with distinct tubercle on dorsal surface; maxillary palpi longer than labial, last segment large, clavate; last segment of labial palpi narrow, elongate, pointed at apex; antennae black, not annulate, usually not attaining elytral apices, stout, tapering apically, basal segments densely clothed with long, black, suberect hairs, remaining segments sparsely ciliate internally, all segments densely clothed with short, appressed, black pubescence; scape longest, second segment as broad as long, third subequal to fourth, shorter than scape, segments five to ten gradually decreasing in length, eleventh longer than tenth, pointed at apex. Pronotum reddish, broader than long; umbone scarcely elevated, not sharply delimited, finely, sparsely punctate, rather sparsely pubescent; four black, densely pubescent spots present on sides of broad discal area; sides moderately densely clothed with long, dark, suberect hairs and very fine, short, pale, appressed pubescence which does not obscure surface; sides abruptly inflated at middle, forming an obtuse tubercle, base and apex broadly constricted; prosternum very finely, densely pubescent; meso- and metasternum finely, sparsely punctate, finely, densely, cinereously pubescent. Elytra reddish, less than twice as long as broad; anterior black chevron usually posteriorly oblique, never expanded to lateral elytral margins, often confluent with apical dark band along suture, usually heart-shaped; apical dark chevron usually posteriorly oblique behind two black spots, extending to elytral apices but usually not laterally expanded to margins of elytra; humeri black; punctures moderately dense, largest slightly smaller than those behind eyes, becoming finer apically; pubescence dense, consisting of longer, suberect, light and dark hairs and fine, pale, appressed hairs; apices rounded to suture; scutellum black, apically rounded, densely pubescent. Legs short, stout, sparsely punctate, finely, densely cinereo-pubescent with longer dark suberect hairs numerously interspersed; tibiae densely clothed with dark brushlike hairs. Abdomen rather finely, sparsely punctate, punctures becoming larger and more numerous apically; densely cinereo-pubescent with long dark suberect hairs rising out of punctures; apex of fifth sternite sinuate truncate. Length, 7.5-11 mm.

Female. Form more robust, larger; antennae barely attaining third abdominal segment; apex of fifth sternite subtruncate, slightly emarginate, impressed at middle; mandibles usually without dorsal tubercle. Length, 7.5-12 mm.

Type locality. Of *melanura*, America borealis; *canteriator*, Savannah, Georgia; *arator*, America septentrionalis.

Range. Length of Atlantic states to midwestern United States (Figure 47).

Flight period. May to August.

Host plants. Asclepias tuberosa.

The habits are not known, except that the adults occur on milkweed and are seldom abundant. This species is quite variable in color pattern throughout its range. The elytral chevrons are often quite reduced, particularly the anterior one. There is also a striking size variation.

T. melanurus differs from the other members of its species group by the poorly developed thoracic umbone and tuberculate pronotal sides. It further differs from the remainder of the genus by the dark chevrons of the elytra.

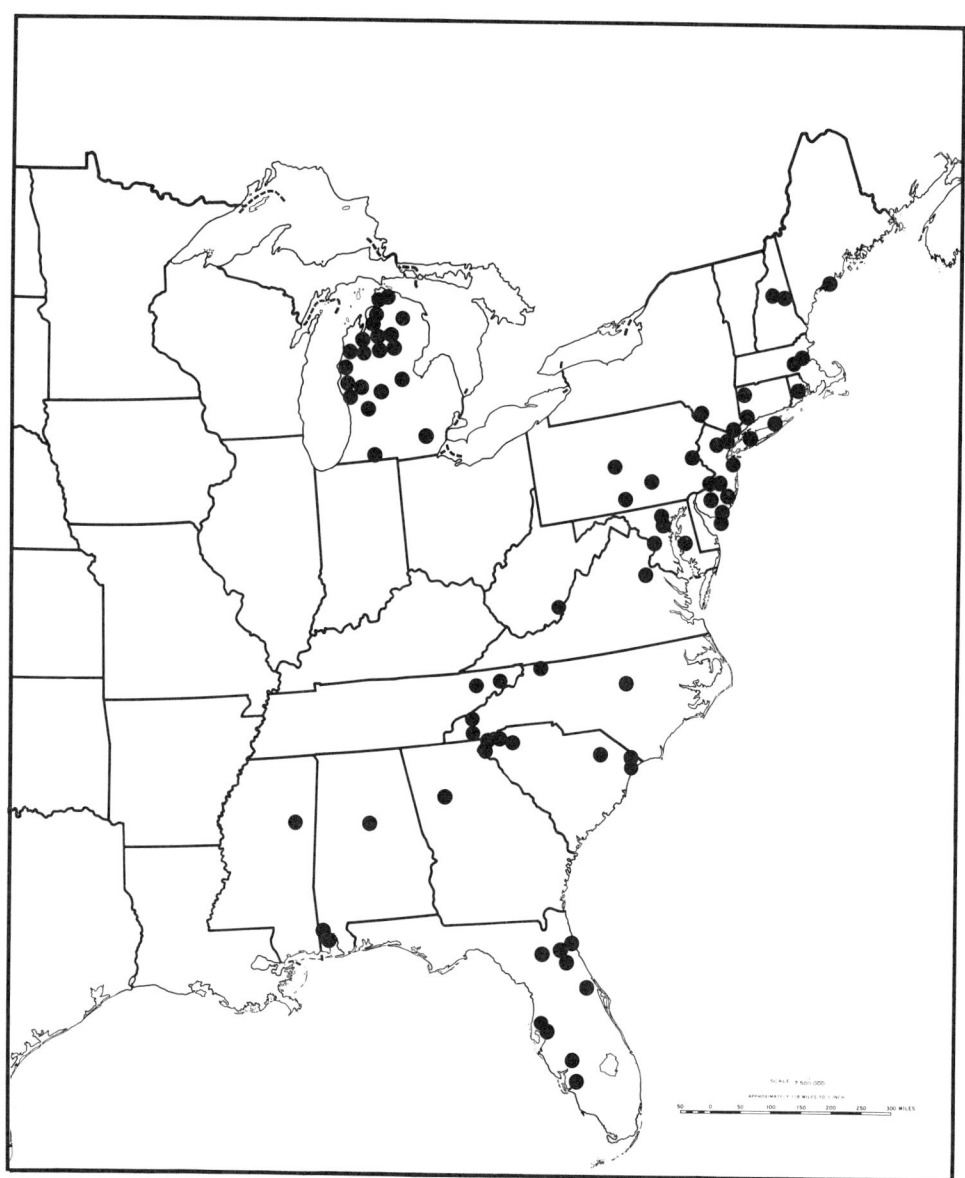

Figure 47. Known geographic range of *Tetraopes melanurus* Schönherr.

Tetraopes linsleyi Chemsak
(Figure 48)

Tetraopes n.sp.: Linsley, Knull, and Statham, 1961, Amer. Mus. Nov., 2050:32.
Tetraopes linsleyi Chemsak, 1963, Univ. Calif. Pub. Entomol., 30:31, pl. 8, fig. 7; Hovore, 1983, Coleop. Bull., 37: 386.

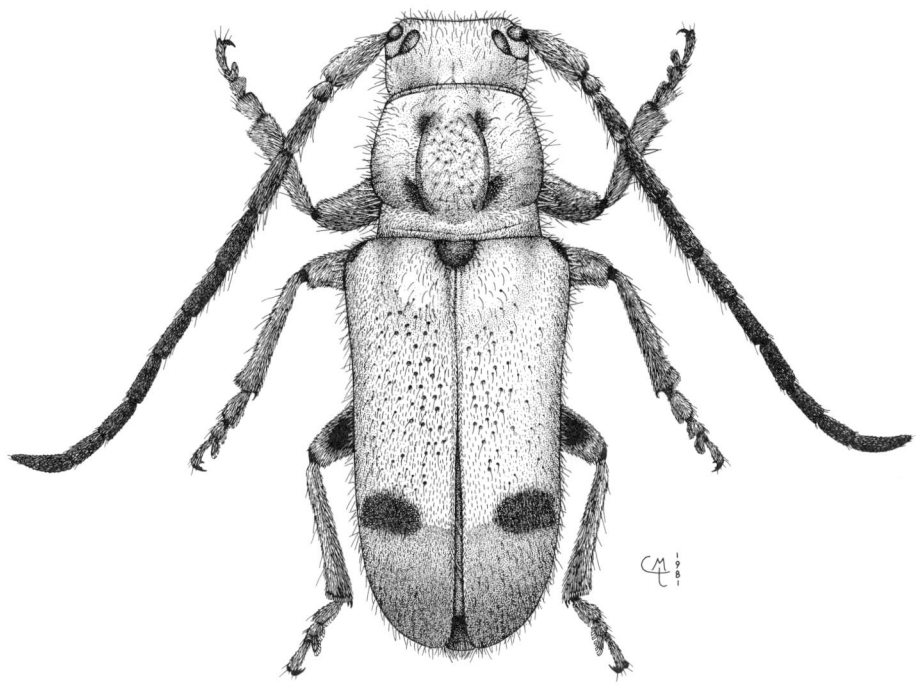

Figure 48. *Tetraopes linsleyi* Chemsak, male.

Male. Form and size of *discoideus* LeConte; color red, underside black; pubescence rather dense, suberect, and appressed. Head red, narrower than prothorax; rather sparsely, moderately coarsely punctate, long dark erect hairs sparse, very short pale appressed pubescence moderate, not obscuring surface; distinct suture extending for entire length to clypeus; labrum heavily sclerotized; maxillary palpi longer than labial, apical segment large, somewhat inflated, not tapering to a sharp point, pubescence moderate; antennae fuscus, stout, extending to about elytral apices; segments three to six subequal in length, shorter than scape, segments seven to ten decreasing in length, eleventh longer than tenth, appendiculate; basal segments moderately clothed with very long suberect hairs; segments from second to about tenth internally clothed with very long suberect hairs; basal segments densely cinereous throughout and remainder beneath, distal segments densely clothed with very short brown appressed pubescence slightly intermingled with cinereous above, somewhat annulate. Pronotum slightly broader than long, red; pronotum with an abruptly raised, slightly elevated, convex, reddish umbone at center, occupying about middle three-fifths of length, about 1-3/4 times longer than broad, rather densely, coarsely punctate, each puncture bearing a long dark erect hair, short pale pubescence sparse; lateral margins of umbone darker red; each side of umbone with two dark pubescent spots at anterior and posterior ends; punctation at sides very sparse, erect hairs sparse, appressed cinereous pubescence sparse; base deeply broadly emarginate, sides slightly inflated at middle; prosternum red, narrow, sparsely punctate and pubescent; meso- and metasternum black, densely cinereously pubescent. Elytra red, twice as long as broad; apical one-third with dark transverse spots present at anterior edge of chevron at about middle of elytra; no anterior chevron present; two faint spots present anteriorly along suture; humeri with dark spots; punctures at base rather sparse, moderately coarse, becoming much coarser posteriorly to about mid elytra, then finer toward apices; dark apical chevron densely cinereously pubescent, hairs obscuring surface, pale appressed pubescence sparse on remainder of elytra, longer suberect hairs somewhat sparse; apices rounded to suture; scutellum black, broadly rounded, densely pubescent. Legs short, robust, densely cinereously pubescent, femora not glabrous internally, mid and hind tibiae with moderate brush of hairs along outside surface. Abdomen densely cinereously pubescent with longer hairs sparsely interespersed; fifth sternite sparsely clothed with suberect dark hairs, apex subtruncate. Length, 5-6 mm.

Female. More robust, antennae almost attaining elytral apices, pygidium exposed; fifth sternite impressed at middle, subtruncate at apex. Length, 5.5-6 mm.

Type locality. Graham Mts., Arizona.
Range. Arizona to Texas.
Flight period. July and August.
Host plants. Asclepias capricornis asperula, A. linaris, A. subverticillata.

This species is similar in size and form to *T. discoideus* LeConte. It differs from that species by the lack of an anterior elytral chevron, completely red prothorax, and reduced cinereous pubescence. Structurally, the shape and size of the apical maxillary palpal segment will also differentiate the two.

Some variation is evident, primarily in coloration. The two anterior elytral spots are often lacking, and a small interior spot at the sides of the pronotum is evident in some specimens. Variation is also present in the cinereous pubescence of the antennae.

Tetraopes discoideus LeConte

Tetraopes discoideus LeConte, 1858, J. Acad. Nat. Sci. Philadelphia, (2)4:26; Horn, 1878, Trans. Amer. Entomol. Soc., 7:49; Horn, 1886, Trans. Amer. Entomol. Soc., 13:13; Hamilton, 1896, Trans. Amer. Entomol. Soc., 23:158; Casey, 1913, Memoirs on the Coleoptera, 4:375; Linsley, Knull, and Statham, 1961, Amer. Mus. Nov., 2050:32, fig. 23 (habits); Chemsak, 1963,

Univ. Calif. Pub. Entomol., 30:33, pl. 8, figs. 2,7,8,9; Hovore et al., 1987, Proc. Calif. Acad. Sci., 44:321.
Tetraopes rubrocinereous Thomson, 1860, Class. ceramb., p. 67; Bates, 1881, Biol. Centr.-Amer., Coleop., 5:202; Lacordaire, 1872, Genera des coléoptères, Atlas, pl. 110, fig. 2, 2a; Horn, 1886, Trans. Amer. Entomol. Soc., 13:13.
Tetraopes nanulus Casey, 1913, Memoirs on the Coleoptera, 4:376.
Tetraopes nigricollis Casey, 1913, Memoirs on the Coleoptera, 4:376.

Male. Form small, robust, color reddish to almost black; pubescence dense, suberect and thickly appressed. Head black or red, slightly narrower than prothorax, punctures fairly large, moderately dense, each puncture giving rise to a long dark erect seta; suture down middle extending from occiput to fronto-clypeal suture; fairly densely clothed with short, pale, appressed, pubescence which arises from very fine punctures in areas between large ones; labrum heavily sclerotized, dark brown, coarsely confluently punctate along middle one-third, or more, the punctures becoming larger basally, apical and basal bands glabrous, the apical portion densely clothed with short golden suberect hairs, which become very long in the punctate area; palpi moderately densely pubescent, apical segment of maxillary palpi elongate, narrow, sharply pointed at apex; antennae black, usually annulated, fairly stout, barely attaining elytral apices, second segment small, segments three to six subequal in length, slightly shorter than scape, segments seven to ten decreasing in length, eleventh longer than tenth, basal segments densely clothed with long erect dark hairs, decreasing in number apically, basal segments densely cinereously pubescent, at least some of the distal segments usually cinereous beneath. Prothorax broader than long; often reddish on dorsal surface on each side of middle at apical margin or completely black, base and middle black; pronotum with an oblong slightly raised umbone at center extending from a little before the base to a little before the apex, umbone almost 1-1/2 times as long as broad, moderately densely punctate, punctures much smaller than basal elytral ones, each puncture bearing a long erect black seta; densely clothed with appressed cinereous pubescence which arises from minute punctures on areas between large ones; four black spots present, two each side of umbone; punctation at sides of umbone sparser, pubescence similar; base of pronotum deeply, broadly emarginate, less so at base of umbone; sides slightly, abruptly inflated at middle; prosternum fairly narrow, sparsely punctate with fine transverse folds, densely clothed with fine, cinereous, appressed pubescence, prosternal coxal process narrow, becoming expanded at apex; meso- and metasternum sparsely punctate, densely pubescent. Elytra reddish, almost twice as long as broad, usually with an oblique black chevron at basal half beginning behind humeri and joining at the suture or broadly expanded into a large diamond pattern which joins apical dark band at suture and anteriorly extends to the scutellum; apical black pattern beginning behind spots at about apical one-third and extending to apex, usually not expanded to elytral edge; punctures large, dense, separated by at least their own diameters at base, much larger than pronotal punctures, becoming finer apically; densely clothed with long, suberect, pale and dark hairs arising from punctures, surface almost obscured by short appressed cinereous pubescence; apices round to suture; scutellum concave, broadly rounded at apex, densely clothed with dark pubescence. Legs short, stout, robust, densely clothed with long, cinereous, appressed pubescence, femora with outside surface glabrous for at least half their lengths, long suberect hairs interspersed on femora, tibiae with dense brush of bristle-like hairs along outside and inside surfaces and also with long dark suberect hairs interspersed throughout. Abdomen finely, densely, shallowly punctate with a few larger punctures sparsely interspersed; pubescence long, dense, appressed, cinereous; fifth sternite with numerous long dark suberect setae arising out of large punctures, apex broadly rounded to subtruncate. Length, 6-10 mm.

Female. More robust, antennae not attaining elytral apices; fifth sternite slightly impressed at apex at middle; pygidium usually exposed. Length, 6.5-10 mm.

Type locality. Of *discoideus*, Llano Estacado, Texas; *rubrocinereous*, Mexico; *nanulus*, Arizona; *nigricollis*, Fort Wingate, New Mexico.

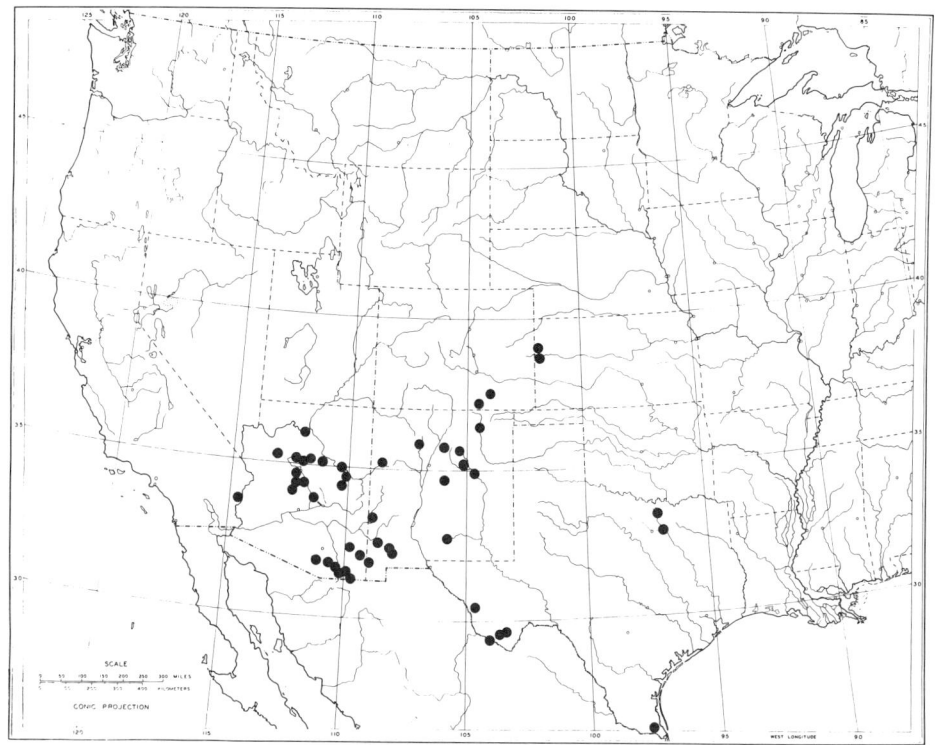

Figure 49. Known geographic range of *Tetraopes discoideus* LeConte north of Mexico.

Range. Colorado to El Salvador, western Arizona to western Kansas down to west Texas (Figure 49).

Flight period. July to October.

Host plants. Asclepias auriculata, A. curassavica, A. glaucescens, A. linaria, A. subverticillata.

This species occurs from Colorado southward to at least Honduras. Eastward it extends into western Kansas, across New Mexico, and down along western Texas. To the west, the limit appears to be western Arizona.

T. discoideus is quite variable in size, coloration, and elytral pattern. Judging from the available series representing the wide range of distribution, evidence for clinal variation is apparent, expressed chiefly in the color of the head and to a lesser degree the extent of the anterior elytral chevrons. Without exception the entire northern populations have red heads and somewhat reduced elytral chevrons. This characteristic continues down to southern Arizona, where populations occur having both red and black heads. From this point southward, every specimen seen has not only a completely black head but also an expanded dark elytral chevron. The black-headed populations also extend into Texas from the eastern side of the Chiracahua Mountains. Within the range of red-headed examples only one small series has been seen with dark heads. This population occurs in the Sierra Ancha Mountains, northwest of the Chiricahua Mountains.

Since the two major types discussed above are sympatric, possess similar habits and host plants, and are morphologically indistinguishable, there appears to be little doubt that they are conspecific.

Although a number of hosts are listed for *discoideus*, all but *A. subverticillata* and *curassavica* are probably incidental. These two are doubtless the primary and preferred hosts, *subverticillata* in the southwestern United States and *curassavica* throughout Mexico.

Tetraopes annulatus LeConte

Tetraopes annulatus LeConte, 1847, J. Acad. Nat. Sci. Philadelphia, (2)1:93; LeConte, 1852, J. Acad. Nat. Sci. Philadelphia, (2)2:157; Thomson, 1857, Arch. Entomol., 1:64; Leng and Hamilton, 1896, Trans. Amer. Entomol. Soc., 23:161; Chemsak, 1963, Univ. Calif. Pub. Entomol., 30:37, fig. 10; Bayer and Shenefelt, 1969, Univ. Wis. Res. Bull. 275:33, fig. 41; Penrose and Westcott, 1974, Coleop. Bull., 28:236; Stein and Tagestad, 1976, USDA For. Ser. Res. Pap. RM-171:42.

Tetraopes canescens LeConte, 1852, J. Acad. Nat. Sci. Philadelphia, (2)2:158; Thomson, 1857, Arch. Entomol., 1:64; Horn, 1878, Trans. Amer. Entomol. Soc., 7:50; Leng and Hamilton, 1896, Trans. Amer. Entomol. Soc., 23:161; Casey, 1913, Memoirs on the Coleoptera, 4:378.

Tetraopes canescens fontinalis Casey, 1913, Memoirs on the Coleoptera, 4:378.

Tetraopes uteanus Casey, 1913, Memoirs on the Coleoptera, 4:378.

Tetraopes vestitus Casey, 1913, Memoirs on the Coleoptera, 4:378.

Male. Form small to moderate-sized, somewhat elongate; color reddish, antennae and usually underside black; pubescence dense, pale appressed, somewhat obscuring surface. Head reddish; shallowly concave between antennae; vertex moderately coarsely, rather sparsely punctate, front more densely; densely clothed with short, pale, appressed pubescence, hairs appearing denser on cheeks, longer erect hairs fairly dense; mandibles not prominently tuberculate on dorsal surface; palpi unequal, apical segment of maxillary somewhat elongate, not broadened excessively; antennae shorter than elytra, black scape often reddish or partially so, some basal segments often rufo-piceous, distinctly annulate with cinereous pubescence, first three segments entirely and often segments six or seven densely, cinereously pubescent beneath, scape and segments two and three bristling with suberect black hairs, remaining segments rather sparsely clothed internally with longer suberect dark hairs which diminish in number apically, eleventh segment not noticeably appendiculate, rather bluntly tapering at apex. Pronotum reddish; less than 1-1/3 wider than long; umbone not elevated prominently nor distinctly delimited, subhexagonal in shape, usually slightly broader than long, densely clothed with short, pale, appressed pubescence with short suberect hairs often sparsely interspersed or shining and glabrous, slightly coarsely, sparsely punctate; four black spots prominent, usually only slightly shining; sides abruptly but not prominently inflated at middle, forming an obtuse tubercle, densely pubescent with few long dark erect hairs interspersed, punctures moderately coarse, sparse; basal margin often dark at sides, broadly shallowly impressed; prosternum densely clothed with pale appressed pubescence; meso- and metasternum sparsely punctate, densely pubescent. Elytra reddish; varying in length from less than twice to more than twice as long as broad; two black sutural spots or at least traces of spots present at about basal one-fifth; two larger spots present at mid elytra a little behind apical half; humeri black but occasionally reddish; apices often narrowly black; basal punctures moderately coarse and dense, becoming finer and denser apically; pubescence short, pale, appressed, obscuring surface, short suberect hairs evident; apices rounded; scutellem black, rounded. Legs infuscated or red, very sparsely punctate, densely clothed with pale appressed pubescence, few longer erect hairs present. Abdomen dark, often reddish, densely clothed with short pale appressed pubescence; apex of fifth sternite truncate, slightly notched at middle. Length, 8-13 mm.

Female. More robust; antennae usually not exceeding third abdominal segment; apex of fifth sternite subtruncate, notched and impressed at middle, segment distinctly longer than fourth; pygidium usually prominently visible dorsally. Length, 10-16 mm.

Type locality. Of *annulatus*, Platte River toward the mountains; *vestitus*, Colorado; *canescens*, New Mexico; *fontinalis*, Las Vegas Hot Springs, New Mexico; *uteanus*, Marysvale, Utah.

Range. Alberta to Arizona, Texas, and South Dakota (Figure 51).

Flight period. May to August.

Host plants. Asclepias subverticillata, A. tuberosus, A. verticillata, A. viridiflorus var. *lanceolatus*.

The distribution and characteristics of this species suggest two largely allopatric populations . *T. annulatus* occurs primarily in the western Great Plains region and near the eastern parts of the Rocky Mountains to southern Canada and Texas and *T. canescens* is found in the Great Basin areas of Utah, western Colorado, Arizona, and New Mexico. Although morphological data do not warrant their separation as species, additional material and biological information will probably confirm their status as subspecies. The Great Basin group has a tendency toward dark appendages and somewhat reduced pubescence, while the others appear more densely pubescent with reddish legs and, often, antennal scape.

Tetraopes pilosus Chemsak
(Figure 50)

Tetraopes pilosus Chemsak, 1963, Univ. Calif. Pub. Entomol., 30:40; Rice et al., 1985, Coleop. Bull., 39:23.

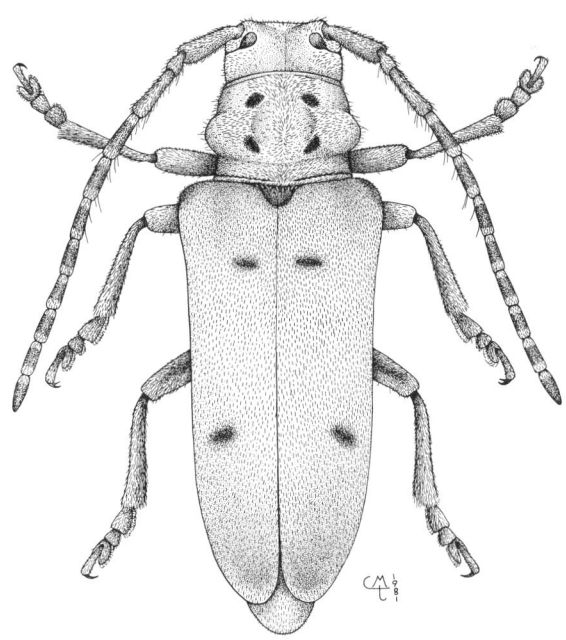

Figure 50. *Tetraopes pilosus* Chemsak, female.

Male. Form rather large, robust; color reddish, opaque; antennae except basal segments dark; very densely, thickly covered with pale, yellowish, short, appressed pubescence which completely obliterates surface. Head finely, sparsely punctate on vertex, more densely on front; antennal tubercles rather abruptly arising, with area between somewhat flat; densely clothed with short, pale, appressed pubescence, thicker on cheeks; short suberect pale hairs sparse, longer dark suberect hairs present on apices of antennal tubercles; mandibles robust, broadly arcuate dorsally, not distinctly tuberculate; apical segment of maxillary palpi slightly swollen, tapering apically; antennae not reaching elytral apices, scape reddish, segments two and three rufo-piceous, remaining segments brownish, segments from fourth distinctly annulate, basal segments densely clothed with pale appressed pubescence, scape moderately densely clothed with longer suberect dark hairs, segments two to four sparsely so internally, and remaining segments with a few long hairs at apices, eleventh segment rather abruptly tapering from apical one-fourth. Pronotum slightly broader than long; umbone scarcely elevated, not prominent, obscurely delimited at sides, subhexagonal, densely clothed with pale appressed pubescence which obscures surface, erect hairs absent, punctures coarse, rather sparse and shallow; sides of pronotum abruptly inflated, obtusely tuberculate, very sparsely coarsely punctate, densely clothed with pale appressed pubescence, longer dark erect hairs very sparse, more numerous on lateral tubercles; four dark spots prominent; base broadly impressed, apex slightly impressed at sides; apical margin narrowly piceous, basal more broadly; prosternum extremely densely clothed with ivory-colored, appressed pubescence; meso- and metasternum extremely densely clothed with rather long, ivory-colored, appressed pubescence, surface completely obscured. Elytra less than twice as long as broad; two dark glabrous sutural spots present at about basal one-fifth; two dark pubescent spots subequal in size to sutural ones present at mid elytra a little before apical one-third; humeri black internally; extremely densely clothed with short, pale appressed pubescence which completely obscures surface, erect hairs totally absent; punctures rather coarse, dense, finer apically, apices slightly dehiscent, rounded to subtruncate; scutellum black, rounded. Legs reddish, short, stout; densely clothed with pale appressed pubescence, femora with few short erect hairs internally, tibiae with longer suberect hairs numerous. Abdomen densely covered with appressed ivory pubescence, long erect hairs absent; apex of fifth sternite broadly concave. Length, 11-16 mm.

Female. Form stouter, more robust; antennae barely extending beyond third abdominal segment; pygidium prominently exposed; fifth sternite elongate, apex truncate, longitudinally impressed down middle. Length, 14-17 mm.

Type locality. Medora, Reno Co., Kansas.
Range. Nebraska and Colorado to Oklahoma and western Texas.
Flight period. June to August.
Host plants. Asclepias arenaria, A. tuberosus.

This species is easily recognizable by the extremely dense, pale, appressed pubescence which completely obscures the surface of the body, and also by its large size. Some variation in color is evident, with the antennal scape, legs, and underside often being dark. In very few cases, the tips of the elytra are vaguely tinged with black.

The very dense, dirty white to yellowish pubescence which completely obliterates the surface distinguishes *T. pilosus* from *T. annulatus*. Additionally, the average size of *T. pilosus* is greater.

Tetraopes tetrophthalmus (Forster)

Cerambyx tetrophthalmus, Forster, 1771, Novae Species Insectorum, Centuria I., p. 41.

Tetraopes tetrophthalma: Haldeman, 1847, Trans. Amer. Phil. Soc., (2)10:53 (habits); Emmons, 1854, Nat. Hist. N.Y., Agric. Rept., 5:124, pl. 5, fig. 11 (habits).

Tetraopes tetraophthalmus: Provancher, 1877, Petite faune entomol. Canada, 1:637; Horn, 1878, Trans. Amer. Entomol. Soc., 7:49; Forbes, 1880, Ill. St. Lab. Nat. Hist. Bull., 4:141; Riley, 1880, Amer. Entomol., (n.s.) 3:271 habits); Robertson, 1891, Trans. St. Louis Acad. Sci., 5:141; Wolcott, 1895, Entomol. News, 6:309; Leng and Hamilton, 1896, Trans. Amer. Entomol. Soc., 23:159; Beutenmuller, 1896, J. N.Y. Entomol. Soc., 4:81 (habits); Harrington, 1899, Ottawa Nat., 13:63; Smith 1900, N.J. St. Bd. 27th Ann. Rep., Agr. Suppl., 1899, p. 297; Chagnon, 1905, Nat. Can., 32:44; Blatchley, 1910, Coleoptera in Indiana, p. 1094; Kellogg, 1914, Amer. Insects, 3rd ed., p. 284, pl. 2, fig. 10 (habits); Adams, 1915, Ill. St. Lab. Nat. Hist. Bull., 11:177 (habits); Weiss and Dickerson, 1921, J. N.Y. Entomol. Soc., 29:124 (habits); Clench, 1923, Papers Mich. Acad. Sci., 3:367 (habits); Craighead, 1923, Can. Dept. Agric. Bull. (n.s.) 27:135, pl. 5, fig. 4, pl. 12, fig. 2, pl. 16, fig. 11, pl. 32, fig. 1 (larva, habits); Jones, 1932, Trans. Entomol. Soc. London, 80:357, pl. 18, fig. 3, pls. 20-22, App. A, p. 372, 374-377; Chagnon, 1933- 40, Coleop. Prov. Quebec, p. 279; Judd, 1949, Can. J. Res., D, 27:195, 7 figs. (morphology); Shiraki, 1952, Cat. Inj. Ins. Japan, 5:16 (habits); Duffy, 1953, Mon. British Timber Beetles, p. 41, 45 (larva, habits); Alexander, 1957, Ohio J. Sci., 57:108 (habits); Edgren and Calhoun, 1958, Bull. Ecol. Soc. Amer., 39:91 (habits); Dillon and Dillon, 1961, Man. Beetles East. North America, p. 657, pl. 65, no. 14; Alexander et al., 1963, Anim. Behav., 11:114, pl. 10, fig. 9; Mason, 1964, Syst. Zool., 13:161; Stein and Tagestad, 1976, USDA For. Serv. Res. Pap., RM-171:43; Scheiring, 1977, Evolution, 31:447; McCauley, 1979, Heredity, 42:143; Davis, 1980, Envir. Entomol., 9:432; Davis, 1980, J. Ins. Phys., 26:403 (habits); Davis, 1981, Ann. Entomol. Soc. Amer., 74:385; Rice and Enns, 1981, Trans. Mo. Acad. Sci., 15:106; McCauley, 1982, Anim. Behav., 30:23; McCauley, 1983, Evolution, 37:701; McCauley, 1983, Evolution, 37:1239; McCauley and Reilley, 1984, Ann. Entomol. Soc. Amer., 77:526.

Tetraopes tetrophthalmus: LeConte, 1852, J. Acad. Nat. Sci. Philadelphia, (2)2:157; Thomson, 1857, Arch. Entomol., 1:62; Thomson, 1868, Mem. Soc. Roy. Sci. Liege, 19:125; Deveraux, 1878, Can. Entomol., 10:143 (habits); Casey, 1913, Memoirs on the Coleoptera, 4:379, 385; Saalas, 1936, Ann. Zool. Soc. Vanamo, 4:165, pl. 16, fig. 258 (morphology); Williams, 1941, Can. Entomol., 73:137 (habits); Smith, 1943, Rep. Kansas Bd. Agric., p. 314; Knull, 1946, Ohio Biol. Surv. Bull., 39:284; Duffy, 1960, Mon. Neotrop. Timber Beetles, p. 280; Gardiner, 1961, Can. Entomol., 93:678, figs. 1, 2 (habits); Chemsak, 1963, Univ. Calif. Pub. Entomol., 30:41, figs. 1, 11; Gardiner, 1966, Can. J. Zool., 44:204, figs. 32, 53; Bayer and Shenefelt, 1969, Univ. Wis. Res. Bull. 275:34, fig. 41; Gardiner, 1969, Can. Dep. Fish. For. Intern. Rept. 0-14:107 (larva); Price and Willson, 1976, Oecologia, 25:331; Laliberte et al., 1977, Fabreries, 3:99; Lawson, 1977, J. Kansas Entomol. Soc., 50:172, figs. 2, 4, 8, 9, 11; Gosling and Gosling, 1977, Gr. Lakes Entomol., 10:36. fig. 181; Headstrom, 1977, Beetles of America, p. 385, fig. 536; Dailey et al., 1978, Coleop. Bull., 32:224; Price and Willson, 1979, Amer. Mid. Nat., 101:76; Ode, 1980, Melsheimer Entomol. Ser., 29:43; Gosling, 1984, Gr. Lakes Entomol., 17:73; Dussourd and Eisner, 1987, Science, 237:898 (habits).

Lamia tornator Fabricius, 1775, Syst. Entomol., p. 176; Fabricius, 1781, Species Insect., 1:223; Oliver, 1792, Encycl. Meth., Hist. Nat. Ins., 7:469; Fabricius, 1793, Entomol. Syst., 1(2):287; Fabricius, 1801, Syst. Eleuth., 2:301; Tigny, 1802, Hist. Nat. Insectes, 7:282; Schonherr, 1817, Synon. Insect. 1(3):401; Latrielle, 1830, Dict. Class. d'Hist. Nat., 16:190; Thon and Reichenbach, 1838, Die Insekten, Krebsund Spinnenthiere, p. 411, pl. 99, fig. 648; Berge, 1844, Kaferbuch, p. 123, pl. 11, fig. 9; Chenu, 1860, Encycl. d'Hist. Nat., 3:325.

Lamia fornator: Fabricius, 1787, Mant. Ins., 1:141.

Cerambyx tornator: Olivier, 1795, Entomol., 4:103, pl. 8, fig. 52.

Tetraopes tornator: Say, 1835, Boston J. Nat. Hist., 1(2) 665; Castelnau, 1840, Hist. Nat. Insectes, Coleop., 2:487, pl. 53, fig. 3.

Lamia 13-punctata Drapiez, 1820, Ann. Gen. Sci. Phys., 5:121, pl. 74, fig. 6.

Tetraopes tetrophthalmus iowensis Casey, 1913, Memoirs on the Coleoptera, 4:386.

Tetraopes humeralis Casey, 1913, Memoirs on the Coleoptera, 4:379.

Male. Form moderate-sized, rather stout, shining; color reddish, underside and appendages black; pubescence moderate. Head reddish, narrower than prothorax; punctures usually coarse, dense; fine, very short pubescence dense but not obscuring surface, longer erect dark hairs moderately dense; mandibles prominent, robust, dorsal surface with prominent tubercles at base giving sharply arcuate appearance, apices bluntly pointed; antennae black, not annulated, extending to about third abdominal segment, first three segments shining, remainder opaque; scape moderately densely clothed with short, appressed, cinereous pubescence and longer, suberect, dark, bristle-like hairs, second and third segments similar to scape; segments four to eleven densely clothed with very short, brown, appressed pubescence; segments two to about six or eight sparsely clothed internally with suberect dark hairs, all segments with a few erect long hairs at apices; segments from third with short erect pale hairs sparsely interspersed; eleventh segment appendiculate. Pronotum reddish, broader than long; umbone scarcely elevated, rather distinctly delimited laterally, subhexagonal, usually finely, rather sparsely punctate, sparsely clothed with long erect dark hairs and very fine pale pubescence which does not obscure surface; four dark spots prominent, shining; sides of pronotum rather finely sparsely punctate, densely clothed with very fine short pale pubescence with long erect dark hairs sparsely interspersed; sides abruptly inflated, lateral tubercles prominent; anterior edge slightly emarginate at middle; basal edge narrowly black, black spot usually present on sides at anterior end; prosternum almost impunctate, densely clothed with appressed cinereous pubescence. Elytra reddish, less than twice as long as wide; humeri black; two black sutural spots at about mid elytra at basal half, these spots frequently enlarged and elongated, occasionally confluent with sutural spots; two other black spots present behind post-humeral at about apical one-third; punctation at basal one-fourth rather fine and sparse, becoming coarser and denser at middle half and again finer at apical one-fourth; very short pale pubescence dense but not obscuring surface, longer suberect dark hairs moderately dense, longer basally and becoming very short at apex; scutellum black, rounded apically; apices rounded to suture with small vague black stripes at extreme apices. Legs short, robust, very sparsely, finely punctate, moderately densely, cinereously pubescent; mid and hind tibiae with external dark brush of bristle-like hairs. Abdomen finely sparsely punctate, densely cinereously pubescent, with longer erect hairs moderately densely interspersed; apex of fifth sternite to sinuate truncate. Length, 8-14 mm.

Female. Form more robust, stouter; antennae usually subequal to male in length; apex of fifth sternite slightly emarginate at middle with median suture extending length of segment; pygidium almost always exposed. Length, 8-15 mm.

Type locality. Of *tetrophthalmus*, America septentrionalis, noveboracensi; *tornator*, America; *13-punctata*, America meridionale; *iowensis*, not given; *humeralis*, Dakota.

Range. Length of Atlantic states, southern Canada west to Rocky Mts.

Flight period. May to September.

Host plants. Acerates viridiflora, Apocynum cannabinum, Asclepias incarnata, A. perennis, A. syriaca.

This species is easily recognizable by the flat, sub-hexagonal thoracic umbone (Figure 46), unannulated antennae, and usually by the distinctive elytral markings.

It is abundant over most of its range and is probably the commonest species in the genus.

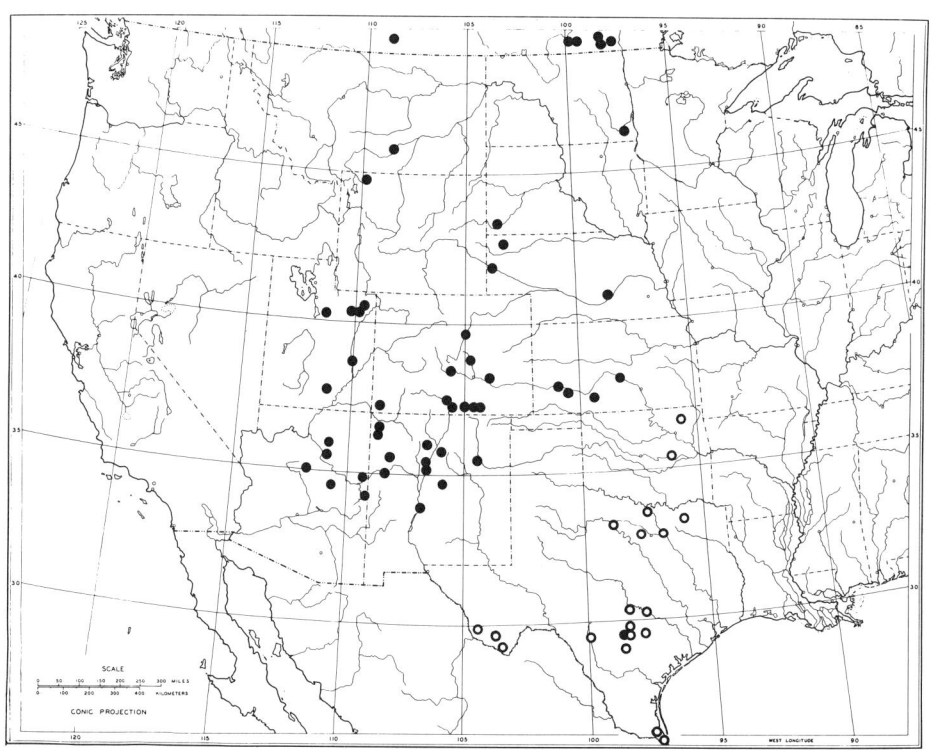

Figure 51. *Known geographic ranges of: Tetraopes texanus* Horn, open circles; *Tetraopes annulatus* LeConte, closed circles.

Tetraopes texanus Horn

Tetraopes quinquemaculatus var. *texana* Horn, 1878, Trans. Amer. Entomol. Soc., 7:49.
Tetraopes quadrimaculatus var. *texanus*: Leng and Hamilton, 1896, Trans. Amer. Entomol. Soc., 23:160.
Tetraopes texanus: Casey, 1913, Memoirs on the Coleoptera, 4:385; Chemsak, 1963, Univ. Calif. Pub. Entomol., 30:46, fig. 10; Hovore et al., 1987, Proc. Calif. Acad. Sci., 44:321.

Male. Form moderate-sized, robust; color reddish, antennae except scape, tibiae, and underside black; pubescence moderate, punctation moderately coarse, dense. Head reddish; moderately coarsely, sparsely punctate on vertex, more densely on front; very short pale pubescence dense but not obscuring surface, long black suberect hairs sparse on vertex, denser on front; mandibles stout, slightly tuberculate basally on dorsal surface, excavated basal area densely clothed with pale appressed pubescence; last segment of maxillary palpi elongate, large, not tapering to sharp point; antennae black, usually distinctly cinereously annulate, scape except apex usually red; shorter than elytra; basal segments shining, rather densely clothed with moderate-sized dark, bristle-like hairs, cinereous beneath; segments from fourth opaque, densely clothed with very short appressed brownish pubescence; apices of segments from fourth with plume of extremely long erect dark hairs beneath which radiate out to sides; eleventh segment rather abruptly tapering at about midpoint, then very narrowly extending apically, slightly arcuate. Pronotum reddish, shining, broader than long; umbone not abruptly elevated nor distinctly delimited, surface broadly convex, somewhat hexagonal in form, surface finely, sparsely punctate, usually sparsely clothed with short pale appressed pubescence and long dark suberect hairs; four dark spots prominent; sides of pronotum abruptly inflated at middle, distinctly obtusely tuberculate, base distinctly deeply emarginate, apex somewhat less; puncture rather fine, sparse, pale appressed pubescence dense, denser at sides; apical margin and base at sides usually narrowly infuscated; prosternum narrow, sparsely, shallowly punctate, slightly transversely rugous, cinereous pubescence moderately dense; meso- and metasternum densely shallowly punctate, densely clothed with appressed cinereous pubescence. Elytra reddish, about twice as long as wide; humeri black; two small black spots usually present near suture at about basal one-fifth, two larger black spots also present at midelytra at about apical two-fifths; punctures coarser than pronotal ones at base, fairly dense, becoming finer and denser apically; short pale pubescence dense, not obscuring surface except partially at apex; long, fine suberect, dark hairs dense; scutellum black, subtruncate at apex. Legs short, robust; femora red, apices black, very finely, shallowly, densely punctate, densely, cinereously pubescent; tibiae black, front tibiae with fine, dense brush of golden hairs internally, mid and hind tibiae with brush externally. Abdomen moderately densely, shallowly punctate, densely cinereously pubescent; apex of fifth sternite concavely truncate. Length, 11-15 mm.

Female. More robust; antennae shorter, usually extending to about third abdominal segment, eleventh segment similar in shape to male; apex of fifth sternite impressed at middle. Length, 12.5-16.5 mm.

Type locality. Texas.

Range. Oklahoma to Texas and northern Mexico (Figure 51).

Flight period. April to September.

Host plants. Asclepias. One female specimen bears the label, "taken on gaillardia."

This species is readily recognizable by the shape of the eleventh antennal segments and fringes of very long hairs at the apices of the antennal segments. These two characteristics will at once separate it from *T.*

quinquemaculatus. It further differs by the reddish femora and antennal scape and the usual presence of anterior sutural spots.

Tetraopes mandibularis Chemsak

Tetraopes mandibularis Chemsak, 1963, Univ. Calif. Pub. Entomol., 30:47; Rice et al., 1984, Coleop. Bull., 39:23.

Male. Form moderate-sized, stout, robust; color reddish, antennae black, tibiae, tarsi, and underside black or partially infuscated; pubescence dense, short, pale and suberect and long, dark, and erect. Head reddish, finely, densely punctate; densely clothed with short, pale, suberect and depressed pubescence with long, black erect hairs numerously interspersed, surface not obscured; palpi unequal, apical segment not inflated; genae prominent, apices diverging from head to beyond outer margins of eyes when viewed head on; mandibles prominent, dorsal surface with very large, elongate tubercle which extends about half the length of mandible; antennae not extending beyond apex of third abdominal segment, first three segments densely cinereously pubescent, longer erect and suberect hairs numerous, segments from third completely clothed with very short, appressed, brownish pubescence, thinly annulate at apex, and base with pale pubescence, long, dark erect hairs diminishing in number apically. Pronotum reddish, about 1-1/2 times broader than long; umbone convex, not distinctly delimited at sides, about as long as broad, finely, moderately densely punctate, short pale hairs not obscuring surface, long, black, erect hairs numerous; sides abruptly inflated, obtusely tuberculate, punctures fine, dense, depressed and suberect, four black spots prominent; basal margin broadly emarginate; prosternum narrow, densely pubescent; meso- and metasternum densely pubescent, metepisternum finely, densely punctate. Elytra reddish, less than twice as long as broad; humeri black; two small, oblique, sutural spots present at basal one-fourth, two present at apical third in middle of elytra, elongate post-humeral spots often present near outer margins; punctation rather fine, dense; very short, pale, subdepressed and suberect pubescence dense, obscuring surface partially, long, erect dark hairs numerous basally; apices rounded; scutellum black, densely pubescent. Legs short, stout; femora reddish, very sparsely punctate, moderately densely pubescent. Abdomen finely punctate, densely pubescent; apex of fifth sternite broadly rounded, vaguely emarginate. Length, 13-15 mm.

Female. Antennae extending to about second abdominal segment; mandibles without prominent dorsal tubercles; apex of fifth abdominal sternite emarginate, longitudinally impressed down middle. Length, 12-15 mm.

Type locality. Childress, Childress Co., Texas.
Range. Texas and Oklahoma.
Flight period. June to August.
Host plants. Asclepias latifolia.

This species is related to *T. texanus* and *T. quinquemaculatus,* as indicated by the broad, convex, poorly defined thoracic umbone. The structure of the male mandibles separates it from both of those species. The spot pattern of the elytra and terminal antennal segment further differentiate *mandibularis* from both.

Tetraopes quinquemaculatus Haldeman

Tetraopes 5-maculata Haldeman, 1847, Trans. Amer. Phil. Soc., (2) 10:53.
Tetraopes 5-maculatus: LeConte, 1852, J. Acad. Nat. Sci. Philadelphia, (2) 2:157; Thomson, 1857, Arch. Entomol., 1:63.
Tetraopes quinquemaculatus: Horn, 1878, Trans. Amer. Entomol. Soc., 7:49; Blatchley, 1910, Coleoptera in Indiana, p. 1094; Casey, 1913, Memoirs on the Coleoptera, 4:374, 377, 385; Knull, 1946, Ohio Biol. Surv. Bull., 39:283; Chemsak, 1963, Univ. Calif. Pub. Entomol., 30:48; Bayer and Shenefelt, 1969, Univ. Wis. Res. Bull. 275:34, fig. 41; Gosling and Gosling, 1977, Gr. Lakes Entomol., 10:36, fig. 182; Price and Willson, 1977, Amer. Mid. Nat., 101:76; Ode, 1980, Melsheimer Entomol. Ser., 29:43; Rice and Enns, 1981, Trans. Mo. Acad. Sci., 15:106; Gosling, 1984, Gr. Lakes Entomol., 17:73.
Tetraopes quadrimaculatus: Leng and Hamilton, 1896, Trans. Amer. Entomol. Soc., 23:160.

Male. Form small, short, robust; color reddish orange, appendages and underside black; pubescence moderately dense. Head reddish; broadly concave between antennae; rather finely, moderately densely punctate, more so on front; very short pale pubescence dense but not obscuring surface, long black erect hairs fairly dense; mandibles moderately sinuate at base, rugous concavity on dorsal surface occupying about basal half; maxillary palpi much longer than labial, last segment large, clavate, rounded at apex; antennae black, may have traces of cinereous annulations, not quite attaining elytral apices, basal segments shining, densely clothed with long dark suberect hairs and cinereous beneath; segments from fourth opaque, very densely clothed with very short, appressed, brownish pubescence, sparsely clothed with long suberect fine hairs internally; scape longest, eleventh segment tapering somewhat gradually to point. Pronotum reddish, about 1-1/3 times wider than long; umbone not elevated nor distinctly delimited, broadly convex, subhexagonal in shape, finely sparsely punctate, shining, rather densely clothed with very fine short pale pubescence which does not obscure surface, long dark suberect hairs sparse; four black spots prominent, densely pubescent; sides of pronotum abruptly inflated, obtusely tuberculate, punctation fine, sparse, long erect hairs sparse, longer at sides, fine pale pubescence dense; basal margin distinctly broadly emarginate; prosternum narrow, almost impunctate, densely pubescent; meso- and metasternum almost impunctate except for episternum of metathorax, densely cinereously pubescent. Elytra reddish, slightly less than twice as long as broad; humeri black; two moderate-sized black spots present at about mid elytra at apical one-third; apices narrowly black; punctation moderately coarse, punctures a great deal larger than those of umbone, dense; short fine pale appressed pubescence dense, usually not obscuring surface except at apex, long dark erect hairs fairly sparse, more abundant at base; scutellum black, broadly rounded apically; apices rounded. Legs short, robust; femora almost impunctate, fairly densely, cinereously pubescent; tibiae with suberect bristle-like hairs dense on outside edge; abdomen almost impunctate, densely cinereously pubescent; fifth segment abundantly clothed with long suberect dark hairs, apex subtruncate. Length, 9-11 mm.

Female. More robust, antennae a little shorter; apex of fifth sternite truncate, impressed at middle. Length, 10.5-12 mm.

Type locality. Sault Ste. Marie.

Range. Ontario, midwestern United States south into Texas.

Flight period. June to August.

Host plants. Asclepias syriaca.

This species may be distinguished from *T. tetrophthalmus* by the lack of anterior elytral spots, the slightly annulated antennae, and the coarse elytral punctures. The non-elevated thoracic umbone will at once separate it from *femoratus* and the distinctive antennae of *texanus* differentiate it from *quinquemaculatus.*

Tetraopes sublaevis Casey

Tetraopes sublaevis Casey, 1913, Memoirs on the Coleoptra, 4:382; Chemsak, 1963, Univ. Calif. Pub. Entomol., 30:52 pls. 1-3, 6, 7, figs. 13, 14.

Tetraopes sandix Casey, 1914, Memoirs on the Coleoptera, 5:370.

Male. Form moderate-sized to large; color reddish, shining, pubescence sparse, punctation fine, sparse, head reddish; area between antennal tubercles distinctly, concavely impressed; punctures on vertex and front moderately coarse, sparse; densely clothed with very short pale appressed pubescence which does not obscure the surface, denser around eyes and on cheeks, fine erect hairs rather sparse, not long, longer and more numerous on cheeks; mandibles with prominent dorsal tubercle at basal half; apical segment of maxillary palpi rather slender, not swollen; antennae dark, annulate, rarely as long as elytra; scape usually reddish, apex black; scape and usually segments to fourth densely cinereously pubescent, segments five to six or seven usually cinereously pubescent beneath, distal segments densely clothed with very short dark appressed pubescence except at apex and base; scape with few short suberect hairs above, more numerous and longer beneath, remaining segments with long erect black hairs beneath becoming less numerous apically; eleventh segment arcuate, tapering. Pronotum reddish, less than 1-1/2 times broader than long; four black dorsal spots present; a black spot usually present laterally on each side of apical margin; umbone abruptly elevated, usually distinctly delimited, sides slightly arcuate to subparallel, surface rather finely, sparsely punctate away from edges, usually sparsely clothed with pale pubescence and longer dark suberect hairs; sides of pronotum abruptly inflated, prominently tuberculate, broadly deeply impressed at base and apex, dorsal edge of tubercles usually extending with short pale appressed pubescence which often slightly obscures the surface, long erect dark hairs sparse, more numerous on lateral tubercles; punctures fine, sparse, coarser on lateral tubercles; prosternum almost impunctate, fairly densely pubescent; meso- and metasternum black, moderately densely, rather finely punctate, densely cinereously pubescent. Elytra reddish; less than twice as long as broad; humeri black; two black sutural spots always present at about basal one-fourth, post-humeral dark spot usually absent but occasionally a very small pair is present near lateral margins, two larger dark spots also present at mid elytra at about apical third; basal punctures very fine and sparse, becoming coarser and denser in central third, than finer apically; short appressed cinereous pubescence usually dense, particularly at apex, longer suberect hairs fairly abundant, longer and more erect basally; apices rounded; scutellum black, densely pubescent. Legs short, robust; femora except apices usually reddish, front and middle tibiae usually reddish, hind tibiae usually black, femora finely, sparsely punctate, densely cinereously pubescent, longer erect hairs sparse. Abdomen black; densely cinereously pubescent, erect hairs sparse, short; apex of fifth sternite concavely truncate. Length, 11.5-16 mm.

Female. Form larger, more robust; antennae rarely extending beyond third abdominal segment; mandibles without prominent dorsal tubercles; apex of fifth sternite rounded, notched at middle, longitudinally impressed. Length, 13-18 mm.

Type locality. Of *sublaevis*, California; *sandix*, Witch Creek, San Diego County, California.

Range. San Diego Co. to Kern Co., California, east to Arizona.

Flight period. April to July.

Host plants. Asclepias erosa.

This species is primarily restricted to southern California, extending only as far north as Kern County. In the large series available, specimens appear to be fairly constant in coloration and size. However, some variation is evident by the presence of a small post-humeral dark spot on the elytra. This is usually absent, but occurs in a relatively few individuals in any given population. The antennal scape is often infuscated, and the femora and tibiae vary slightly in the degree of redness or infuscation. The sutural and post-median spots are constant in their occurrence, varying only in size.

T. sublaevis may be readily separated from the other Pacific Coast forms by the very fine, sparse, basal elytral punctures. It differs from *basalis* by the abruptly, highly elevated thoracic umbone.

Tetraopes thermophilus Chevrolat

Tetraopes thermophilus Chevrolat, 1861, J. Entomol., 1(3):190, 254; Bates, 1881, Biol. Cent.-Amer. Insecta, Coleoptera, 5:201; Chemsak, 1963, Univ. Calif. Pub. Entomol., 30:54; Hovore et al., 1987, Proc. Calif. Acad. Sci., 44:321.

Male. Form small to moderate-sized; color reddish, antennae black, barely annulated, all femora reddish. Head reddish; area between antennae shallowly concave; vertex moderately coarsely, rather sparsely punctate, punctures more abundant on front; short pale pubescence denser on cheeks and around eyes; erect black hairs moderately dense, approximately equal in length to width of last antennal segment; mandibles with distinct dorsal tubercles at base; apical segment of maxillary palpi stout, somewhat tapering; antennae black, thinly annulated, shorter than body, three basal segments somewhat shining, densely clothed with suberect black hairs with pale appressed pubescence densely interspersed, segments from fourth opaque, densely clothed with short dark appressed pubescence, antennae beneath densely cinereo-pubescent, long, black, erect, internal hairs sparse, diminishing in number apically; eleventh segment slightly tapering apically. Pronotum reddish, at least 1.4 times wider than long; four black spots prominent; umbone abruptly but not highly elevated, sides distinctly arcuate and delimited, surface convex; umbone elongate, about three-fourths as wide as long, almost three-fourths as long as pronotum; disk rather finely sparsely punctate, moderately densely pubescent; meso- and metasternum black, densely clothed with pale appressed pubescence. Elytra reddish, about twice as long as wide; humeri black; two black sutural spots usually present at about basal one-fifth, two other black spots occasionally present a little behind the sutural toward the lateral margins; two larger spots also present at about apical one-third; basal punctures fine, rather sparse, becoming coarse and large to about apical one-third, then fine apically, large punctures well separated; short pale pubescence moderately dense, denser toward apex where surface is partially obscured, long erect hairs abundant, longer and denser toward base; scutellum black, densely pubescent. Legs short, stout, femora reddish except narrowly at base and apex; femora rather densely clothed with fine pale appressed pubescence which does not obscure the surface. Abdomen black, densely cinereously pubescent; apex of fifth sternite broadly subtruncate to rounded. Length, 9-13 mm.

Female. Usually more robust, antennae not extending beyond second or third abdominal segment; mandibles without prominent dorsal tubercles; apex of fifth sternite rounded, segment longitudinally impressed and notched at middle. Length, 8-14 mm.

Type locality. Veracruz, Mexico.
Range. Southeastern Texas south to El Salvador along the tropical belt.
Flight period. April to October.
Host plants. Unknown.

This species is readily separable from *thoreyi* by the reddish femora, annulated antennae, and longer, narrower thoracic umbone. This latter character will at once distinguish it from *varicornis*.

Tetraopes basalis LeConte

Tetraopes basalis LeConte, 1852, J. Acad. Nat. Sci. Philadelphia (2) 2:158; Thomson, 1857, Arch. Entomol., 1:63; Hatch, 1961, Univ. Wash. Pub. Biol., 16:157; Chemsak, 1963, Univ. Calif. Pub. Entomol., 30:55, pls. 4-7, figs. 1, 15, 16.
Tetraopes femoratus var. *basalis*: Horn, 1878, Trans. Amer. Entomol. Soc., 7:49; Leng and Hamilton, 1896, Trans. Amer. Entomol. Soc., 23:160.
Tetraopes mancus LeConte, 1859, Proc. Acad. Nat. Sci. Philadelphia, 11:81.
Tetraopes femoratus var. *mancus*: Horn, 1878, Trans. Amer. Entomol. Soc., 7:49; Leng and Hamilton, 1896, Trans. Amer. Entomol. Soc., 23:160.
Tetraopes omissus Casey, 1913, Memoirs on the Coleoptera, 4:377.
Tetraopes coccineus Casey, 1913, Memoirs on the Coleoptera, 4:382.
Tetraopes obsoletus Casey, 1913, Memoirs on the Coleoptera, 4:382.
Tetraopes junctus Casey, 1913, Memoirs on the Coleoptera, 4:380.
Tetraopes latior Casey, 1924, Memoirs on the Coleoptera, 11:296.

Male. Form moderate-sized to large; color reddish, shining, pubescence sparse to moderately dense; appendages black, scape and femora often reddish. Head reddish; area between antennal tubercles flattened, often concave; punctures on vertex moderately coarse, sparse, denser and finer on front; pale appressed pubescence usually sparse except on cheeks and around eyes, long black erect hairs numerous; mandibles with an unusually prominent dorsal tubercle near base; apical segment of maxillary palpi not prominently swollen, rather slender, only slightly broader than apical segment of labial palpi; antennae dark, usually distinctly annulated, usually shorter than body; scape reddish or black, scape and usually segments to third densely cinereously pubescent, densely clothed with long dark suberect hairs, segments from fourth opaque, densely clothed with very short dark appressed pubescence; segments from third with numerous long erect hairs internally which diminish in number apically. Pronotum reddish, less than 1 1/2 times wider than long; four dorsal black spots prominent; umbone somewhat abruptly but not highly elevated, often not distinctly delimited at sides, sides usually arcuate; surface of umbone with a narrow longitudinal glabrous line; disk shining, rather finely, sparsely punctate, very sparsely clothed with short pale pubescence, long erect hairs sparse; sides of pronotum rather abruptly inflated medially, distinctly but not prominently tuberculate, broadly impressed at base and apex; lateral tubercles coarsely, densely punctate, remainder of pronotum sparsely to densely punctate; short pale pubescence sparse, long erect black hairs abundant to moderately numerous; a black spot occasionally present laterally at apical margin; prosternum moderately densely punctate, densely pubescent, meso- and metasternum finely, rather sparsely punctate, densely pubescent. Elytra reddish, shining; slightly less than twice as long as broad; humeri black; two black sutural spots present or not; post-humeral spot always absent; two larger black spots always present at mid elytra at about apical one-third; elytral punctures moderately coarse, not dense, becoming much finer toward apical one-third; short pale pubescence sparse but dense at apical one-third, long erect hairs fairly dense; apices rounded; scutellum black, densely pubescent, often with a small glabrous spot at middle. Legs short, stout; femora varying from all red except apices to all black; tibiae and tarsi black, densely pubescent. Abdomen black, densely cinereously pubescent; apex of fifth sternite shallowly, concavely truncate. Length, 8-15 mm.

Female. Form larger, more robust; mandibles not dorsally tuberculate; antennae seldom extending much beyond apical spots; apex of fifth sternite concavely truncate, impressed at middle. Length, 10-17 mm.

Type locality. Of *basalis*, Sierra Nevada; *mancus*, Tejon; *omissus*, California; *coccineus*, California; *obsoletus*, Siskiyou Co., California; *junctus*, not indicated; *latior*, Mt. Wilson, Los Angeles Co., California.

Range. Southwestern Oregon to southern California along the west side of the Sierra Nevada.

Flight period. April to August.

Host plants. Asclepias eriocarpa, A. fascicularis, A. speciosa.

T. basalis is an extremely variable species, particularly when specimens from the extremes of its range are compared. The long series available for study reveals a north-south cline in color and size. At the northern limits of the range, the specimens have the antennal scape and femora reddish. The sutural spot is almost always present but usually reduced. Along with the broad and poorly defined thoracic umbone, the smaller size range is characteristic of populations in southwestern Oregon and northern California. Progressing southward, there is a marked increase in size and elytral spot development and also a slight tendency toward infuscation of the scape and femora. The reddish coloration of these parts is evident in examples from the foothill areas of the Sierra Nevada as far south as Yosemite National Park. However, south of this region, the scape and femora are chiefly black with only traces of redness in some individuals. This dark aspect is encountered farther north near the coast at about Napa County, and continues southward and inland down the San Joaquin Valley to southern California. The San Joaquin Valley appears to be the transition zone between the red and dark forms. The chief characteristics of the valley, coastal, and transverse-range specimens are the dark scape and femora, intermediate size range, and usual absence of all but the large apical elytral spot.

The extreme southern forms are readily discernible by their large size, prominently expanded sutural, humeral, and apical elytral spots, dark scape and femora, and somewhat longer, denser erect pubescence.

This species may be readily distinguished from *T. sublaevis* by the coarser, more numerous elytral punctures and the broader less elevated thoracic umbone. The lateral tubercles of the thorax are also much more prominent in *T. sublaevis*. *T. basalis* differs from *T. femoratus* by the broader, vaguely defined umbone and the narrow non-inflated apical segment of the maxillary palpi. In *T. femoratus*, this segment is inflated and much broader than the apical segment of the labial palpi.

Tetraopes femoratus LeConte
(Figure 52)

Tetraopes femoratus LeConte, 1847, J. Acad. Nat. Sci. Philadelphia, (2)1:93; LeConte, 1852, J. Acad. Nat. Sci. Philadelphia, (2)2:157; Thomson, 1857, Arch. Entomol., 1:63; Horn, 1878, Trans. Amer. Entomol. Soc., 7:49; Gahan, 1892, Trans. Entomol. Soc. London, 1892, p. 267; Leng and Hamilton, 1896, Trans. Amer. Entomol. Soc., 23:160; Casey, 1913, Memoirs on the Coleoptera, 4:383; Graham, 1922, Ann. Entomol. Soc. Amer., 15:196, fig. 3, no. 2; Saalas, 1936, Ann. Zool.-Soc. Zool.-Bot. Fenn. Vanamo, 4: 168; Williams, 1941, Psyche, 48:169 (biology); Knull, 1946, Ohio Biol. Surv. Bull., 39:284; Edgren and Calhoun, 1958, Bull. Ecol. Soc. Amer., 39:91 (habits); Linsley, Knull and Statham, 1961, Amer. Mus. Nov., 2050:32 (habits); Dillon and Dillon, 1961, Man. Beetles East. North America, p. 657, pl. D, pl. 65, no. 15; Alexander et al., 1963, Anim. Behav., 11:114, pl. 10, fig. 10; Chemsak, 1963, Univ. Calif. Pub. Entomol., 30:61, pls. 5, 6, 9; figs. 1, 17; Bayer and Shenefelt, 1969, Univ. Wis. Res. Bull. 275:33, fig. 41; Hatch, 1971, Univ. Wash. Pub. Biol., 16:157, pl. 18, fig. 8; Stein and Tagestad, 1976, USDA For. Serv. Res. Pap. RM-171:43; Headstrom, 1977, Beetles of America, p. 385; Lawson, 1977, J. Kansas Entomol. Soc., 50:172, figs. 1, 3, 5, 6, 7, 10, 13, 14, 15; Gosling and Gosling, 1977, Gr. Lakes Entomol., 10:36, fig. 183; Darley et al., 1978, Coleop. Bull., 32:224; Price and Willson, 1979, Amer. Med. Nat., 101:76; Ode, 1980, Melsheimer Entomol. Ser., 29:43; Rice and Enns, 1981, Trans. Mo. Acad. Sci., 15:106; Gosling, 1984, Gr. Lakes Entomol., 17:73; Hovore et al., 1987, Proc. Calif. Acad. Sci., 44:322; MacKay, Zak and Havore, 1987, Coleop. Bull., 41: 366.

Tetraopes femoratus var. *femoratus*: Horn, 1878, Trans. Amer. Entomol. Soc., 7:49; Leng and Hamilton, 1896, Trans. Amer. Entomol. Soc., 23:160.

Tetraopes oregonensis LeConte, 1854, Proc. Acad. Nat. Sci. Philadelphia, 7:19, LeConte, 1860, Rep. Exp. Surv. Mississippi Pacific, 9:65.

Tetraopes femoratus var. *oregonesis*: Horn, 1878, Trans. Amer. Entomol. Soc., 7:49; Leng and Hamilton, 1896, Trans. Amer. Entomol. Soc., 23:160.

Tetraopes collaris Horn, 1878, Trans. Amer. Entomol. Soc., 7:49; Leng and Hamilton, 1896, Trans. Amer. Entomol. Soc., 23:160; Casey, 1913, Memoirs on the Coleoptera, 4:385.

Tetraopes nigripes Casey, 1913, Memoirs on the Coleoptera, 4:377.

Tetraopes vegasensis Casey, 1913, Memoirs on the Coleoptera, 4:380.

Tetraopes velutinus Casey, 1913, Memoirs on the Coleoptera, 4:380.

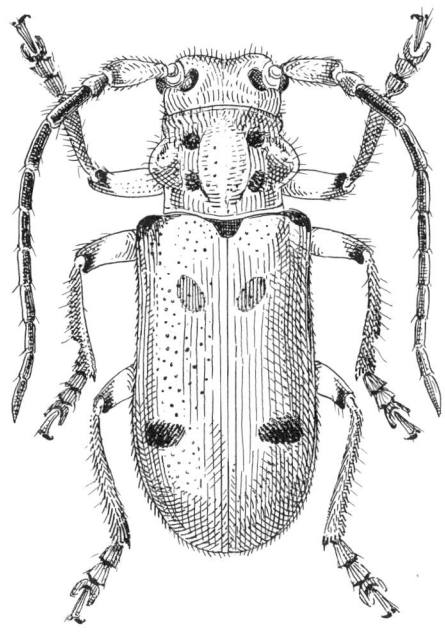

Figure 52. *Tetraopes femoratus* LeConte, male.

Tetraopes robustus Casey, 1913, Memoirs on the Coleoptera, 4:381.
Tetraopes brevisetosus Casey, 1913, Memoirs on the Coleoptera, 4:381.
Tetraopes punctipennis Casey, 1913, Memoirs on the Coleoptera, 4:383.
Tetraopes atrisetosus Casey, 1913, Memoirs on the Coleoptera, 4:383.
Tetraopes femoratus amnicola Casey, 1913, Memoirs on the Coleoptera, 4:384.
Tetraopes femoratus monticola Casey, 1913, Memoirs on the Coleoptera, 4:384.
Tetraopes spissicornis Casey, 1913, Memoirs on the Coleoptera, 4:384.
Tetraopes ruber Casey, 1913, Memoirs on the Coleoptera, 4:384.
Tetraopes fortis Casey, 1913, Memoirs on the Coleoptera, 4:384.
Tetraopes caseyi Aurivillius, 1923, Coleop. Cat., 74:575.

Male. Form small to large, color reddish, shining or dull; pubescence dense to sparse; legs reddish or black, antennal scape red or black. Head reddish; area between antennal tubercles flattened to deeply impressed, usually shallowly concave; punctures on vertex fine to moderately coarse, sparse, finer and denser on front; short cinereous pubescence usually dense, denser on front and around eyes, long erect dark hairs numerous, usually longer and bristling on cheeks; mandibles with a prominent dorsal tubercle near base; apical segment of maxillary palpi swollen, not elongate, definitely broader than apical segment of labial palpi; antennae black, distinctly annulated, usually as long as body; scape occasionally infuscated; scape and usually segments to third densely cinereously pubescent, densely clothed with

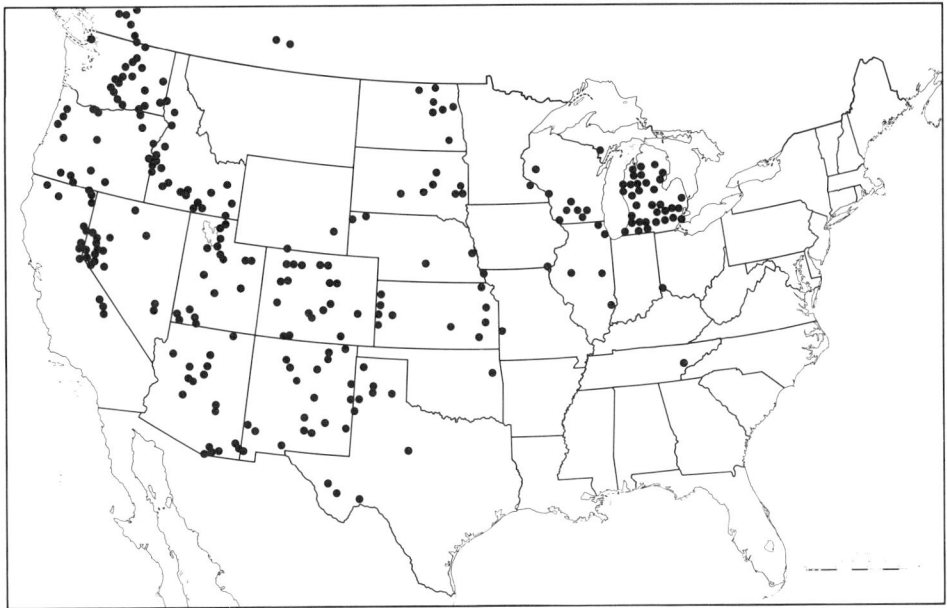

Figure 53. Known geographic range of *Tetraopes femoratus* LeConte north of Mexico.

long suberect black hairs, segments from fourth opaque, densely clothed with very short dark appressed pubescence; segments from third with numerous long erect hairs internally which diminish in number apically. Pronotum reddish, varying from about 1-1/3 to 1-1/2 times wider than long; four black dorsal spots almost always distinct; umbone abruptly prominently elevated, sides distinctly delimited; disk usually shining, convex, moderately coarsely, sparsely punctate, usually sparsely clothed with short pale pubescence, long dark erect hairs sparse; umbone usually longer than wide; sides of pronotum abruptly inflated medially, prominently tuberculate, basal and apical margins impressed transversely; lateral tubercles usually densely, coarsely punctate, remainder of pronotum less coarsely, sparsely; short pale pubescence usually dense, long erect black hairs sparse; a black spot usually present laterally at apical margin; prosternum densely to moderately densely pubescent; meso- and metasternum sparsely, finely punctate, densely cinereously pubescent. Elytra reddish; slightly less than twice as long as broad; humeri black or almost always so; two sutural spots usually present, but may be absent or greatly reduced; post-humeral spots usually present, but often lacking; apical dark spots always present, but varying greatly in size; punctures moderately coarse, usually only moderately dense and well separated; short pale pubescence variable, often quite dense but usually only moderately so, longer suberect hairs fairly numerous; scutellum black, densely pubescent; apices rounded. Legs short, stout; varying in color from all red to only femora red with tibiae black or some tibiae red, also some femora black; densely pubescent. Abdomen usually black, densely cinereously pubescent; apex of fifth sternite shallowly, concavely subtruncate to rounded. Length, 8-16 mm.

Female. Form larger, more robust; mandibles not dorsally tuberculate; antennae shorter than body; apex of fifth sternite rounded, impressed at middle. Length, 9-19 mm.

Type locality. Of *femoratus*, "toward the Arkansas River near the mountains"; *oregonensis*, Wenass River to Fort Coleville (Washington); *collaris*, New Mexico; *nigripes*, Colonia Garcia, Chihuahua; *vegasensis*, Las Vegas, New Mexico; *velutinus*, Finney and Hamilton Cos., Kansas; *robustus*, not indicated; *brevisetosus*, Kansas; *punctipennis*, Texas; *atrisetosus*, Fort Wingate, New Mexico; *amnicola*, Keokuk (Iowa); *monticola*, Utah; *spissicornis*, Oak Creek Canyon, Arizona; *ruber*, southern Arizona; *fortis*, Durango City, Mexico.

Range. Great Basin, Great Plains to Ohio, Arizona to Texas south to central Mexico (Figure 53).

Flight period. May to October.

Host plants. Asclepias fascicularis, A. hallii, A. hirtella, A. latifolia, A. lemmonii, A. speciosa, A. syriaca, A. viridis.

T. femoratus is without doubt the most variable and probably the most widespread species in the genus. Some difficulty is encountered in defining the species as a whole, since the large geographical areas occupied give rise to different phenotypes. Four major areas, the Great Basin, the Great Plains, montane Arizona, and Mexico, contain distinct populations which are usually readily recognizable, particularly in a series of specimens. Chemsak (1963) has characterized each of these populations.

Genus *Phaea* Newman

Phaea Newman, 1840, Entomol., 1:13; Thomson, 1860, Class. ceramb., p. 67; Chevrolat, 1861, J. Entomol., 1:190 (synonymy); Thomson, 1864, Syst. ceramb., p. 121; Bates, 1866, Ann. Mag. Nat. Hist., 3(17): 366 (238);

Lacordaire, 1872, Genera des coléoptères, 9(2):878; LeConte, 1873, Smithson. Misc. Coll., 11:(265): 347; Bates, 1881, Biol. Centr.-Amer., Coleop., 5: 199; LeConte and Horn, 1883, Smithson. Misc. Coll., 507: 332; Casey, 1913, Memoirs on the Coleoptera, 4: 386; Chemsak, 1977, Pan-Pac. Entomol., 53: 269; Rice and Enns, 1981, Trans. Mo. Acad. Sci., 15: 106.

Tetrops: LeConte, 1852 (not Stephens, 1831), J. Acad. Nat. Sci. Philadelphia, 2: 155; Thomson, 1857, Arch. Entomol., 1: 65 (part); Thomson, 1864, Syst. ceramb., p. 115 (part); Lacordaire, 1872, Genera des coléoptères, 9(2): 880 (part); LeConte, 1873, Smithson. Misc. Coll., 11:265): 347; Horn, 1878, Trans. Amer. Entomol. Soc., 7: 50; Bates, 1881, Biol. Centr.-Amer., Coleop., 5: 95 (part); LeConte and Horn, 1883, Smithson. Misc. Coll., 507: 332; Leng and Hamilton, 1896, Trans. Amer. Entomol. Soc., 23: 157; Blatchley, 1910, Coleoptera in Indiana, p. 1093; Bradley, 1930, Man. Genera Beetles, p. 247; Knull, 1946, Ohio Biol. Surv. Bull., 39: 282; Arnett, 1962, Beetles U.S., 103: 873.

Lamprocleptes Thomson, 1857, Arch. Entomol., 1: 64. (Type species: *Lamprocleptes entomologorum* Thomson, monobasic.)

Oberopa Haldeman, 1847, Proc. Amer. Philos. Soc., 4: 373. (Type species: *Oberea monostigma* Haldeman, monobasic.)

Form small to moderate-sized, subparallel. Head with front convex, interantennal area rather deeply impressed; eyes finely faceted, completely divided; genae about as long as lower eye lobes; antennae often stout, about as long as body in males, shorter in females, scape often asperate apically, segments with flying hairs beneath, third segment longer than first and fourth. Pronotum usually broader than long, sides rounded; apex and base broadly impressed transversely; dorsal umbone usually well developed; prosternum with intercoxal process narrow, expanded at apex, coxal cavities closed behind; mesosternum with coxal cavities open to epimeron; metasternum with episternum a little broader in front. Elytra with sides usually expanding slightly toward apex; apices rounded. Legs short; intermediate tibiae with external sinus; tarsal claws appendiculate. Abdomen of females inflated, longitudinally linearly impressed.

Type species. *Phaea saperda* Newman (monobasic).

The genus *Phaea* is a large Mexican and Central American group which extends northward into the United States and southward to northern South America. Two species occur in our fauna, one rather widespread in the eastern United States and the other apparently restricted to some of the midwestern states.

Phaea may be easily separated from the closely related genus *Tetraopes* by the appendiculate rather than bifid claws.

KEY TO THE NORTH AMERICAN SPECIES OF *PHAEA*.

Elytra very densely clothed with grayish appressed pubescence which partially obscures the surface;

form broad, robust. Length, 9-12 mm. Kansas,
Colorado, New Mexico .... *canescens*
Elytra moderately densely clothed with dark
appressed pubescence, surface not obscured; form
slender. Length, 7-10 mm. Eastern United States
to Texas .... *monostigma*

Phaea canescens (LeConte)
(Figure 54)

Tetrops canescens LeConte, 1852, J. Acad. Nat. Sci. Philadelphia, 2: 156; Thomson, 1857, Arch. Entomol., 1: 66; Lacordaire, 1872, Genera des coléoptères, 9(2): 881, fn; Horn, 1878, Trans. Amer. Entomol. Soc., 7: 50; Leng and Hamilton, 1896, Trans. Amer. Entomol. Soc., 23: 157; Smith, 1910, N.J. St. Mus. Rept., 1909: 337; Casey, 1913, Memoirs on the Coleoptera, 4: 386.

Phaea canescens: LeConte, 1873, Smithson. Misc. Col., 11(265): 347; Chemsak and Linsley, 1974, Coleop. Bull., 28: 184.

Male. Form moderate-sized, parallel; integument black, head and pronotum orange except for tips of mandibles, eyes, and umbone, outer antennal segments narrowly pale annulate at bases; pubescence dense, grayish, appressed. Head with front convex, moderately coarsely, separately punctate, densely clothed with pale appressed pubescence and longer erect hairs; interantennal depression deep, vertex irregularly punctate; antennae stout, slightly shorter than body, scape not apically asperate, segments from about fourth narrowly pale annulate at bases, long erect hairs beneath becoming sparser toward apex, third segment longer than first, fourth equal to first. Pronotum broader than long, sides rounded; apex and base broadly, deeply impressed transversely; umbone not prominent, distinctly delineated only on apical half, sparsely, irregularly punctate; sides sparsely, irregularly punctate; pubescence fairly sparse, mostly long and erect; prosternum transversely rugulose, densely pubescent; meso- and metasternum sparsely punctate, densely clothed with very short, pale, appressed pubescence. Elytra about 2-1/2 times as long as broad, sides vaguely impressed at middle; basal punctures coarse, subserial toward middle, becoming finer toward apex; pubescence dense, grayish, appressed, partially obscuring surface, longer suberect hairs numerous; apices broadly rounded. Legs short; femora sparsely punctate. Abdomen sparsely punctate, moderately densely pubescent; last sternite broadly truncate at apex. Length, 9-11 mm.

Female. Form more robust, broader. Antennae extending to about second abdominal segment. Abdomen with last sternite convex, linearly impressed longitudinally at middle, apex triangularly impressed. Length, 9-12 mm.

Type locality. Missouri Territory.
Range. Kansas, Colorado, New Mexico.

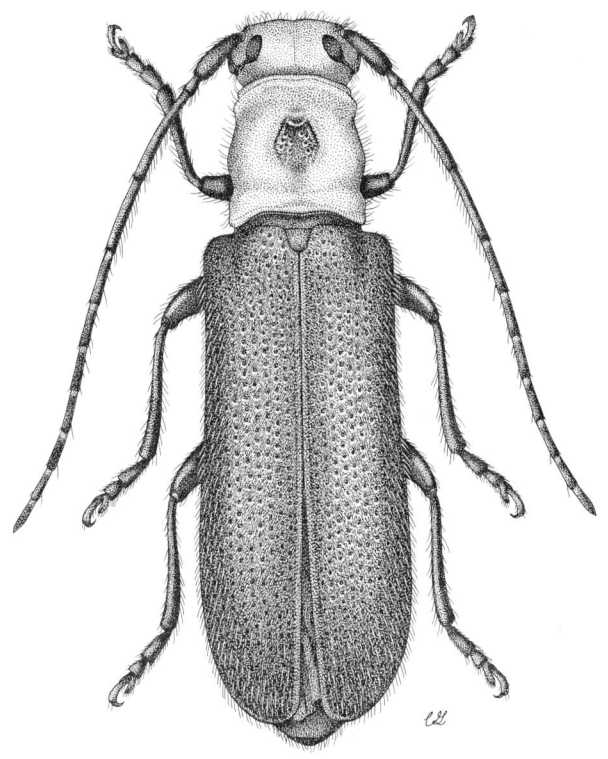

Figure 54. *Phaea canescens* (LeConte), female.

Flight period. June and July.
Host plants. Alnus.
The broader body form and dense grayish pubescence of the elytra readily separate this species from *monostigma*.

Phaea monostigma (Haldeman)

Oberea monostigma Haldeman, 1847, Trans. Amer. Philos. Soc., (2)10: 57.
Oberopa monostigma: Haldeman, 1847, Proc. Amer. Philos. Soc., 4: 373.
Tetrops monostigma: LeConte, 1852, J. Acad. Nat. Sci. Philadelphia, 2: 156; Thomson, 1857, Arch. Entomol., 1: 66; Lacordaire, 1872, Genera des coléoptères, 9(2): 881, fn; Horn, 1878, Trans. Amer. Entomol. Soc., 7: 50;

Leng and Hamilton, 1896, Trans. Amer. Entomol. Soc., 23: 158; Blatchley, 1910, Coleoptera in Indiana, p. 1093; Casey, 1913, Memoirs on the Coleoptera, 4: 386; Knull, 1946, Ohio Biol. Surv. Bull., 39: 282, pl. 26, fig. 111; Fattig, 1947, Emory Univ. Mus. Bull., 5: 45; Kirk, 1970, S.C. Agr. Exp. Sta. Bull., 1038: 84.

Phaea monostigma: LeConte, 1873, Smithson. Misc. Coll., 11(265): 347; Chemsak and Linsley, 1974, Coleop. Bull., 28: 184; Rice and Enns, 1981, Trans. Mo. Acad. Sci., 15: 106.

Tetrops jucunda LeConte, 1962, Proc. Acad. Nat. Sci. Philadelphia, 1862: 40; Horn, 1978, Trans. Amer. Entomol. Soc., 7: 50; Leng and Hamilton, 1896, Trans. Amer. Entomol. Soc., 23: 157; Blatchley, 1910, Coleoptera in Indiana, p. 1093 (synonymy); Casey, 1913, Memoirs on the Coleoptera, 4: 386; Craighead, 1923, Can. Dept. Agr. Bull., (n.s.) 27: 143; Kirk, 1969, S.C. Agr. Exp. Sta. Bull., 1033: 88.

Tetrops expurgata Casey, 1913, Memoirs on the Coleoptera, 4: 386.

Male. Form small, subparallel; integument black, head except apices of mandibles and eyes reddish, pronotum except umbone reddish, antennae with outer segments narrowly to broadly pale annulate, middle and front femora pale, front tibiae partially pale; pubescence dark, dense, short and depressed and long and erect. Head with front convex, moderately coarsely, separately punctate, interspaces micropunctate, moderately densely clothed with pale depressed pubescence, long, dark, suberect hairs numerous; area between antennal tubercles deeply impressed, vertex sparsely punctate and pubescent; antennae slender, almost as long as elytra, scape not apically asperate, segments from fourth narrowly to broadly pale annulate basally, segments with numerous long flying hairs beneath, becoming sparser toward apex, third segment longer than first, fourth slightly shorter than third, longer than first. Pronotum slightly broader than long, sides rounded; base and apex broadly, deeply impressed transversely; umbone not distinctly delineated except at apical half; middle rather sparsely punctate, each puncture giving rise to a long, erect, black seta; sides sparsely punctate, long erect hairs numerous; prosternum shallowly impressed, transversely rugose, moderately densely pubescent; meso- and metasternum sparsely punctate, densely clothed with pale depressed pubescence. Elytra about 2-1/2 times as long as broad, sides vaguely impressed at middle; punctures coarse, subserial, becoming finer toward apex; pubescence fine, dark, short, subappressed, long suberect hairs numerous; apices broadly rounded. Legs short; femora finely punctate, moderately densely pubescent. Abdomen finely punctate, moderately densely clothed with long, suberect pubescence; last sternite broadly subtruncate at apex. Length, 7-9.5 mm.

Female. Form more robust. Antennae reaching to about second abdominal segment. Abdomen with last sternite convex, linearly impressed at middle, apex shallowly, triangularly impressed. Length, 8-10 mm.

Type locality. Of *monostigma*, Pennsylvania; *jucunda*, Middle States; *expurgata*, Indiana.

Range. Eastern United States to Texas.

Flight period. May to July.
Host plants. Ipomoea.

The infuscation of the umbone and legs varies, as well as the extent of the pale annulations of the antennae.

The small form and sparser appressed pubescence of the elytra readily separate this species from *P. canescens* (LeConte). The Mexican *P. acromela* Bates is similar in size and coloration but the much less elevated thoracic umbone of *monostigma* separate the two.

TRIBE HEMILOPHINI THOMSON

Thomson, 1868, Physis, 2: 189 (Hemilophitae).
Lacordaire, 1872, Genera des coléoptères, 9(2): 881 (Amphionychides).
LeConte, 1873, Smithson. Misc. Coll., 11(265): 347 (Phytoeciini, part).
Bates, 1881, Biol. Centr.-Amer., Coleoptera, 5: 201 (Phytoeciini, part).
LeConte and Horn, 1883, Smithson. Misc. Coll., 507: 332 (Phytoeciini, part).
Leng and Hamilton, 1896, Trans. Amer. Entomol. Soc., 23: 161 (Phytoeciini, part).
Aurivillius, 1923, Coleop. Cat., 74: 584.
Bradley, 1930, Man. Genera Beetles, p. 247.
Knull, 1946, Ohio Biol. Surv. Bull., 39: 285.
Duffy, 1960, Mon. Imm. Neotrop. Timber Beetles, p. 276.
Villiers, 1980, Ann. Soc. Entomol. Fr., (n.s.) 16: 119.
Rice and Enns, 1981, Trans. Mo. Acad. Sci., 15: 90.

Form small to moderate-sized. Head with front short, transverse, convex; eyes finely faceted, deeply emarginate; antennal tubercles depressed, widely separated; antennae slender, usually at least as long as body, scape slender, subcylindrical, basal segments fringed with long flying hairs beneath. Pronotum rounded, apex often narrower than base, disk convex; prosternum narrow, intercoxal process narrow, expanded at apex, coxal cavities closed behind; mesosternum with coxal cavities open to epimeron; metasternum with episternum moderately broad, slightly broader anteriorly. Elytra often with vertical epipleura; wings present. Legs short; femora linear; intermediate tibiae lacking or with a vague tarsal sinus; tarsal claws bifid. Abdomen with sternites two to four shorter than others.

The bifid tarsal claws, fairly narrow metepisternum, and proportions of the abdominal sternites characterize this tribe. The Tetraopini differ by the completely divided eyes.

This group attains its greatest development in the Neotropical region. Most of the species are mimetic, utilizing Lampyridae, Lycidae, Cantharidae, and Chrysomelidae as models.

Two genera occur in the United States. We are retaining the historical nomenclature for them until such time as the genera of Mexico and Central America can be revised. These will no doubt fall into one of the Mexican groups such as *Alampyris* when that fauna becomes better known.

KEY TO THE NORTH AMERICAN GENERA OF HEMILOPHINI

Elytra with posthumeral carinae well developed, sides
    steeply declivous below carina .... *Cathetopteron*
Elytra lacking post-humeral carinae, sides less
    abrupt .... *Hemierana*

Genus *Cathetopteron* Hamilton

Cathetopteron Hamilton, 1896, Trans. Amer. Entomol. Soc., 23: 162; Bradley, 1930, Man. Genera Beetles, p. 247.
Cathstopteron: Arnett, 1962, Beetles U.S., 103: 873 (error).

    Form small, slightly tapering posteriorly. Head with front convex, short, broad, interantennal depression shallow; eyes deeply emarginate, upper lobes small; genae about as long as lower eye lobes; antennae slender, about as long as body in males, long, flying hairs abundant on basal segments, outer segments short, scape slender, third segment longer than first, fourth shorter than first. Pronotum broader than long, sides vaguely sinuate; base deeply impressed at each side of middle; prosternum very short; mesosternum with intercoxal process rather abruptly declivous anteriorly. Elytra with strong carinae extending over humeri to about apical one-fifth, epipleura vertical with strong carinae between margins and dorsal carinae; apices rounded. Legs short; femora linear. Abdomen with last sternite of females linearly impressed longitudinally.
    Type species. Amphionycha amoena Hamilton (monobasic).
    The prominent carinae and vertical epipleura of the elytra characterize this genus in the North American fauna.
    A single species is known.

Cathetopteron amoena Hamilton

Cathetopteron amoena Hamilton, 1896, Trans. Amer. Entomol. Soc., 23:161; Linsley and Martin, 1933, Entomol. News, 44:183; Turnbow and Wappes, 1978, Coleop. Bull., 32:370; Hovore et al., 1987, Proc. Calif. Acad. Sci., 44:322.
Amphionycha amoena: Townsend, 1902, Trans. Texas Acad. Sci., 5:80.

    Male. Form small, slightly tapering posteriorly; integument black, head and pronotum partially pale yellowish; pubescence dense, appressed, long, erect hairs numerous. Head with front strongly convex, separately punctate, densely clothed with short, pale, appressed pubescence, long, erect hairs numerous; each side behind eyes with a broad, brownish vitta which extends onto apical one-fourth of pronotum; antennae slightly larger than body, segments three and four narrowly pale annulate at bases, long flying hairs numerous, becoming sparser toward apex, outer segments vaguely annulate with grayish pubescence at bases. Pronotum broader than long, sides rounded; disk convex, each side of middle of apex with a shallow brownish impression, base with longer, deeper brownish impressions at sides of middle; punctures moderately coarse, dense, obscured by pubescence; pubescence creamy, very dense, short, long, pale, erect hairs sparsely interspersed; prosternum very narrow, finely pubescent; meso- and metasternum finely punctate, very fine pubescent. Elytra about 2-1/2 times as long as basal width; lateral carinae distinct; punctures at base moderately coarse, separated, becoming finer toward apex; pubescence dense, dark, very short, appressed, long suberect hairs numerous; apices broadly rounded. Legs finely punctate, finely pubescent. Abdomen densely, minutely punctate, finely pubescent; last sternite with apex broad, vaguely emarginate. Length, 6-7 mm.

Female. Form more robust. Antennae slightly shorter than body. Abdomen with last sternite convex; linearly impressed down middle. Length, 6.5-9 mm.

Type locality. near Brownsville, Texas.
Range. Southern Rio Grande Valley.
Flight period. April to October.
Host plants. Celtis.

This species is distinctive by the vertical epipleura of the elytra and four dark spots of the pronotum.

Genus *Hemierana* Aurivillius

Hemierana Aurivillius, 1923, Coleop. Cat., 74:594 (new name for *Amphionycha* Haldeman); Bradley, 1930, Man. Genera Beetles, p. 247; Knull, 1946, Ohio Biol. Surv. Bull., 39:285; Arnett, 1962, Beetles U.S., 103:874.

Amphionycha Haldeman, 1847 (not Dejean, 1837), Trans. Amer. Philos. Soc., (2)10:57; LeConte, 1852 (not Leseleuc, 1844), J. Acad. Nat. Sci. Philadelphia, 2:154; Lacordaire, 1872, Genera des coléoptères, 9(2):890 (part); LeConte, 1873, Smithson. Misc. Coll., 11(265):347; Horn, 1875, Trans. Amer. Entomol. Soc., 4:150; Horn, 1878, Trans. Amer. Entomol. Soc., 7:50; LeConte and Horn, 1883, Smithson. Misc. Coll., 507:332; Horn, 1886, Trans. Amer. Entomol. Soc., 13:x; Leng and Hamilton, 1896, Trans. Amer. Entomol. Soc., 23:161; Wickham, 1897, Can. Entomol., 29:203; Wickham, 1898, Can. Entomol., 30:40; Blatchley, 1910, Coleoptera in Indiana, p. 1095.

Form small to moderate-sized, subparallel. Head with front convex; eyes deeply emarginate, upper lobes small; genae about as long as lower eye lobes; antennae slender, about as long as body, scape slender, subcylindrical, basal segments densely fringed with long, flying hairs, segments from fifth short, third segment longer than first, fourth subequal to first. Pronotum broader than long, sides rounded; apex narrower than base; base shallowly, broadly impressed at sides of middle; apex broadly, shallowly impressed; prosternum short, mesosternum with intercoxal process shallowly arcuate. Elytra lacking carinae, sides rounded to epipleura; apices rounded. Legs short; femora sublinear; intermediate tibiae with a vague external sinus. Abdomen with last sternite of females not impressed.

Type species. Saperda marginata Fabricius (by present designation).

The lack of carinae on the elytra separates this genus from *Cathetopteron*.

Aurivillius (1923) proposed *Hemierana* as a new name for *Amphionycha* LeConte (not Leseleuc 1844). Actually, Dejean (1837) is the author of the name, having included names of a number of valid species in his list, among them *marginata* Fabricius. Haldeman (1847) listed *Amphionycha marginata* Fabricius as a valid North American species, and this concept of *Amphionycha* was followed by LeConte in 1852. Various interpretations of *Amphionycha* by subsequent authors is not now at issue, and the name *Hemierana* appears to be properly used in the North American fauna.

A single, apparently polytypic, species is known.

Hemierana marginata (Fabricius)

Saperda marginata Fabricius, 1798, Suppl. Entomol. Syst., p. 148.
Amphionycha marginata: Haldeman, 1847, Trans. Amer. Philos. Soc., (2)10:57.
Hemierana marginata: Aurivillius, 1923, Coleop. Cat., 74:594.

Male. Form small to moderate-sized, subparallel; integument black, head and pronotum partially yellowish, elytra often pale on humeri; pubescence dense, fine, appressed, long and short, pale and dark. Head with front strongly convex, moderately coarsely, separately punctate, densely clothed with yellowish, appressed pubescence, long, erect hairs moderately dense; vertex with a brownish, triangular vitta which extends onto pronotum; antennae slightly longer than body, segments often pale annulate at bases, long, flying hairs numerous, becoming sparser toward apex, segments often vaguely grayish pubescent beneath and at bases; third segment longer than first, fourth subequal to first. Pronotum broader than long, sides rounded, apex narrower than base, barely impressed transversely; base broadly, shallowly impressed on each side of middle; disk with a broad brownish, longitudinal vitta, sides with broad, brownish vittae, punctures moderately coarse, contiguous to separated; pale, appressed pubescence very dense, obscuring surface, long, erect hairs moderately dense; prosternum finely, densely pubescent; meso- and metasternum densely micropunctate, finely, densely pubescent. Elytra about 2-1/2 times longer than broad, subparallel; punctures coarse to moderately coarse, separated to subconfluent; pubescence fine, short, appressed, dark to pale, suture and lateral margins often with narrow bands of pale pubescence, longer hairs suberect to erect; apices rather narrowly rounded. Legs with femora densely, minutely punctate, finely, densely pubescent. Abdomen densely, minutely punctate, finely, densely pubescent; last sternite broadly rounded to subtruncate at apex. Length, 6-9 mm.

Female. Form more robust. Antennae slightly shorter than body. Abdomen with last sternite convex, broadly rounded at apex. Length, 8-10.5 mm.

Type locality. Carolina.

Range. Eastern United States to Texas and Kansas.

The annulate antennae and yellowish pubescent sides of the pronotum make this species distinctive.

There are three recognizable populations of *H. marginata* within the boundaries of the United States.

Hemierana marginata marginata (Fabricius)

Saperda marginata Fabricius, 1798, Suppl. Entomol., System., p. 148; Fabricius, 1801, Syst. Eleuth., 2:331; LeConte, 1852, J. Acad. Nat. Sci. Philadelphia, 2:165.

Amphionycha marginata: Haldeman, 1847, Trans. Amer. Philos. Soc., (2)10:57; Haldeman, 1847, Proc. Amer. Philos. Soc., 4:373 (synonymy).

Hemierana marginata: Aurivillius, 1923, Coleop. Cat., 74:594 (part); Fattig, 1947, Emory Univ. Mus. Bull., 5:45.

Saperda flammata Newman, 1840, Entomol., 1:13.

Amphionycha flammata: Haldeman, 1847, Proc. Amer. Philos. Soc., 4:373; LeConte, 1852, J. Acad. Nat. Sci. Philadelphia, 2:154; Lacordaire, 1872, Genera des coléoptères, 9(2):892, fn.; Horn, 1878, Trans. Amer. Entomol. Soc., 7:50 (part); Leng and Hamilton, 1896, Trans. Amer. Entomol. Soc., 23:161 (part).

Hemierana flammata: Kirk, 1970, S.C. Agr. Exp. Sta. Tech. Bull., 1038:84.

Pubescence of elytra grayish, suture and lateral margins with narrow bands of dense yellowish to grayish pubescence. Pronotum moderately coarsely, densely punctate. Elytra moderately coarsely, subcontiguously punctate, punctures larger than those of pronotum. Antenna with segments vaguely, narrowly grayish annulate at bases. Length, 6-9 mm.

Type locality. Of marginata, Carolina; *flammata*, St. John's Bluff, Florida.

Range. Southeastern United States.

Flight period. May to August.

Host plants. Not known.

The dense pubescence of the elytral suture and margins and less coarsely punctate elytra separate this subspecies from *H.m. ardens*. It differs from *H.m. suturalis* by the broader brownish vitta and denser punctures of the pronotum.

The extent of the range of this subspecies outside of Florida is not known at present.

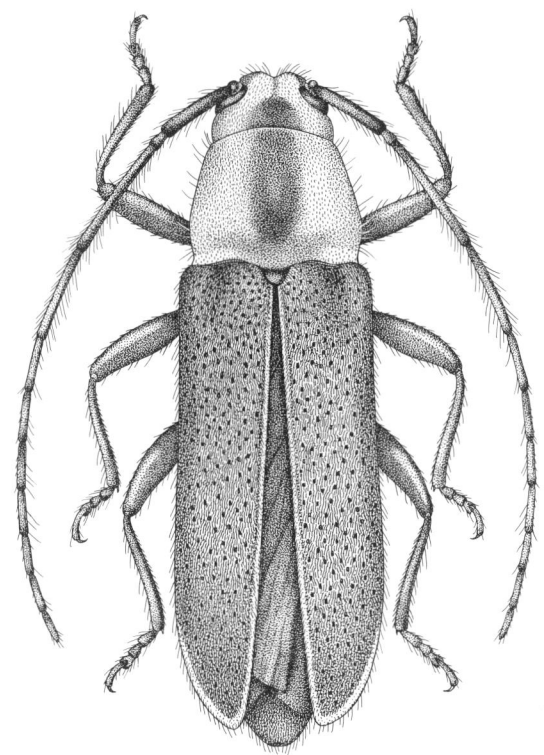

Figure 55. *Hemierana marginata ardens* (LeConte), female.

Hemierana marginata ardens (LeConte)
(Figure 55)

Amphionycha ardens LeConte, 1859, Smithson. Contr. Knowl., 11:22.
Amphionycha flammata var. *ardens*: Leng and Hamilton, 1896, Trans. Amer. Entomol. Soc., 23:161.
Hemierana marginata ab. *ardens*: Aurivillius, 1923, Coleop. Cat., 74:595.
Hemierana marginata: Aurivillius, 1923, Coleop. Cat., 74:595 (part); Knull, 1937, Entomol. News, 48:42; Schwitzgebel and Wilbur, 1942, J. Kansas Entomol. Soc., 15:37 (habits); Knull, 1946, Ohio Biol. Surv. Bull., 39:285, pl. 26, fig. 114; Kirk and Balsbaugh, 1975, S.D. Agr. Exp. Sta. Tech. Bull., 42:101; Rice and Enns, 1981, Trans. Mo. Acad. Sci., 15:107: Hovore et al., 1987, Proc. Calif. Acad. Sci. 44:322.
Amphionycha flammata: Horn, 1878, Trans. Amer. Entomol. Soc., 7:50 (part); Leng and Hamilton, 1896, Trans. Amer. Entomol. Soc., 23:161; Wickham, 1898, Can. Entomol., 30:40; Blatchley, 1910, Coleoptera in Indiana, p. 1095;

Smith, 1910, N.J. St. Mus. Rept., 1909: 337; Leonard, 1928; Cornell Agr. Exp. Sta. Mem., 101:457.

Pubescence of elytra dark, very short, not obscuring surface. Pronotum densely punctate, median dark vitta broad, broader posteriorly. Elytra coarsely, usually subconfluently punctate. Antenna with segments usually narrowly pale annulate at bases. Length, 6-10.5 mm.

Type locality. Fort Riley, Kansas.
Range. Eastern United States to Texas and Kansas.
Flight period. April to August.
Host plants. Vernonia.

This is the most common and widespread subspecies and may extend its range into northern Mexico. The lack of pubescent vittae on the elytra, basally broader, dark vitta of the pronotum, and coarsely, subcontiguously punctate elytra characterize it.

Hemierana marginata suturalis (Linell)

Amphionycha suturalis Linell, 1896, Proc. U.S. Nat. Mus., 19:398; Townsend, 1902, Trans. Tex. Acad. Sci., 5:80.

Hemierana suturalis: Aurivillius, 1923, Coleop. Cat., 74:595; Linsley and Martin, 1933, Entomol. News, 44:183; Hovore et al., 1987, Proc. Calif. Acad. Sci., 44:322.

Pubescence of elytra dense, grayish, partially obscuring surface, sutural and lateral vittae well developed. Pronotum sparsely, rather finely punctate, median brownish vitta narrower, usually narrowing posteriorly. Elytra moderately coarsely, separately punctate. Antennae with basal segments often narrowly pale annulate at bases. Length, 7-8 mm.

Type locality. Brownsville, Texas.
Range. Southern Rio Grande Valley.
Flight period. May to July.
Host plants. Bernardia.

H. m. suturalis probably ranges into at least northeastern Mexico, but we have seen no specimens resembling the Texas forms. This subspecies is distinctive by the denser pubescence of the elytra, narrower dark vitta of the pronotum, and sparse punctures of the pronotum.